FUNDAMENTALS OF MICROMECHANICS OF SOLIDS

FUNDAMENTALS OF MICROMECHANICS OF SOLIDS

Jianmin Qu

Mohammed Cherkaoui

WILEY

John Wiley & Sons, Inc.

This book is printed on acid-free paper. ♾

Published by John Wiley & Sons, Inc., Hoboken, New Jersey
Published simultaneously in Canada

Library of Congress Cataloging-in-Publication Data:
Qu, Jianmin.
Fundamentals of micromechanics of solids/Jianmin Qu, Mohammed Cherkaoui.
p. cm.
Includes bibliographical references.
ISBN-13: 978-0-471-46451-8 (cloth)
ISBN-10: 0-471-46451-1 (cloth)
1. Nanostructured materials—Mechanical properties—Textbooks.
2. Micromechanics—Textbooks. 3. Microelectromechanical systems—Materials—Textbooks. I. Cherkaoui, Mohammed. II. Title.
TA418.9.N35Q22 2006
620.1′05—dc22

2005030781

10 9 8 7 6 5 4 3 2 1

To Qianqian, Bouchra, Lisa, Jalal, Rayan and Hajja-Fetouma.

CONTENTS

PREFACE

For many years, we have taught courses on micromechanics at our respective institutions (JQ at the Georgia Institute of Technology and MC at the University of Metz). We both felt the need for a textbook that (1) provides an integrated approach to the various topics of micromechanics, (2) covers the basic ideas of homogenization, (3) is written pedagogically for graduate and upper-level undergraduate students, (4) contains exercises problems, and (5) can be taught in a semester course. It was such a need that motivated us to start this project. We hope the book in front of you fulfills these objectives.

Micromechanics had flourished over the past half-century. The literature is rich and diverse. Holding the book to a reasonable length prevented us from including many of the elegant results published over the years. Although we used our best judgment to cite and compile a list of references that we believe are most relevant to the text, listing such a limited number of references does not do justice to the breadth and depth of the field. We sincerely apologize to those authors whose works were not given proper credit or were not cited.

The book was developed primarily from our notes for the micromechanics courses that we have taught over the years. Our students, through their questions and comments, provided valuable input to the book. We are very grateful for their contributions. We would like to thank our colleagues at Georgia Tech for their encouragement and collaboration. It has been a great pleasure to be among such an intellectually stimulating group of people.

The book was written mainly during evenings and on weekends. At times, we had to neglect our duties as husbands and fathers. We want to acknowledge that without the support of our wives and families, this project would not have been completed. Their understanding and sacrifices are greatly appreciated. This book is dedicated to them.

JIANMIN QU
Atlanta, Georgia

MOHAMMED CHERKAOUI
Metz, France

1

INTRODUCTION

Mechanics studies the theory of forces and their interactions. It is the pillar of modern sciences. Many of the great scientists in history have been mechanicians including, for example, Aristotle, Archimedes, Leonardo da Vinci, Kepler, and Newton.

Continuum mechanics is a relatively young branch of mechanics. It studies the deformation of bodies (solid or fluid) under forces or stresses. Although the basic foundation of continuum mechanics was laid by Galileo, it was during the late eighteenth and the early nineteenth centuries that modern theories of continuum mechanics were gradually developed by Laplace, Fourier, Coriolis, Lagrange, Hamilton, Navier, and Cauchy, among others. It was during this period and up until the early 1900s that continuum mechanics enjoyed its most rapid development into maturity. By the mid-1900s, theories of continuum mechanics had been established upon vigorous mathematics.

One of the most successful stories of continuum mechanics is the development and application of fracture mechanics. In 1913, C.E. Inglis looked at a thin plate of glass with an elliptical hole in the middle in a new and different way (Fig. 1.1). The plate was pulled at both ends perpendicular to the ellipse. He found that the stress at point A is given by

$$\sigma_A = \sigma\left(1 + \frac{2a}{b}\right).$$

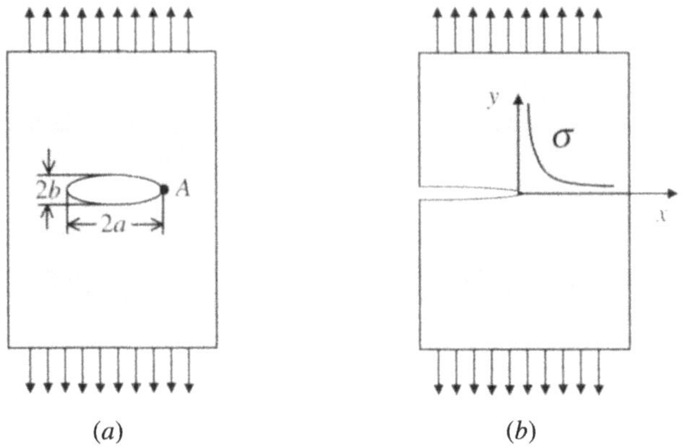

Figure 1.1 (*a*) Inglis' work on stress concentration near an elliptical hole. (*b*) Griffith's work on stress concentration near a sharp crack tip.

In other words, the stress at the tip of the elliptical hole can be much larger than the stress applied. In the 1920s, A.A. Griffith extended Inglis' work. He found that the stress at the ends of the crack approaches infinity:

$$\sigma = \frac{K}{\sqrt{2\pi x}}.$$

Griffith also introduced the notion of energy. He said that for a crack to grow, it was necessary for there to be enough potential energy in the system to create the new surface area of the crack. Although he did not know that it takes more than this for a crack to grow, Griffith's idea of fracture criterion laid the foundation for a brand new theory called fracture mechanics, which is one of the most celebrated branches of continuum mechanics in modern history.

More importantly, fracture mechanics brought continuum mechanics and material science together and opened up new opportunities for studying not only deformation of solid bodies but also the failure behavior of solid materials under load. Thus a new field called mechanics of materials emerged. In mechanics of material, we use the vigorous continuum mechanics theories to investigate and study how materials with certain microstructures deform and eventually fail under given loads or stresses.

Micromechanics is a branch of mechanics of materials. It is the most recent development in applying continuum mechanics theories to real materials. The beginning of micromechanics may be traced back to Eshelby's seminal study (Eshelby, 1957). But, the theory of micromechanics was not fully developed into a subject area of its own until the early 1980s. Even though it is still an actively researched area now, the theory of micromechanics has matured enough that several books have been published. Among them, the books by Mura (1986), Nemat-Nasser and Hori (1993), and Krajcinovic (1996) are probably the most comprehensive and influential ones.

The theory of micromechanics of solids describes the scientific concepts, principles, and methodologies for the study of thermomechanical behavior of heterogeneous materials. Although the fundamental equations of micromechanics are based on mechanics of continuum, its applications cover a broad range of thermomechanical behavior of materials including plasticity, fracture, and fatigue, constitutive equations of composites, and polycrystalline materials. For example, it applies the theories of elasticity and plasticity to study imperfections in crystals, inclusions, and inhomogeneities in alloys and composite materials. The objective is to study the macroscopic mechanical behavior of materials from an understanding of their microstructure. This involves the application of continuum mechanics to identifiable small-scale structures and the use of analytical and numerical methods to compute the macroscopic responses. This science-based approach enables us to predict the behavior of new materials without the need for physical experimentation. It provides a powerful tool for engineering design, fabrication, and analysis of a wide range of materials including polycrystalline, composite, geotechnical, biological, and electronic materials. Optimum microstructures can be forecasted rather than found by trial and error. Fracture and fatigue of solids and structures, martensitic transformations, interphases in composites, and dispersion hardening of alloys are examples of the phenomena that are being elucidated and qualified by micromechanics.

1.1 BACKGROUND AND MOTIVATION

As the theory of micromechanics matures, many universities around the world are offering courses on this subject. For the past 15 years, the authors have taught micromechanics classes in their respective institutions. We have always been frustrated by not being able to find an appropriate textbook for the course. Most existing books on this subject are research monographs, primarily for experts and researchers. They can be excellent research tools but not convenient to use as textbooks. Because the theory of micromechanics is still in its infancy, results were obtained by individual researchers using, sometime, very different approaches/methodologies. In order for the students (or first-time learners) to understand the intrinsic connections among different concepts and approaches, a unified approach (including the use of notations) is needed to develop the micromechanics theory. This will allow both the instructor and the students to fully grasp the essence of the theory. Furthermore, instead of collecting and compiling existing results from the literature, we should identify a set of topics that convey the fundamental ideas of micromechanics and focus on these topics. Related topics not covered in the text should be referenced for those who wish to learn more. Exercise problems should be provided for the convenience of the instructor, as well as for those who wish to study the subject on their own. These are the major considerations that motivated us to write this textbook.

1.2 OBJECTIVES

The intent of this book is not to provide a comprehensive collection of results of micromechanics in the literature, nor is it to be a research reference book for relevant publications. Instead, it is intended to be a textbook for graduate and possibly upper-level undergraduate students. It is to provide a teaching tool for an instructor to teach and a learning aid for a beginner to learn (what we believe) the most fundamental ideas and approaches, the basic concepts, principles, and methodologies of micromechanics. To this end, a unified mathematical framework is introduced early on in the book. The rest of the theories will be developed based on this framework in a logical and easily understandable approach. In addition to some new results from the authors' own research, many of the available results in the literature will be derived or re-derived based on this unified mathematical framework. This approach enables the students to follow the various developments of the micromechanics theories. It also helps the students to quickly comprehend and appreciate the wide range of applications of micromechanics.

1.3 ORGANIZATION OF BOOK

The book is organized into 13 chapters. References and/or Suggested Readings are included at the end of each chapter. Some of these references contain certain results not derived but used in the text. Others are listed because they present either alternative approaches for the same problem or provide additional topics related to those discussed in the chapter. We have made great effort to make the book somewhat self-contained. The goal was that a reader with the basic knowledge of continuum mechanics should be able to follow this book without consulting other publications.

Each chapter also contains a set of problems. The students, as well as the instructor, may find these exercises useful. The level of difficulty varies significantly among the problems. The students should not feel discouraged if they cannot solve some of the problems on their first attempt.

The reminder of this chapter presents the most frequently used notations and notation conventions used in this book. In the next chapter, a brief summary of the basic theories of continuum mechanics is presented. The rest of the chapters are grouped into linear theories (Chapters 3–10) and nonlinear theories (Chapter 11–13).

1.4 NOTATION CONVENTIONS

One of the difficulties many people encounter in studying micromechanics is the different kinds of notations used in the literature. For consistency, we will use the following conventions throughout this book, unless otherwise noted.

Index notation for vectors and tensors will be used extensively. Whenever possible, the base letter for a vector (first-order tensor) will be a lowercase italic letter, for a second-order tensor it will be a lowercase Greek letter, and for a fourth-order tensor it will be an uppercase italic letter. For example, u_i represents a vector, ε_{ij} represents a second-order tensor, and L_{ijkl} represents a fourth-order tensor. Exceptions to these rules are certain letters conventionally used for specific physical entities. For example, G_{ij} is for the Green function, which is a second-order tensor.

The summary convention will be used:

$$L_{ijkl}\varepsilon_{kl} \equiv \sum_{l=1}^{3}\sum_{k=1}^{3} L_{ijkl}\varepsilon_{kl}. \tag{1.4.1}$$

Alternately, when it is convenient, the direct (or matrix) notation of vectors and tensors will be used as well. Boldface letters will be used for this purpose. For example, $\mathbf{u}$ represents a vector, $\boldsymbol{\varepsilon}$ represents a second-order tensor, and $\mathbf{L}$ represents a fourth-order tensor. To distinguish tensors with the same base letter a subscript will be used. For example, $\mathbf{L}_1$ and $\mathbf{L}_2$ are used to denote two different fourth-order tensors. Note that these subscripts are entirely different from the subscripts used to represent the components of a tensor. Subscripts on matrix notations do not follow the summation convention, therefore, when a tensor in the matrix notation with a subscript is written in index forms, we will change the subscript to superscript with parenthesis, for example,

$$\mathbf{L}_1 \Leftrightarrow L_{ijkl}^{(1)} \quad \text{and} \quad \mathbf{L}_2 \Leftrightarrow L_{ijkl}^{(2)}.$$

Throughout the book, index and matrix notations will be used interchangeably based on whichever is convenient.

To avoid confusion, we use the following notations to represent the tensor algebraic operations.

Dot product:	$\sigma_{ij}n_j \Leftrightarrow \boldsymbol{\sigma} \cdot \mathbf{n}$	
Double-dot product:	$L_{ijkl}\varepsilon_{kl} \Leftrightarrow \mathbf{L}:\boldsymbol{\varepsilon}$,	$L_{ijkl}T_{klmn} \Leftrightarrow \mathbf{L}:\mathbf{T}$
Dyad:	$m_in_j \Leftrightarrow \mathbf{m} \otimes \mathbf{n}$,	$n_im_j \Leftrightarrow \mathbf{n} \otimes \mathbf{m}$

Since dot and double-dot operations will be used extensively, we will, when there is no ambiguity, neglect the dot(s) and simply write, for example, $\mathbf{L}:\boldsymbol{\varepsilon} \Leftrightarrow \mathbf{L}\boldsymbol{\varepsilon}$ and $\sigma \cdot \mathbf{n} \Leftrightarrow \sigma\mathbf{n}$.

A fourth-order tensor, $\mathbf{A}$, is nonsingular if and only there exists a fourth-order tensor, for example, $\mathbf{B}$, such that

$$\mathbf{AB} = \mathbf{BA} = \mathbf{I}. \tag{1.4.2}$$

In this case, $\mathbf{B}$ is the inverse of $\mathbf{A}$, or $\mathbf{A}$ is the inverse of $\mathbf{B}$, that is,

$$\mathbf{B} = \mathbf{A}^{-1} \quad \text{or} \quad \mathbf{A} = \mathbf{B}^{-1}. \tag{1.4.3}$$

An equivalent definition can be given as follows: $\mathbf{A}$ is singular if and only if there exists a second-order tensor $\sigma \neq \mathbf{0}$, such that $\mathbf{A}\sigma = \mathbf{0}$.

A fourth-order isotropic tensor can be written as

$$A_{ijkl} = a\delta_{ij}\delta_{kl} + b(\delta_{ik}\delta_{jl} + \delta_{il}\delta_{jk} - \tfrac{2}{3}\delta_{ij}\delta_{kl}). \tag{1.4.4}$$

For convenience, we introduce the following two fourth order tensors:

$$I^h_{ijkl} = \tfrac{1}{3}\delta_{ij}\delta_{kl}, \qquad I^d_{ijkl} = \tfrac{1}{2}(\delta_{ik}\delta_{jl} + \delta_{il}\delta_{jk} - \tfrac{2}{3}\delta_{ij}\delta_{kl}), \tag{1.4.5}$$

so that the fourth-order isotropic tensor given in (1.4.4) can be written as

$$\mathbf{A} = 3a\mathbf{I}^h + 2b\mathbf{I}^d. \tag{1.4.6}$$

It can be easily shown that, if $\boldsymbol{\sigma}$ is a second-order tensor, then

$$\mathbf{A}\boldsymbol{\sigma} = 3a\sigma\mathbf{I} + 2b\boldsymbol{\sigma}', \tag{1.4.7}$$

where $\sigma = \sigma_{kk}$ is the spherical part of $\boldsymbol{\sigma}$, and

$$\sigma'_{ij} = \sigma_{ij} - \tfrac{1}{3}\sigma_{kk}\delta_{ij} \tag{1.4.8}$$

is the deviatoric part of $\boldsymbol{\sigma}$. Furthermore, the following statements can be easily proven:

1. $\mathbf{A} = 3a\mathbf{I}^h + 2b\mathbf{I}^d$ is positive definite if and only if $a > 0$, $b > 0$.
2. $\mathbf{A}^{-1} = \dfrac{1}{3a}\mathbf{I}^h + \dfrac{1}{2b}\mathbf{I}^d$.
3. $\mathbf{I} = \mathbf{I}^h + \mathbf{I}^d$ is the identity tensor, that is, $\mathbf{I} = \mathbf{A}\mathbf{A}^{-1} = \mathbf{A}^{-1}\mathbf{A}$.
4. If $\mathbf{B} = 3c\mathbf{I}^h + 2d\mathbf{I}^d$, then $\mathbf{A} + \mathbf{B} = 3(a + c)\mathbf{I}^h + 2(b + d)\mathbf{I}^d$.
5. $\mathbf{AB} = \mathbf{BA} = 9ac\mathbf{I}^h + 4bd\mathbf{I}^d$.

Additionally, we introduce another symbolic notation for fourth-order isotropic tensors:

$$\mathbf{A} = (3a, 2b). \tag{1.4.9}$$

The following statements can be easily proven:

1. $\mathbf{A} = (3a, 2b)$ is positive definite if and only if $a > 0$, $b > 0$.
2. $\mathbf{A}^{-1} = \left(\dfrac{1}{3a}, \dfrac{1}{2b}\right)$.
3. $\mathbf{I} = (1,1)$ is the identity tensor, that is, $\mathbf{I} = \mathbf{A}\mathbf{A}^{-1} = \mathbf{A}^{-1}\mathbf{A} = (1,1)$.
4. If $\mathbf{B} = (3c, 2d)$, then $\mathbf{A} + \mathbf{B} = (3a + 3c, 2b + 2d)$.
5. $\mathbf{AB} = \mathbf{BA} = (9ac, 4bd)$.

For tensor calculus, the most often used operations are the gradient and divergence. The gradient of a scalar function results in a first-order tensor:

$$\mathbf{b} = \nabla f \Leftrightarrow b_j = \frac{\partial f}{\partial x_j} \equiv f_{,j}. \tag{1.4.10}$$

The divergence of a vector is a scalar:

$$a = \text{div}[\mathbf{b}] = \frac{\partial b_1}{\partial x_1} + \frac{\partial b_2}{\partial x_2} + \frac{\partial b_3}{\partial x_3} \Leftrightarrow a = b_{j,j}. \tag{1.4.11}$$

In the above, the notation $(\bullet)_{,j}$ indicates the derivative with respect to the independent variable of the function. If the function has more than one independent variable, ambiguity may arise. In this case, instead of using, for example, $f_{,j}(x, y)$, we will use the standard notation to explicitly indicate partial derivatives, for example,

$$\frac{\partial f(x, y)}{\partial x}. \tag{1.4.12}$$

The divergence of a second-order tensor becomes a vector:

$$\mathbf{p} = \nabla \cdot \boldsymbol{\sigma} \Leftrightarrow p_i = \frac{\partial \sigma_{ij}}{\partial x_j} = \sigma_{ij,j}. \tag{1.4.13}$$

For integrals, a single integral sign will be used. For line (one-dimensional) integrals, the integration variable will be used for the infinitesimal line element, for example,

$$y = \int_L f(x)\, dx,$$

where L is the line of integration. For surface (two-dimensional) integrals, we will typically use dS for the infinitesimal area element, for example,

$$y = \int_S f(\mathbf{x})\, dS,$$

where S is the area of integration. Here, it is implied that $\mathbf{x}$ is the integration variable since the integrand depends on $\mathbf{x}$ only. If the integrand depends on more than one independent variable, we will explicitly indicate which variable is being integrated, for example,

$$y(\mathbf{z}) = \int_S f(\mathbf{x}, \mathbf{z})\, dS(\mathbf{x}).$$

Similarly, for volume (three-dimensional) integrals, we will use either one of the following two forms:

$$y = \int_V f(\mathbf{x})\, dV, \qquad y(\mathbf{z}) = \int_V f(\mathbf{x}, \mathbf{z})\, dV(\mathbf{x}).$$

Finally, we introduce some special tensors. The Kronecker delta δ_{ij} is defined by

$$\delta_{ij} = \begin{cases} 1 & \text{for } i = j \\ 0 & \text{for } i \neq j \end{cases} \tag{1.4.14}$$

And the permutation tensor ε_{ijk} is defined by

$$\varepsilon_{ijk} = \begin{cases} 1 & \text{when } i, j, k \text{ are an even permutation of } 1, 2, 3 \\ -1 & \text{when } i, j, k \text{ are an odd permutation of } 1, 2, 3 \; . \\ 0 & \text{when any two indices are equal} \end{cases} \tag{1.4.15}$$

For example, $\varepsilon_{123} = \varepsilon_{231} = \varepsilon_{312} = 1$, $\varepsilon_{213} = \varepsilon_{132} = \varepsilon_{321} = -1$, and others are zero.

The permutation tensor and the Kronecker delta δ_{ij} are related through the $\varepsilon - \delta$ relationship:

$$\varepsilon_{ijk}\varepsilon_{mnk} = \delta_{im}\delta_{jn} - \delta_{in}\delta_{jm}. \tag{1.4.16}$$

REFERENCES

Eshelby, J. D. (1957). The Determination of Elastic Field of an Ellipsoidal Inclusion, and Related Problems, *Proc. Roy. Soc. London,* Vol. A241, pp. 376–396.

Krajcinovic, D. (1996). *Damage Mechanics,* North-Holland, New York.

Mura, T. (1987). *Micromechanics of Defects in Solids,* Martinus Nijhoff Pub., Boston.

Nemat-Nasser, S. and M. Hori (1993). *Micromechanics: Overall Properties of Heterogeneous Materials,* North-Holland, New York.

2

BASIC EQUATIONS OF CONTINUUM MECHANICS

In this chapter, we will introduce some basic equations of continuum mechanics. It is assumed that the readers already have had some prior knowledge of the subject. Therefore, lengthly derivations are omitted. Only equations that are needed later for deriving the micromechanics theories are outlined. For more detailed and in-depth knowledge of continuum mechanics, we refer to the References and Suggested Readings at the end of this chapter.

The fundamental equations governing the deformation of continuum can be roughly classified into four categories: (1) kinematic equations that deal with the description of motion and deformation, (2) kinetic equations that deal with forces, stresses, and their equilibrium, (3) constitutive equations that describe the material's response under load, and (4) boundary and initial conditions that are necessary to uniquely define the boundary/initial value problems. We will discuss these classes of equations separately.

2.1 DISPLACEMENT AND DEFORMATION

At any given instant of time t, a continuum having a volume V and bounding surface S will occupy a certain region R of the physical space; see Figure 2.1. When the continuum undergoes deformation, the particles (material points) within the continuum may be displaced from one spatial position to another along various paths. The displacement of a particle can be written as

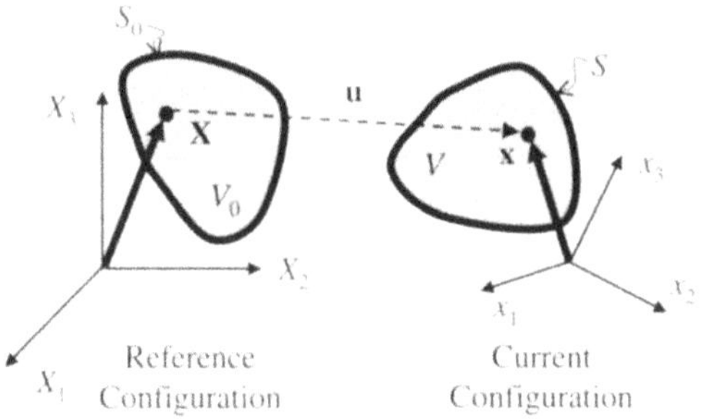

Figure 2.1 Reference and current configurations.

$$\mathbf{u} = \mathbf{x} - \mathbf{X}, \tag{2.1.1}$$

where $\mathbf{x}$ is the position occupied at current time t by the particle that occupied position $\mathbf{X}$ in the initial configuration ($t = 0$). In some literature, $\mathbf{x}$ is referred to as the spatial (Eulerian) coordinate, and $\mathbf{X}$ is referred to as the material coordinate. This is because $\mathbf{X}$ can be viewed as a label attached to a material particle while $\mathbf{x}$ indicates where the particle is located at a give time.

The current position $\mathbf{x}$ of a particle originally located at $\mathbf{X}$ in the initial configuration can be expressed by

$$x_i = x_i(\mathbf{X}, t). \tag{2.1.2}$$

Clearly, the above equation provides a way to find the new location of each particle after the deformation. This way of describing the deformation is called the Lagrangian description. On the other hand, the Eulerian description uses the following equation:

$$X_I = X_I(\mathbf{x}, t) \tag{2.1.3}$$

to keep track of which particle happens to be at the location x_i at the current time t. For clarity, capital letters are used for the subscript of a tensor component associated with the Lagrangian coordinate system (the initial configuration), and the lowercase letters are used for the subscript of a tensor component associated with the Eulerian coordinate system (the current or deformed configuration). Such a distinction is unnecessary if both the Lagrangian and Eulerian configurations use the same coordinate system.

Equations (2.1.2)–(2.1.3) may be interpreted as a mapping between the initial configuration and the current configuration. The continuum mechanics theory assumes that such mapping is sufficiently smooth (differentiable) and one to one, that is, the Jacobian $J = |\partial x_i/\partial X_J|$ should not become zero at any time.

Partial differentiation of (2.1.2) with respect to X_J and (2.1.3) with respect to x_j yield, respectively,

$$dx_i = F_{iJ}\, dX_J, \qquad dX_I = F_{Ij}^{-1}\, dx_j, \tag{2.1.4}$$

where

$$F_{iJ} = \frac{\partial x_i}{\partial X_J} \quad \text{and} \quad F_{Ij}^{-1} = \frac{\partial X_I}{\partial x_j} \tag{2.1.5}$$

are called the material deformation gradient tensor and the spatial deformation gradient tensor, respectively.

Through the polar decomposition (Malvern, 1969), the material deformation gradient tensor can be written as

$$F_{iJ} = R_{iK} U_{KJ} = V_{ik} R_{kJ} \tag{2.1.6}$$

where R_{ij} is the (orthogonal) rotation tensor, and U_{ij} and V_{ij} are symmetric, positive definite tensors known as the right- and left-stretch tensors, respectively.

Based on their definitions, a length segment dX_I connecting two neighboring particles in the undeformed configuration becomes a length segment dx_i connecting the same two particles now in the deformed configuration. The difference between the lengths (square) of these two segments, $dx_i\, dx_i - dX_I\, dX_I$, can be used as a measure of deformation occurring in the neighborhood of these particles when the continuum is deformed from the initial configuration to the current configuration, that is,

$$dx_i\, dx_i - dX_I\, dX_I = 2E_{IJ}\, dX_I\, dX_J = 2E^*_{ij}\, dx_i\, dx_j, \tag{2.1.7}$$

where

$$E_{IJ} = \frac{1}{2}\left(\frac{\partial x_k}{\partial X_I}\frac{\partial x_k}{\partial X_J} - \delta_{IJ}\right) = \frac{1}{2}\left(\frac{\partial u_i}{\partial X_J} + \frac{\partial u_j}{\partial X_I} + \frac{\partial u_k}{\partial X_I}\frac{\partial u_k}{\partial X_J}\right) \tag{2.1.8}$$

$$E^*_{ij} = \frac{1}{2}\left(\delta_{ij} - \frac{\partial X_K}{\partial x_i}\frac{\partial X_K}{\partial x_j}\right) = \frac{1}{2}\left(\frac{\partial u_I}{\partial x_j} + \frac{\partial u_J}{\partial x_i} - \frac{\partial u_K}{\partial x_i}\frac{\partial u_K}{\partial x_j}\right) \tag{2.1.9}$$

are called the Lagrangian (Green) finite strain tensor and the Eulerian (or Cauchy) finite strain tensor, respectively.

When the same coordinate system is used for both the initial reference configuration and the current configurations, the strain tensors can be further written as

$$E_{ij} = \frac{1}{2}\left(\frac{\partial u_i}{\partial X_i} + \frac{\partial u_j}{\partial X_i} + \frac{\partial u_k}{\partial X_i}\frac{\partial u_k}{\partial X_j}\right), \qquad E^*_{ij} = \frac{1}{2}\left(\frac{\partial u_i}{\partial x_i} + \frac{\partial u_j}{\partial x_i} - \frac{\partial u_k}{\partial x_i}\frac{\partial u_k}{\partial x_j}\right). \tag{2.1.10}$$

When the displacement gradients are small, that is, $|\partial u_i/\partial X_j| << 1$ and $|\partial u_i/\partial x_j| << 1$, neglecting the higher order terms leads to

$$\frac{\partial u_i}{\partial x_j} = \frac{\partial u_i}{\partial X_k}\frac{\partial X_k}{\partial x_j} = \frac{\partial u_i}{\partial X_k}\left(\delta_{kj} - \frac{\partial X_k}{\partial x_j}\right) = \frac{\partial u_i}{\partial X_j}. \tag{2.1.11}$$

Thus one has, from (2.1.8) and (2.1.9), that

$$E_{ij} = \frac{1}{2}\left(\frac{\partial u_i}{\partial X_i} + \frac{\partial u_j}{\partial X_i}\right) = E^*_{ij} = \frac{1}{2}\left(\frac{\partial u_i}{\partial x_i} + \frac{\partial u_j}{\partial x_i}\right) = \varepsilon_{ij}, \tag{2.1.12}$$

where ε_{ij} is called the infinitesimal strain tensor.

Other kinematic tensors often used in viscoelastic and plastic theories are the rate of deformation tensor and the spin tensor:

$$D_{ij} = \frac{1}{2}\left(\frac{\partial v_i}{\partial x_j} + \frac{\partial v_j}{\partial x_i}\right), \qquad W_{ij} = \frac{1}{2}\left(\frac{\partial v_i}{\partial x_j} - \frac{\partial v_j}{\partial x_i}\right), \tag{2.1.13}$$

where $v_i(x_1, x_2, x_3, t)$ is the velocity field

$$v_i = \frac{du_i}{dt} = \frac{dx_i}{dt} \tag{2.1.14}$$

expressed in terms of the Eulerian (spatial) coordinates x_i.

For small strain deformation,

$$\frac{d\varepsilon_{ij}}{dt} = \frac{d}{dt}\left[\frac{1}{2}\left(\frac{\partial u_i}{\partial x_j} + \frac{\partial u_j}{\partial x_i}\right)\right] = \frac{1}{2}\left[\frac{\partial(du_i/dt)}{\partial x_i} + \frac{\partial(du_j/dt)}{\partial x_i}\right]$$
$$= \frac{1}{2}\left(\frac{\partial v_i}{\partial x_j} + \frac{\partial v_j}{\partial x_i}\right),$$

that is,

$$\frac{d\varepsilon_{ij}}{dt} = D_{ij}. \tag{2.1.15}$$

This is no longer true for finite deformation. Instead, we have (Malvern, 1969)

$$\frac{dE_{IJ}}{dt} = F_{mI} D_{mn} F_{nJ}. \tag{2.1.16}$$

2.2 STRESSES AND EQUILIBRIUM

The state of stress at a point in a continuum can be represented by the Cauchy stress tensor σ_{ij} in the current (or deformed) configuration. Note that the components of the Cauchy stress tensor are defined as force per unit area in the deformed configuration. Balance of the moment of momentum (Cauchy's second law of motion) dictates the symmetry of the Cauchy stress tensor, that is, $\sigma_{ij} = \sigma_{ji}$. Furthermore, in the deformed configuration, Cauchy's first law of motion yields the equilibrium equation

$$\frac{\partial \sigma_{ji}}{\partial x_j} + f_i = 0 \quad \text{or} \quad \nabla \cdot \boldsymbol{\sigma} + \mathbf{f} = 0, \tag{2.2.1}$$

where f_i is the body force per unit volume in the deformed configuration.

Equation (2.2.1) is valid for any material point within a continuum body. For a material point on the surface (boundary) of the a continuum (see Fig. 2.2), the following Cauchy formula applies:

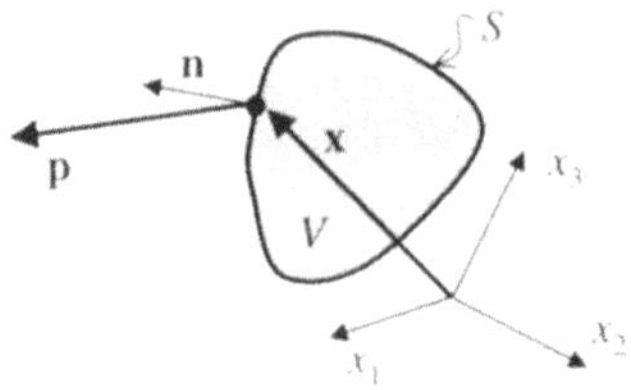

Figure 2.2 Cauchy formula on the deformed (current) surface.

$$\sigma_{ij}n_j|_S = p_i \quad \text{or} \quad \boldsymbol{\sigma} \cdot \mathbf{n}|_S = \mathbf{p}, \tag{2.2.2}$$

where, as shown in Figure 2.2, n_i is the unit outward normal vector of the surface S and p_i is the traction vector applied on S. Note that p_i is measured as force per unit deformed area.

Note that (2.2.1) describes the equilibrium of an infinitesimal material element in the deformed configuration. The Cauchy stress tensor is typically written as a function of the Eulerian spatial coordinates x_i. To describe the equilibrium in the initial (undeformed) configuration, the Piola–Kirchhoff stress tensors may be introduced.

The first Piola–Kirchhoff stress tensor, σ^0_{Ij}, is defined as the actual force in the deformed configuration per unit undeformed area in the undeformed configuration, while the second Piola–Kirchhoff stress tensor, $\tilde{\sigma}_{IJ}$, is defined as a fictitious force in the undeformed configuration per unit undeformed area in the undeformed configuration. The fictitious force here is obtained by transforming the actual force in the deformed configuration back to the undeformed configuration, which is equivalent to premultiplying the actual force in the deformed configuration by the spatial deformation gradient tensor, F^{-1}_{Ij}. Hence, the two Piola–Kirchhoff stresses are related by the deformation gradient tensors:

$$\tilde{\sigma}_{IJ} = F^{-1}_{Jk}\sigma^0_{Ik} \quad \text{or} \quad \sigma^0_{Ij} = \tilde{\sigma}_{IK}F_{jK}. \tag{2.2.3}$$

The Piola–Kirchhoff stresses are typically written as functions of the Lagrangian material coordinates X_i. They can be related to the Cauchy stress tensor by

$$\sigma^0_{Ij} = JF^{-1}_{Ik}\sigma_{kj}, \qquad \tilde{\sigma}_{IJ} = JF^{-1}_{Ik}F^{-1}_{Jm}\sigma_{km} \tag{2.2.4}$$

or

$$\sigma_{ij} = \frac{1}{J} F_{iK} F_{jM} = \tilde{\sigma}_{KM} = \frac{1}{J} F_{iK} \sigma^0_{Kj}, \tag{2.2.5}$$

where J is the Jacobian given by $J = |\partial x_i / \partial X_j| = \rho_0 / \rho$, and ρ_0 and ρ are the mass density measured in the initial reference state and the current deformed state, respectively. It is seen from (2.2.4) that $\tilde{\sigma}_{IJ}$ is symmetric while σ^0_{Ij} is generally nonsymmetric. It is important to keep in mind that the Piola–Kirchhoff stress tensors are not actual stresses, that is, they do not exist in the continuum under consideration. They are introduced to simplify the equilibrium equation in the initial configuration, namely,

$$\frac{\partial \sigma^0_{Ji}}{\partial X_J} + f_0 = 0 \quad \text{or} \quad \frac{\partial (\tilde{\sigma}_{JK} F_{iK})}{\partial X_J} + f_0 = 0, \tag{2.2.6}$$

where f_0 is the body force per unit volume in the initial configuration.

The Cauchy formula given by (2.2.2) can also be expressed in the initial configuration,

$$\sigma^0_{Ji} N_J = F_{iJ} \tilde{p}_J, \qquad \tilde{\sigma}_{JI} N_j = F^{-1}_{Ij} \, p^0_j, \tag{2.2.7}$$

where

$$p^0_j = \frac{dS}{dS_0} p_j, \qquad \tilde{p}_J = \frac{dS}{dS_0} F^{-1}_{Jk} \, p_k, \tag{2.2.8}$$

and dS_0 is an area element in the initial configuration, and dS is the same area element in the deformed configuration. Note that p_j is the force acting on the deformed surface per unit deformed area. Therefore, p^0_j is the force acting on the deformed surface per unit undeformed area, while $\tilde{p}_J$ is a fictitious force acting on the deformed surface per unit undeformed area. This fictitious force is the actual force acting on the deformed surface transformed by F^{-1}_{Jk}. Neither of p^0_j and $\tilde{p}_J$ is a real traction vector. They are called pseudotraction vectors introduced to accommodate the pseudostress tensors.

For small deformation, that is, $|\partial u_i / \partial X_J| << 1$ and $|\partial u_i / \partial x_j| << 1$, one can show from (2.1.5) that when the same coordinate system is used for both the Lagrangian and Eulerian configurations,

$$F_{ij} \approx \delta_{ij} + \frac{\partial u_i}{\partial X_j} \quad \text{and} \quad J = \left| \frac{\partial x_i}{\partial X_j} \right| \approx 1 + \frac{\partial u_i}{\partial X_i}. \tag{2.2.9}$$

Thus, it follows from (2.2.4) that Piola–Kirchhoff stresses reduce to the Cauchy stress,

$$\sigma_{ij} = \tilde{\sigma}_{ij} = \sigma^0_{ij} \tag{2.2.10}$$

for small deformation. Furthermore, the distinction between n_i and N_i and between dS and dS_0 can also be neglected. Thus, the pseudotraction vectors reduce to the Cauchy traction vector as well.

2.3 ENERGY, WORK, AND THERMODYNAMIC POTENTIALS

We first review some of the integral identities from mathematical physics. Consider region D bounded by the surface S. The outward unit normal vector of S is denoted by n_i; see Figure 2.3.

Let $f(x_1, x_2, x_3)$ be a continuously differentiable scalar function defined in D. The *Green's theorem* states that

$$\int_D \frac{\partial f}{\partial x_i}\, dV = \int_S f n_i\, dS. \tag{2.3.1}$$

Let $u_i(x_1, x_2, x_3)$ be a continuously differentiable vector function defined in D. The *divergence theorem* states that

$$\int_D \frac{\partial u_i}{\partial x_i}\, dV = \int_S u_i n_i\, dS \quad \text{or} \quad \int_D \nabla \mathbf{u}\, dV = \int_S \mathbf{u} \cdot \mathbf{n}\, dS \tag{2.3.2}$$

Next consider a simple, closed, and smooth curve L. Let S be any smooth surface spanned across the closed curve. The tangential direction of the curve and the normal direction of the surface are denoted

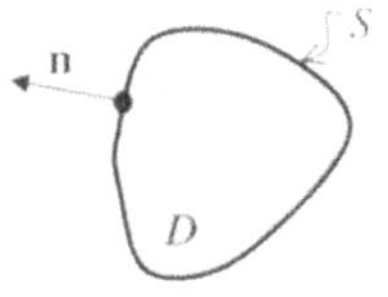

Figure 2.3 Region D bounded by the surface S.

by t_i and n_i, respectively. To get the positive orientation of L think of yourself as walking along the curve. While you are walking along the curve if your head is pointing in the same direction as the unit normal vectors of the surface while the surface is on the left, then you are walking in the positive direction on L; see Figure 2.4. Now, let $u_i(x_1, x_2, x_3)$ be a continuously differentiable scalar function defined on the surface S. The Stokes' theorem states that

$$\int_S \varepsilon_{ijk} u_{k,j} n_i \, dS = \int_L u_i t_i \, dL \quad \text{or} \quad \int_S (\nabla \times \mathbf{u}) \bullet \mathbf{n} \, dS = \int_L \mathbf{u} \bullet \mathbf{t} \, dL, \tag{2.3.3}$$

where ε_{ijk} is the permutation symbol.

Note that (2.3.1)–(2.3.3) are mathematical identities. They are true regardless of the physical interpretation of the functions involved. Nevertheless, when properly interpreted, these integral identifies may be used to state the conservation of certain physical entities.

Now, let us consider the energy of a continuum. Generally speaking, the total energy of a continuum is comprised of two parts, the kinetic energy and the internal energy. The kinetic energy is associated with the macroscopically observable motion of the continuum. Usually, the kinetic energy of the random thermal agitation of the atoms is considered part of the internal energy associated with the temperature of the continuum. The other part of the internal energy may include stored strain energy and possibly other forms of energy, such as chemical, electrical, and optical energies.

Let ρ be the mass density and ϕ be the internal energy per unit mass or specific internal energy for a continuum in the current (deformed) configuration. The first law of thermodynamics in conjunction with the divergence theorem leads to the following energy equation (Malvern, 1969):

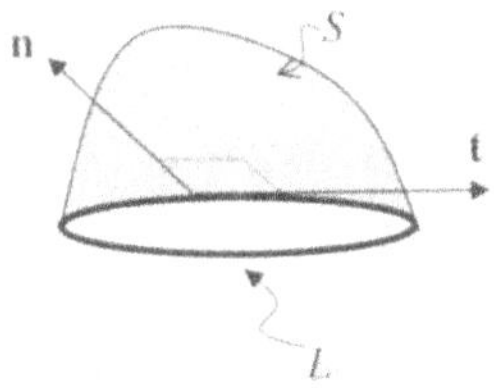

Figure 2.4 Stokes' theorem.

$$\rho \frac{d\phi}{dt} = \sigma_{ij} D_{ij} + \rho r - \frac{\partial q_i}{\partial x_i}, \tag{2.3.4}$$

where σ_{ij} is the Cauchy stress tensor, D_{ij} is the rate of deformation tensor, r is the distributed internal heat source per unit mass, and q_i is the (outward) heat flux vector. The first term on the right-hand side of (2.3.4) is called stress power. It represents the power input per unit deformed volume provided by the stress field. The last two terms of the right-hand side represent the total heat input from internal heat source and external heat flux. Together, the right-hand side of (2.3.4) gives the total energy input, while the left-hand side shows the time rate of change of internal energy. Therefore, (2.3.4) is a mathematical statement of conservation of energy.

The quantity of particular interest to us is the stress power, $\sigma_{ij} D_{ij}$. In a volume V of the deformed configuration, the stress power is given by

$$\int_V \sigma_{ij} D_{ij}\, dV. \tag{2.3.5}$$

This can be converted to the initial reference configuration in terms of the Piola–Kirchhoff stress tensors:

$$\int_V \sigma_{ij} D_{ij}\, dV = \int_{V_0} \sigma^0_{Ij} \frac{dF_{jI}}{dt}\, dV_0 = \int_{V_0} \tilde{\sigma}_{IJ} \frac{dE_{IJ}}{dt}\, dV_0. \tag{2.3.6}$$

For this reason, we call (σ_{ij}, D_{ij}) a conjugate pair in the current deformed configuration, and $(\sigma^0_{Ij}, dF_{jI}/dt)$ and $(\tilde{\sigma}_{IJ}, dE_{IJ}/dt)$ conjugate pairs in the initial reference configuration. In computing the energy, we must use variables that form a conjugate pair in the same configuration.

We are now ready to introduce the thermodynamic potentials. The fundamental assumption here is that the internal energy per unit mass of a continuum can be uniquely determined by the specific entropy s, and some state variables v_i, that is,

$$\phi = \phi(s, v_1, v_2, \ldots, v_n). \tag{2.3.7}$$

This is called the caloric equation of state, and ϕ is, therefore, a thermodynamic potential. From here, one can define the thermodynamic temperature and thermodynamic forces, respectively:

$$T = \left(\frac{\partial \phi}{\partial s}\right)_{v_i} \quad \text{and} \quad \tau_i = \left(\frac{\partial \phi}{\partial v_i}\right)_s. \tag{2.3.8}$$

These yield

$$d\phi = \left(\frac{\partial \phi}{\partial v_i}\right)_s dv_i + \left(\frac{\partial \phi}{\partial s}\right)_{v_i} ds = \tau_i \, dv_i + T \, ds. \tag{2.3.9}$$

For adiabatic ($\rho r - q_{i,i} = 0$) and isentropic ($ds = 0$) deformation, and for isothermal ($dT = 0$) deformation with reversible heat transfer ($\rho r - q_{i,i} = \rho T \, ds/dt$), the work of the thermodynamic forces is recoverable, and the rate of work of the thermodynamic forces equals the stress power:

$$\rho \tau_i \frac{dv_i}{dt} = \sigma_{ij} D_{ij}. \tag{2.3.10}$$

Thus, in these cases, the rate of change of the internal energy is given by

$$\rho \frac{d\phi}{dt} = \sigma_{ij} D_{ij} + \rho T \frac{ds}{dt}. \tag{2.3.11}$$

In the reference configuration, this takes the form

$$\frac{d\phi}{dt} = \frac{1}{\rho_0} \tilde{\sigma}_{IJ} \frac{dE_{IJ}}{dt} + T \frac{ds}{dt}. \tag{2.3.12}$$

Other thermodynamic potentials used often in continuum mechanics are:

Enthalpy: $$\eta = \phi - \tau_i v_i. \tag{2.3.13}$$

Helmholtz free energy: $$h = \phi - sT. \tag{2.3.14}$$

Gibbs free energy: $$g = \phi - sT - \tau_i v_i. \tag{2.3.15}$$

2.4 CONSTITUTIVE LAWS

Constitutive laws are used to describe material behavior when subjected to applied thermomechanical loads. One of the major areas of micromechanics is to develop constitutive laws for heterogeneous materials under various thermomechanical loading conditions. For real materials, their thermomechanical behavior can be rather complex. It is usually not feasible to write down one or a set of equations to describe the entire range of material behavior. Instead, we formulate separately constitutive equations describing various kinds of idealized material response, each of which is a mathematical formulation designed to approximate physical observations of a real material's response over a suitably restricted range. Specifically, constitutive laws for ideal materials are prescribed by equations that establish the relationship between deformation and the internal stresses in the material. For example,

$$G(\sigma_{ij}, D_{ij}) = 0 \tag{2.4.1}$$

describes a relationship between the Cauchy stress tensor and the rate of deformation tensor. In what follows, we will describe some commonly used constitutive laws governing certain idealized material behavior.

Elasticity

If there exists a potential function W of strain for a given material such that the stress in the material is related to the strain by

$$\tilde{\sigma}_{IJ} = \frac{\partial W}{\partial E_{IJ}}, \tag{2.4.2}$$

then the material is called an elastic (or hyperelastic) material, and such material behavior is called elasticity. Equation (2.4.2) is the constitutive equation for elastic materials. For elastic material under isothermal and isentropic conditions, the Helmholtz free energy can be taken as the elastic potential $W = \rho_0 h$, that is,

$$\tilde{\sigma}_{IJ} = \frac{\partial W}{\partial E_{IJ}} = \rho_0 \frac{\partial h}{\partial E_{IJ}} = \rho_0 \frac{\partial \phi}{\partial E_{IJ}}. \tag{2.4.3}$$

Thus, $\tilde{\sigma}_{IJ}/\rho_0$ is the thermodynamic force conjugate to E_{IJ}.

Another consequence of the elastic potential is that the state of deformation is uniquely determined by the state variables. In the isothermal case, the state variables are the strain components. The stress field corresponding to a strain field is unique regardless how that state of strain is arrived. In other words, the deformation is history independent. Consequently, upon unloading, an elastic material recovers its original state or shape and size.

Since W is assumed to be a function of strain only, it can also be interpreted as the strain energy per unit undeformed volume. In the initial reference configuration, the Taylor expansion of W can be written as

$$W = W_0 + c_{IJ}E_{IJ} + \tfrac{1}{2} L_{IJKL}E_{IJ}E_{KL} + \tfrac{1}{6} c_{IJKLMN}E_{IJ}E_{KL}E_{MN} + \cdots, \tag{2.4.4}$$

where, traditionally, L_{IJKL} are called the second-order elastic constants and c_{IJKLMN} are called the third-order elastic constants. The constants W_0 and c_{IJ} are typically zero unless, for example, when residual stress exists in the initial reference configuration.

A material is said to behave linear elastically if the higher order terms in (2.4.4) can be neglected and the stresses are linearly related to the strains. In the undeformed configuration, the constitutive equation for linear elastic materials can be written in terms of the second Piola–Kirchhoff stress tensor and the Lagrangian finite strain tensor,

$$\tilde{\sigma}_{IJ} = L_{IJKL}E_{KL} \quad \text{or} \quad E_{IJ} = M_{IJKL}\tilde{\sigma}_{KL}, \tag{2.4.5}$$

where the fourth-order tensor L_{IJKL} is called the elasticity (or stiffness) tensor, and M_{IJKL} is the compliance tensor, which is the inverse of the stiffness tensor, that is,

$$M_{IJMN}L_{MNKL} = I_{IJKL} \equiv \tfrac{1}{2}(\delta_{IK}\delta_{JL} + \delta_{IL}\delta_{JK}), \tag{2.4.6}$$

where δ_{IJ} is the Kronecker delta:

$$\delta_{IJ} = \begin{cases} 1 & \text{for } I = J \\ 0 & \text{for } I \neq J \end{cases}. \tag{2.4.7}$$

Most engineering materials show linear elastic behavior only when the deformation is very small. In this case, the small-strain constitutive law for linear elastic materials can be written in terms of the Cauchy stress tensor and the infinitesimal strain tensor:

$$\sigma_{ij} = L_{ijkl}\varepsilon_{kl} \quad \text{or} \quad \varepsilon_{ij} = M_{ijkl}\sigma_{kl}. \tag{2.4.8}$$

The above equation is often referred to as the generalized Hooke law.

For linear elastic materials, it is sometime convenient to write the Hooke law in terms of the Voigt elastic constants:

$$\boldsymbol{\sigma} = \mathbf{C}\boldsymbol{\varepsilon}, \tag{2.4.9}$$

where

$$\boldsymbol{\sigma} = \begin{bmatrix} \sigma_{11} \\ \sigma_{22} \\ \sigma_{33} \\ \sigma_{23} \\ \sigma_{31} \\ \sigma_{12} \end{bmatrix}, \quad \boldsymbol{\varepsilon} = \begin{bmatrix} \varepsilon_{11} \\ \varepsilon_{22} \\ \varepsilon_{33} \\ 2\varepsilon_{23} \\ 2\varepsilon_{31} \\ 2\varepsilon_{12} \end{bmatrix}, \tag{2.4.10}$$

and the elements of the 6×6 stiffness matrix $\mathbf{C}$ are called the (Voigt) elastic constants. They are related to the components of stiffness tensor $\mathbf{L}$ through the relationship shown in Table 2.1.

Both the stiffness and compliance tensors are fourth-order tensors. They obey the tensor transformation law,

$$\hat{L}_{ijkl} = \alpha_{im}\alpha_{jn}\alpha_{kp}\alpha_{lq}L_{mnpq}, \qquad \hat{M}_{ijkl} = \alpha_{im}\alpha_{jn}\alpha_{kp}\alpha_{lq}M_{mnpq}, \tag{2.4.11}$$

where $\hat{L}_{ijkl}$ and $\hat{M}_{ijkl}$ are the components of the stiffness and compliance tensors, respectively, in the $\hat{x}_i$ coordinate system, which is related to the original x_i coordinate system by the rotation tensor α_{ij} through

Table 2.1 Relationship between L_{ijkl} and C_{ij}

L_{ijkl}			kl			
ij	11	22	33	23	31	12
11	C_{11}	C_{12}	C_{13}	C_{14}	C_{15}	C_{16}
22	C_{21}	C_{22}	C_{23}	C_{24}	C_{25}	C_{26}
33	C_{31}	C_{32}	C_{33}	C_{34}	C_{35}	C_{36}
23	C_{41}	C_{42}	C_{43}	C_{44}	C_{45}	C_{46}
31	C_{51}	C_{52}	C_{53}	C_{54}	C_{55}	C_{56}
12	C_{61}	C_{62}	C_{63}	C_{64}	C_{65}	C_{66}

$$\hat{x}_i = \alpha_{ij} x_j. \tag{2.4.12}$$

For the Voigt elasticity matrix, if the coordinate transformation is a rotation about the x_3 axis by an angle θ, one can show that (Ting, 1996),

$$\hat{\mathbf{C}} = \mathbf{\Omega C \Omega}^{\mathrm{T}}, \quad \hat{\mathbf{S}} = (\mathbf{\Omega}^{-1})^{\mathrm{T}}\, \mathbf{S \Omega}^{-1}, \tag{2.4.13}$$

where

$$\mathbf{\Omega} = \begin{bmatrix} m^2 & n^2 & 0 & 0 & 0 & 2mn \\ n^2 & m^2 & 0 & 0 & 0 & -2mn \\ 0 & 0 & 1 & 0 & 0 & 0 \\ 0 & 0 & 0 & m & -n & 0 \\ 0 & 0 & 0 & n & m & 0 \\ -mn & mn & 0 & 0 & 0 & m^2 - n^2 \end{bmatrix} \tag{2.4.14}$$

in which $m = \cos\theta$, $n = \sin\theta$.

The stiffness tensor of a linear elastic material is positive definite and possesses the following symmetries:

$$L_{ijkl} = L_{klij} = L_{jikl}. \tag{2.4.15}$$

Because of these symmetries, there are 81 independent components called elastic constants in the most general case. The number of independent components of the stiffness tensor is much less when the material possesses certain symmetries. For example, for materials with cubic symmetry, such as face-center-cubic (FCC) and body-center-cubic (BCC) crystals, the stiffness tensor has only three independent components:

$$L_{ijkl} = \lambda \delta_{ij}\delta_{kl} + \mu(\delta_{ik}\delta_{jl} + \delta_{il}\delta_{jk}) + \chi d_{ijkl}, \tag{2.4.16}$$

where λ, μ, and χ are related to the Voigt constants by

$$\lambda = C_{12}, \qquad \mu = C_{44}, \qquad \chi = C_{11} - C_{12} - 2C_{44}. \tag{2.4.17}$$

The symbol d_{ijkl} is defined by the following:

$$d_{1111} = d_{2222} = d_{3333} = 1, \quad \text{others} = \text{zero}. \tag{2.4.18}$$

For isotropic materials, there are only two independent components in the stiffness tensor, that is,

$$L_{ijkl} = \lambda\delta_{ij}\delta_{kl} + \mu(\delta_{ik}\delta_{jl} + \delta_{il}\delta_{jk}), \tag{2.4.19}$$

where the Lame constants λ and μ are related to the Voigt constants by

$$\lambda = C_{12}, \qquad \mu = C_{44}. \tag{2.4.20}$$

In some cases, it is more convenient to rewrite (2.4.19) as

$$L_{ijkl} = K\delta_{ij}\delta_{kl} + \mu(\delta_{ik}\delta_{jl} + \delta_{il}\delta_{jk} - \tfrac{2}{3}\delta_{ij}\delta_{kl}), \tag{2.4.21}$$

where K and μ are the bulk and shear moduli, respectively. The corresponding compliance tensor is given by

$$M_{ijkl} = \frac{1}{9K}\delta_{ij}\delta_{kl} + \frac{1}{4\mu}(\delta_{ik}\delta_{jl} + \delta_{il}\delta_{jk} - \tfrac{2}{3}\delta_{ij}\delta_{kl}). \tag{2.4.22}$$

Using the symbolic notation (1.4.2), the isotropic stiffness tensor and compliance tensor can be written as

$$\mathbf{L} = (3K, 2\mu), \qquad \mathbf{M} = \left(\frac{1}{3K}, \frac{1}{2\mu}\right). \tag{2.4.23}$$

Other elastic constants such as Young's modulus, Poisson's ratio, the shear modulus, and so forth, are also commonly used for isotropic materials. The relationships among the various elastic constants for isotropic materials are given in Appendix 2.A. The Voigt elastic constants for materials with other types of symmetries are listed in Appendix 2.B.

Thermoelasticity

In addition to mechanical stresses, temperature change may also cause deformation in a continuum. Vice versa, deformation of a continuum may cause the temperature in the continuum to change. Such temperature effects in elastic materials are described by thermoelastic constitutive laws. Some commonly used thermoelastic constitutive laws are presented here.

We assume that, at the temperature T_0, the continuum is stress free in the initial reference state. In the deformed (current) state, let the temperature be T. The Helmholtz free energy is given by (2.3.14)

$$h(E_{IJ}, T) = \phi(E_{IJ}, T) - Ts(E_{IJ}, T), \tag{2.4.24}$$

or

$$\frac{dh}{dt} = \frac{\partial h}{\partial E_{IJ}}\frac{dE_{IJ}}{dt} + \frac{\partial h}{\partial T}\frac{dT}{dt} = \frac{d\phi}{dt} - s\frac{dT}{dt} - T\frac{ds}{dt}. \tag{2.4.25}$$

Making use of (2.3.12) and (2.3.11) in the above, we have

$$\left(s + \frac{\partial h}{\partial T}\right)\frac{dT}{dt} - \left(\frac{1}{\rho_0}\tilde{\sigma}_{IJ} - \frac{\partial h}{\partial E_{IJ}}\right)\frac{dE_{IJ}}{dt} = 0. \tag{2.4.26}$$

Since E_{IJ} and T are independent state variables, satisfaction of the above equation means

$$s = -\frac{\partial h}{\partial T}, \qquad \tilde{\sigma}_{IJ} = \rho_0 \frac{\partial h}{\partial E_{IJ}}. \tag{2.4.27}$$

The difference between the second of (2.4.27) and (2.4.3) is that the Helmholtz free energy in (2.4.27) is a function of both strain and temperature.

The Taylor expansion of the Helmholtz free energy in terms of strains and temperature can be written as

$$\begin{aligned}\rho_0 h = h_0 + c_{IJ}E_{IJ} + \tfrac{1}{2}L_{IJKL}E_{IJ}E_{KL} + \alpha\,\Delta T + \tfrac{1}{2}\beta\,\Delta T^2 \\ - \alpha_{KL}L_{IJKL}E_{IJ}\,\Delta T \cdots .\end{aligned} \tag{2.4.28}$$

For convenience, we have used

$$\Delta T = T - T_0. \tag{2.4.29}$$

It thus follows that

$$s = -\frac{\partial h}{\partial T} = -\frac{1}{\rho_0}(\alpha + \beta\,\Delta T + \alpha_{IJ}E_{IJ}\cdots), \tag{2.4.30}$$

$$\tilde{\sigma}_{IJ} = \rho_0\frac{\partial h}{\partial E_{IJ}} = c_{IJ} + L_{IJKL}E_{KL} - L_{IJKL}\alpha_{KL}\,\Delta T \cdots . \tag{2.4.31}$$

Equation (2.4.31) is the stress–strain relationship in thermoelasticity. The constant term c_{IJ} is usually zero unless initial stresses are present. The fourth-order tensor L_{IJKL} is the second-order elasticity tensor introduced in (2.4.4). The second-order tensor α_{IJ} is called the linear coefficient of thermal expansion. The quantity

$$E^*_{KL} = \alpha_{KL}\, \Delta T \tag{2.4.32}$$

corresponds to the deformation due to temperature change. We will call it thermal strain.

Note that although (2.4.31) provides the stress–strain relationship, it is not a complete description of the constitutive law for thermoelasticity. In thermoelasticity, in addition to stresses and strains, temperature is also an unknown field quantity that needs to be solved simultaneously with the deformation. Therefore, the complete constitutive law of thermoelasticity should also include (2.4.30), which will provide the constitutive relation for heat conduction in the continuum. Such coupled deformation and temperature problems are rather difficult to solve. Fortunately, in most engineering applications, the effect of deformation on the temperature field is negligible. Therefore, one can solve for the temperature field first by solving the heat conduction problem independent of the deformation. Once the temperature field is known, the only constitutive relation needed for solving the deformation problem in thermoelastic materials is (2.4.31). Such approach that decouples the heat conduction and thermoelasticity into two separate problems is often used in engineering applications.

In case of small strain deformation, the linear thermoelastic stress–strain relationship can be written as

$$\sigma_{ij} = L_{ijkl}(\varepsilon_{kl} - \varepsilon^*_{kl}), \tag{2.4.33}$$

where the thermal strain ε^*_{kl} is given by

$$\varepsilon^*_{kl} = \alpha_{kl}\Delta T. \tag{2.4.34}$$

If the material is isotropic (both elastically and thermally), the elastic stiffness tensor L_{ijkl} is given by (2.4.21), and the coefficient of thermal expansion is given by

$$\alpha_{ij} = \alpha\delta_{ij}, \tag{2.4.35}$$

where α is the isotropic coefficient of thermal expansion. Thus, the small-strain constitutive law for isotropic thermoelastic material can be written as

$$\sigma_{ij} = (K - \tfrac{2}{3}\mu)(\varepsilon_{kk} - 3\alpha\,\Delta T)\delta_{ij} + 2\mu\varepsilon_{ij}. \tag{2.4.36}$$

Viscoelasticity

Viscoelastic constitutive laws are used to describe the time-dependent characteristics of material deformation. In this section, we will limit ourselves to small-strain deformation and linear viscoelasticity theories. In this case, a commonly used stress–strain relationship is given by

$$\sigma_{ij}(t) = G_{ijkl}(t)\varepsilon_{kl}(0) + \int_0^t G_{ijkl}(t - s)\,\frac{d\varepsilon_{kl}(s)}{ds}\,ds, \tag{2.4.37}$$

where $\varepsilon_{ij}(t)$ is the time-dependent strain tensor and $G_{ijkl}(t)$ is called the relaxation function. It has been assumed that for $t < 0$, both $\varepsilon_{ij}(t)$ and $G_{ijkl}(t)$ are zero, that is, $\varepsilon_{ij}(t) = 0$ and $G_{ijkl}(t) = 0$ for $t < 0$. Further, the symmetry of the stress and strain tensors implies

$$G_{ijkl}(t) = G_{jikl}(t) = G_{ijlk}(t). \tag{2.4.38}$$

Under constant strain $\varepsilon_{kj}(t) = \varepsilon_{kj}(0)$, we have $d\varepsilon_{kl}(t)/dt = 0$. Thus, (2.4.37) reduces to

$$\sigma_{ij}(t) = G_{ijkl}(t)\varepsilon_{kj}(0). \tag{2.4.39}$$

This is the reason that $G_{ijkl}(t)$ is termed a relaxation function, for it represents the stress relaxation under constant strain.

Alternative to (2.4.37), the viscoelastic stress–strain law can also be written as

$$\varepsilon_{ij}(t) = J_{ijkl}(t)\sigma_{kl}(0) + \int_0^t J_{ijkl}(t - s)\,\frac{d\sigma_{kl}(s)}{ds}\,ds, \tag{2.4.40}$$

where it has been assumed that for $t < 0$, both $\sigma_{ij}(t)$ and $J_{ijkl}(t)$ are zero, that is, $\sigma_{ij}(t) = 0$ and $J_{ijkl}(t) = 0$ for $t < 0$. The fourth-order tensor function $J_{ijkl}(t)$ is called the creep function because under constant stress $\sigma_{kj}(t) = \sigma_{kj}(0)$, we have $d\sigma_{kl}(t)/dt = 0$. Thus, it follows from (2.4.40) that

$$\varepsilon_{ij}(t) = J_{ijkl}(t)\sigma_{kj}(0). \tag{2.4.41}$$

Now, consider the Laplace transform

$$\widehat{f}(s) = \int_0^\infty f(t)e^{-st}\,dt. \tag{2.4.42}$$

Note that

$$\int_0^\infty \left[\int_0^t f(t-s)g(s)\,ds\right]e^{-st}\,dt = \widehat{f}(s)\widehat{g}(s), \tag{2.4.43}$$

$$\int_0^\infty \frac{dg(s)}{ds}e^{-st}\,dt = s\widehat{g}(s) - g(0). \tag{2.4.44}$$

Applying the Laplace transform to both sides of (2.4.37)–(2.4.40) and making use of the above equations, we obtain

$$\widehat{\sigma}_{ij}(t) = s\widehat{G}_{ijkl}(s)\widehat{\varepsilon}_{kl}(s) \quad \text{and} \quad \widehat{\varepsilon}_{ij}(t) = s\widehat{J}_{ijkl}(s)\widehat{\sigma}_{kl}(s). \tag{2.4.45}$$

It thus follows that

$$s\widehat{G}_{ijkl}(s) = [s\widehat{J}_{ijkl}(s)]^{-1}. \tag{2.4.46}$$

We see that, in the Laplace transform space, (2.4.45) is very similar to the linear elastic constitutive law (Hooke's law). The function $s\widehat{G}_{ijkl}(s)$ can be viewed as the "stiffness tensor" of the viscoelastic material in the Laplace transform space, while $s\widehat{J}_{ijkl}(s)$ is the "compliance tensor." Note that it is generally not true that $G_{ijkl}(t) = [J_{ijkl}(t)]^{-1}$, although one can show that (Christensen, 1982)

$$\lim_{t\to 0} G_{ijkl}(t) = \lim_{t\to 0}[J_{ijkl}(t)]^{-1} \quad \text{and} \quad \lim_{t\to\infty} G_{ijkl}(t) = \lim_{t\to\infty}[J_{ijkl}(t)]^{-1}. \tag{2.4.47}$$

When the material is isotropic, the relaxation and creep functions can be written as

$$G_{ijkl}(t) = \tfrac{1}{3}[G_b(t) - G_s(t)] + \tfrac{1}{2}G_s(t)(\delta_{ik}\delta_{jl} + \delta_{il}\delta_{jk}), \quad (2.4.48)$$

$$J_{ijkl}(t) = \tfrac{1}{3}[J_b(t) - J_s(t)] + \tfrac{1}{2}J_s(t)(\delta_{ik}\delta_{jl} + \delta_{il}\delta_{jk}), \quad (2.4.49)$$

where $G_s(t)$ and $G_b(t)$ are, respectively, the shear and bulk relaxation functions, while $J_s(t)$ and $J_b(t)$ are the shear and bulk creep functions, respectively. Making use of the above in (2.4.37) and (2.4.40), respectively, leads to

$$\sigma'_{ij}(t) = G_s(t)\varepsilon'_{ij}(0) + \int_0^t G_s(t - s)\,\frac{d\varepsilon'_{ij}(s)}{ds}\,ds, \quad (2.4.50)$$

$$\sigma_{kk}(t) = G_b(t)\varepsilon_{kk}(0) + \int_0^t G_b(t - s)\,\frac{d\varepsilon_{kk}(s)}{ds}\,ds, \quad (2.4.51)$$

and

$$\varepsilon'_{ij}(t) = J_s(t)\sigma_{ij}(0) + \int_0^t J_s(t - s)\,\frac{d\sigma'_{ij}(s)}{ds}\,ds, \quad (2.4.52)$$

$$\varepsilon_{kk}(t) = J_b(t)\sigma_{kk}(0) + \int_0^t J_b(t - s)\,\frac{d\sigma_{kk}(s)}{ds}\,ds. \quad (2.4.53)$$

In the Laplace transform space, these stress–strain relationships can be written as

$$\widehat{\sigma}'_{ij}(s) = s\widehat{G}_s(s)\widehat{\varepsilon}'_{kl}(s), \qquad \widehat{\sigma}_{kk}(s) = s\widehat{G}_b(s)\widehat{\varepsilon}_{kk}(s) \quad (2.4.54)$$

and

$$\widehat{\varepsilon}'_{ij}(s) = s\widehat{J}_s(s)\widehat{\sigma}'_{ij}(s), \qquad \widehat{\varepsilon}'_{kk}(s) = s\widehat{J}_b(s)\widehat{\sigma}'_{kk}(s), \quad (2.4.55)$$

where $\widehat{\varepsilon}'_{ij}(s)$ and $\widehat{\sigma}'_{ij}(s)$ are the deviatoric strain and deviatoric stress tensors in the Laplace transform space, respectively,

Plasticity

For many engineering materials, particularly, metallic materials, deformation becomes permanent once the strain goes beyond the elastic limit. Although the mechanisms causing such permanent deformation are rather complex, the macroscopic behavior can be approximately represented by certain functions called elastic-plastic constitutive laws. More detailed and in-depth discussions of elastic-plastic constitutive laws will be given in Chapters 11–13. Here, we will only outline the basic ideas of some commonly used constitutive laws for idealized elastic-plastic materials.

For a material to be plastically deformed, certain combinations of the stress components, called the effective stress, must be larger than a threshold value, which is typically called the yield strength of the material. When the effective stress is below the yield strength, the material behaves linear elastically, and the linear elastic constitutive law can be used. When the effective stress is above the yield strength, the material behaves plastically, and elastic-plastic constitutive laws are needed to describe the postyield behavior. Thus, elastic-plastic constitutive laws typically consist of a yield criterion and a function to describe the postyield stress–strain relationship. The latter is called the flow rule; for plastic deformation it is called historically plastic flow.

It is assumed that, for a given state of a material, there exists a function $f(\sigma_{ij})$ called the yield function such that the material behaves elastically if

$$f(\sigma_{ij}) < 0, \quad \text{or} \quad f(\sigma_{ij}) = 0 \quad \text{and} \quad \frac{\partial f}{\partial \sigma_{ij}} \frac{d\sigma_{ij}}{dt} < 0, \qquad (2.4.56)$$

and the material behaves plastically if

$$f(\sigma_{ij}) = 0 \quad \text{and} \quad \frac{\partial f}{\partial \sigma_{ij}} \frac{d\sigma_{ij}}{dt} \geq 0. \qquad (2.4.57)$$

The above yield criterion can be understood as follows. The equation $f(\sigma_{ij}) = 0$ defines a yield surface in the stress space. It is one of the basic assumptions in plasticity that the yield surface is a closed concave surface enclosing the origin of the stress space. Therefore, we can speak of its inside and outside. For a stress state inside the yield surface, we have $f(\sigma_{ij}) < 0$. The stress is not high enough to yield the material yet, and the deformation is elastic. When the stress state reaches the yield surface, that is, $f(\sigma_{ij}) = 0$, the situation is not unique. It depends on whether, at the next moment, the stress state is moving out of or moving back inside the yield surface. Note that $\partial f / \partial \sigma_{ij}$ is an outward (not necessarily unit) vector of the yield surface, and $d\sigma_{ij}/dt$ is the direction of stress increment. Therefore, $(\partial f/\partial\sigma_{ij})(d\sigma_{ij}/dt) \geq 0$ means that the direction of stress increment at the next moment is moving out of the yield surface, thus plastic yielding occurs. On the other hand, $(\partial f/\partial\sigma_{ij})(d\sigma_{ij}/dt) < 0$ means that the stress state is moving back inside the yield surface; thus the deformation is elastic.

The most commonly used yield functions are the von Mises yield function

$$f(\sigma_{ij}) = J_2 - k^2, \tag{2.4.58}$$

and the Tresca yield function

$$f(\sigma_{ij}) = 4J_2^3 - 27J_3^2 - 36J_2^2 + 96k^4J_2 - 64k^6, \tag{2.4.59}$$

where k is a material constant (yield strength in pure shear), and

$$J_2 = \tfrac{1}{2}\sigma'_{ij}\sigma'_{ij}, \qquad J_3 = \tfrac{1}{3}\sigma'_{ij}\sigma'_{jk}\sigma'_{ki}, \tag{2.4.60}$$

are called, respectively, the second and third invariants of the deviatoric stress tensor

$$\sigma'_{ij} = \sigma_{ij} - \tfrac{1}{3}\sigma_{kk}\delta_{ij}. \tag{2.4.61}$$

For other types of yield functions, the readers are referred to the References and Suggested Reading at the end of this chapter.

Among the many flow rules for postyield behavior, the simplest one is the Levy–Mises perfectly plastic constitutive law. It assumes that postyield deformation is incompressible and fully characterized by the rate of deformation tensor, which is related to the deviatoric Cauchy stress tensor through

$$\sigma'_{ij} = \frac{k\sqrt{2}}{\sqrt{D_{mn}D_{mn}}}D_{ij}, \qquad D_{kk} = 0. \tag{2.4.62}$$

Together, the yield conditions (2.4.56) and (2.4.57) and the postyield stress–strain relationship (2.4.62) form a constitutive law to describe mechanical behavior of an elastic–perfectly plastic material. Note that such simple constitutive law is rather limited in its application to real engineering materials because (1) it assumes that plastic deformation does not produce any volume change (incompressibility), and (2) it assumes no work hardening. More realistic constitutive laws for elastic-plastic materials will be discussed in Chapters 10–13.

2.5 BOUNDARY VALUE PROBLEMS FOR SMALL-STRAIN LINEAR ELASTICITY

For the remaining part of this chapter, we will focus on linear elastic material under small-strain deformation. As discussed in previous sections, the distinction between the reference and current configurations can be neglected when the deformation is small. Therefore, for such small deformation, we will not distinguish the different types of stress and strain measures, nor do we need to distinguish the spatial coordinate from the material coordinate. Instead, for convenience, we will simply use σ_{ij} for stress and ε_{ij} for strain, and x_i for coordinates.

Now, consider a continuum of volume V and surface S (see Fig. 2.5). The total surface S may be divided into $S = S_u + S_\sigma$, where it is assumed that the displacements are prescribed on S_u and the traction is prescribed on S_σ, that is,

$$u_i|_{S_u} = u_i^{(0)} \tag{2.5.1}$$

$$\sigma_{ij} n_j|_{S_\sigma} = p_i^{(0)}, \tag{2.5.2}$$

where $u_i^{(0)}$ and $p_i^{(0)}$ are given values of displacement and traction on S_u and S_σ, respectively.

Within the domain V, the stress must satisfy the equilibrium equation (2.2.1),

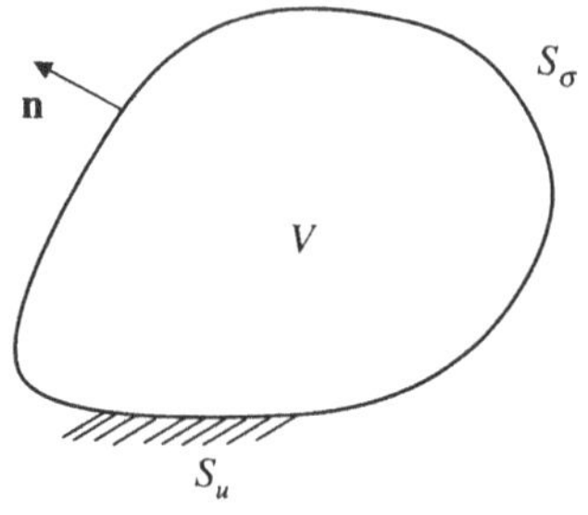

Figure 2.5 An elastic body V in the undeformed configuration.

$$\sigma_{ji,j} + f_i = 0 \quad \text{in } V \tag{2.5.3}$$

where f_i is the body force per unit volume. Combining (2.5.1), (2.5.2), and (2.5.3) in conjunction with Hooke's law, we arrive at the following boundary value problem:

$$\sigma_{ji,j} + f_i = 0 \quad \text{in } V, \tag{2.5.4}$$

$$\sigma_{ij} = L_{ijkl}u_{k,l} \quad \text{in } V, \tag{2.5.5}$$

$$u_i|_{S_u} = u_i^{(0)}, \tag{2.5.6}$$

$$\sigma_{ij}n_j|_{S_\sigma} = p_i^{(0)}. \tag{2.5.7}$$

This is a well-posed boundary value problem with nine unknowns (six stress components plus three displacement components) and nine equations. Theoretically speaking, for a given set of functions $u_i^{(0)}$ and $p_i^{(0)}$ on S, and a given function f_i in V, the stresses and displacements in V can be uniquely determined by solving the above boundary value problem. In most cases, numerical methods by computers are needed to obtain the solutions.

Making use of Hooke's law, the stress tensor can be eliminated from the above boundary value problem. Thus we have alternatively,

$$L_{ijkl}u_{k,lj} + f_i = 0 \quad \text{in } V, \tag{2.5.8}$$

$$u_i|_{S_u} = u_i^{(0)}, \tag{2.5.9}$$

$$L_{ijkl}u_{k,l}n_j|_{S_\sigma} = p_i^{(0)}. \tag{2.5.10}$$

2.6 INTEGRAL REPRESENTATIONS OF ELASTICITY SOLUTIONS

Let us first consider a boundary value problem in an unbounded domain V comprised of a homogeneous linear elastic material with elasticity tensor L_{ijkl}. We assume that the body force f_i is localized, that is,

$$f_i \to 0 \quad \text{as} \quad x_1^2 + x_2^2 + x_3^2 \to \infty. \tag{2.6.1}$$

and the entire boundary (at infinity) is traction free, that is,

$$L_{ijkl}u_{k,l}n_j|_S = p_i^{(0)} \to 0 \quad \text{as} \quad x_1^2 + x_2^2 + x_3^2 \to \infty. \tag{2.6.2}$$

Under these conditions, the displacement u_i should satisfy the following boundary value problem:

$$L_{ijkl}u_{k,lj} + f_i = 0 \quad \text{in } V, \tag{2.6.3}$$

$$u_i \to 0 \quad \text{as} \quad x_1^2 + x_2^2 + x_3^2 \to \infty. \tag{2.6.4}$$

Several approaches can be used to solve this problem. We will use the Fourier transform method. The one-dimensional Fourier transform is defined by the following integrals:

$$\hat{g}(\xi) = \frac{1}{2\pi}\int_{-\infty}^{\infty} g(x)e^{-i\xi x}\, dx, \tag{2.6.5}$$

$$g(x) = \int_{-\infty}^{\infty} \hat{g}(\xi)e^{i\xi x}\, d\xi, \tag{2.6.6}$$

where $i = \sqrt{-1}$ and $\hat{g}(\xi)$ is used to denote the Fourier transform of $g(x)$. In three-dimensional space, the Fourier transform may be defined by the following pair of integrals:

$$\hat{g}(\boldsymbol{\xi}) = \frac{1}{(2\pi)^3}\int_{-\infty}^{\infty} g(\mathbf{x})e^{-i\boldsymbol{\xi}\cdot\mathbf{x}}\, d\mathbf{x} \tag{2.67}$$

$$g(\mathbf{x}) = \int_{-\infty}^{\infty} \hat{g}(\boldsymbol{\xi})e^{i\boldsymbol{\xi}\cdot\mathbf{x}}\, d\boldsymbol{\xi} \tag{2.6.8}$$

where

$$\mathbf{x} = (x_1, x_2, x_3), \qquad \boldsymbol{\xi} = (\xi_1, \xi_2, \xi_3), \tag{2.6.9}$$

and the integrals in (2.6.7) and (2.6.8) are over the entire three-dimensional space, while $d\mathbf{x}$ and $d\boldsymbol{\xi}$ are volume elements in the $\mathbf{x}$ space and the $\boldsymbol{\xi}$ space, respectively. If $g(\mathbf{x}) \to 0$ as $|\mathbf{x}| \to \infty$, then by integration by parts, we have

$$\frac{1}{(2\pi)^3}\int_{-\infty}^{\infty}\frac{\partial g(\mathbf{x})}{\partial x_j}e^{-i\boldsymbol{\xi}\cdot\mathbf{x}}\,d\mathbf{x} = i\xi_j\hat{g}(\boldsymbol{\xi}). \tag{2.6.10}$$

Next, multiply both sides of (2.5.8) by $e^{-i\boldsymbol{\xi}\cdot\mathbf{x}}/(2\pi)^3$ and integrate:

$$L_{ijkl}\left\{\frac{1}{(2\pi)^3}\int_{-\infty}^{\infty}u_{k,lj}(\mathbf{x})e^{-i\boldsymbol{\xi}\cdot\mathbf{x}}\,d\mathbf{x}\right\} + \frac{1}{(2\pi)^3}\int_{-\infty}^{\infty}f_i(\mathbf{x})e^{-i\boldsymbol{\xi}\cdot\mathbf{x}}\,d\mathbf{x} = 0. \tag{2.6.11}$$

Since u_i vanishes at infinity, (2.6.10) can be used in (2.6.11) to yield

$$L_{ijkl}(i\xi_l)(i\xi_j)\hat{u}_k(\boldsymbol{\xi}) + \hat{f}_i(\boldsymbol{\xi}) = 0, \tag{2.6.12}$$

where $\hat{u}_k$ and $\hat{f}_i$ are the Fourier transforms of and u_k and f_i, respectively. Equation (2.6.12) can also be written as

$$K_{ik}\hat{u}_k = \hat{f}_i, \tag{2.6.13}$$

where the dependence of $\hat{u}_k$ and $\hat{f}_i$ on $\boldsymbol{\xi}$ is implied without writing it out explicitly as an argument, and

$$K_{ik} = K_{ik}(\boldsymbol{\xi}) = L_{ijkl}\xi_l\xi_j. \tag{2.6.14}$$

It then follows from (2.6.13) that

$$\hat{u}_i = \hat{f}_j N_{ij}(\boldsymbol{\xi})/D(\boldsymbol{\xi}), \tag{2.6.15}$$

where

$$N_{ij}(\boldsymbol{\xi}) = \tfrac{1}{2}\varepsilon_{ikl}\varepsilon_{jmn}K_{km}(\boldsymbol{\xi})K_{ln}(\boldsymbol{\xi}), \tag{2.6.16}$$

$$D(\boldsymbol{\xi}) = \varepsilon_{mnl}K_{m1}(\boldsymbol{\xi})K_{n2}(\boldsymbol{\xi})K_{l3}(\boldsymbol{\xi}), \tag{2.6.17}$$

and ε_{ijk} is the permutation tensor introduced in (1.4.15). Note that N_{ij} is the cofactor of K_{ij} and D is the determinant of K_{ij}. Therefore, N_{ij}/D is the inverse of K_{ij}. Expressions of N_{ij} and D for isotropic, cubic, and transversely isotropic materials are given in Appendix 2.C.

Equation (2.6.15) means that $\hat{u}_i$ is known explicitly as a function of $\boldsymbol{\xi}$ once the stiffness tensor $\mathbf{L}$ and the body force $\mathbf{f}$ are known. The displacement in the physical space (the $\mathbf{x}$ space) can be obtained by taking the inverse transform of (2.6.15),

$$u_i(\mathbf{x}) = \int_{-\infty}^{\infty} \hat{f}_j(\boldsymbol{\xi}) N_{ij}(\boldsymbol{\xi}) D^{-1}(\boldsymbol{\xi}) e^{i\boldsymbol{\xi}\cdot\mathbf{x}}\, d\boldsymbol{\xi}, \tag{2.6.18}$$

$$\hat{f}_i(\boldsymbol{\xi}) = \frac{1}{(2\pi)^3} \int_{-\infty}^{\infty} f_i(\mathbf{x}) e^{-i\boldsymbol{\xi}\cdot\mathbf{x}}\, d\mathbf{x}. \tag{2.6.19}$$

The corresponding stresses can then be obtained by substituting (2.6.18) into (2.4.8):

$$\sigma_{ij}(\mathbf{x}) = i \int_{-\infty}^{\infty} \hat{f}_m(\boldsymbol{\xi}) L_{ijkl} N_{km}(\boldsymbol{\xi}) \xi_l D^{-1}(\boldsymbol{\xi}) e^{i\boldsymbol{\xi}\cdot\mathbf{x}}\, d\boldsymbol{\xi}. \tag{2.6.20}$$

Equations (2.6.18) and (2.6.20) are integral representations of the displacements and stresses on V.

An alternative form of (2.6.18) can be obtained by substituting (2.6.19) into (2.6.18):

$$u_i(\mathbf{x}) = \int_{-\infty}^{\infty} f_j(\mathbf{y}) G_{ij}^{\infty}(\mathbf{x}, \mathbf{y})\, d\mathbf{y}, \tag{2.6.21}$$

where

$$G_{ij}^{\infty}(\mathbf{x}, \mathbf{y}) = \frac{1}{(2\pi)^3} \int_{-\infty}^{\infty} N_{ij}(\boldsymbol{\xi}) D^{-1}(\boldsymbol{\xi}) e^{i\boldsymbol{\xi}\cdot(\mathbf{x}-\mathbf{y})}\, d\boldsymbol{\xi}. \tag{2.6.22}$$

The second-order tensor function G_{ij}^{∞} given in (2.6.22) is called the Green's function in an unbounded elastic domain. One can easily show that G_{ij}^{∞} is a solution to the following problem:

$$L_{ijkl} \frac{\partial^2 G_{km}^{\infty}(\mathbf{x}, \mathbf{y})}{\partial x_l \partial x_j} + \delta_{im} \delta(\mathbf{x} - \mathbf{y}) = 0 \tag{2.6.23}$$

$$G_{ij}^{\infty}(\mathbf{x}, \mathbf{y}) \rightarrow 0 \quad \text{as} \quad |\mathbf{x}| \rightarrow \infty, \tag{2.6.24}$$

where $\delta(\mathbf{x})$ is the three-dimensional Dirac delta function. Physically, the Green's function is the displacement field generated by a unit force. Since the domain is homogeneous and unbounded, only the difference between the $\mathbf{x}$ point (where the displacement is measured) and the $\mathbf{y}$

point (where the unit force is applied) determines the Green function, that is,

$$G_{ij}^{\infty}(\mathbf{x}, \mathbf{y}) = G_{ij}^{\infty}(\mathbf{x} - \mathbf{y}). \tag{2.6.25}$$

Additionally, by using the reciprocal theorem, it can be easily shown that

$$G_{ij}^{\infty}(\mathbf{x}, \mathbf{y}) = G_{ij}^{\infty}(\mathbf{x} - \mathbf{y}) = G_{ji}^{\infty}(\mathbf{y} - \mathbf{x}) = G_{ji}^{\infty}(\mathbf{y}, \mathbf{x}). \tag{2.6.26}$$

It then follows that

$$\frac{\partial G_{ij}^{\infty}(\mathbf{x}, \mathbf{y})}{\partial x_i} = -\frac{\partial G_{ij}^{\infty}(\mathbf{x}, \mathbf{y})}{\partial y_i}. \tag{2.6.27}$$

Recall that $N_{ij}(\boldsymbol{\xi})$ and $D^{-1}(\boldsymbol{\xi})$ are homogeneous polynomials of degree 4 and 6, respectively, with respect to $\boldsymbol{\xi}$, we have $N_{ij}(-\boldsymbol{\xi})D^{-1}(-\boldsymbol{\xi}) = N_{ij}(\boldsymbol{\xi})D^{-1}(\boldsymbol{\xi})$. Therefore, by changing $\boldsymbol{\xi}$ to $-\boldsymbol{\xi}$, we can write (2.6.22) in a slightly different form:

$$G_{ij}^{\infty}(\mathbf{x}, \mathbf{y}) = \frac{1}{(2\pi)^3} \int_{-\infty}^{\infty} N_{ij}(\boldsymbol{\xi}) D^{-1}(\boldsymbol{\xi}) e^{-i\boldsymbol{\xi}\cdot(\mathbf{x}-\mathbf{y})}\, d\boldsymbol{\xi}. \tag{2.6.28}$$

Furthermore, the volume element in the $\boldsymbol{\xi}$ space can be written as

$$d\boldsymbol{\xi} = d\xi_1\, d\xi_2\, d\xi_3 = \xi^2\, d\xi\, d\hat{S}(\hat{\boldsymbol{\xi}}), \tag{2.6.29}$$

where

$$\xi = |\boldsymbol{\xi}| = \sqrt{\xi_1^2 + \xi_2^2 + \xi_3^2}, \qquad \hat{\boldsymbol{\xi}} = \frac{\boldsymbol{\xi}}{|\boldsymbol{\xi}|}, \tag{2.6.30}$$

and $d\hat{S}(\hat{\boldsymbol{\xi}})$ is a surface element on the unit sphere in the $\boldsymbol{\xi}$ space. Thus, the integral in (2.6.22) can be converted to

$$G_{ij}^{\infty}(\mathbf{x}, \mathbf{y}) = \frac{1}{(2\pi)^3} \int_0^{\infty} \left[\int_S \xi^2 N_{ij}(\boldsymbol{\xi}) D^{-1}(\boldsymbol{\xi}) e^{i\boldsymbol{\xi}\cdot(\mathbf{x}-\mathbf{y})}\, d\hat{S}(\hat{\boldsymbol{\xi}}) \right] d\xi. \tag{2.6.31}$$

The homogeneity of degree zero of $\xi^2 N_{ij}(\boldsymbol{\xi}) D^{-1}(\boldsymbol{\xi})$ with respect to $\boldsymbol{\xi}$ leads to

$$\xi^2 N_{ij}(\xi\hat{\boldsymbol{\xi}})D^{-1}(\xi\hat{\boldsymbol{\xi}}) = N_{ij}(\hat{\boldsymbol{\xi}})D^{-1}(\hat{\boldsymbol{\xi}}). \tag{2.6.32}$$

It thus follows from (2.6.31) that

$$G_{ij}^{\infty}(\mathbf{x}, \mathbf{y}) = \frac{1}{(2\pi)^3}\int_0^{\infty}\left[\int_S N_{ij}(\hat{\boldsymbol{\xi}})D^{-1}(\hat{\boldsymbol{\xi}})e^{i\xi\hat{\boldsymbol{\xi}}\cdot(\mathbf{x}-\mathbf{y})}\, d\hat{S}(\hat{\boldsymbol{\xi}})\right] d\xi. \tag{2.6.33}$$

Following the same procedure, we can rewrite (2.6.28) into

$$G_{ij}^{\infty}(\mathbf{x}, \mathbf{y}) = \frac{1}{(2\pi)^3}\int_{-\infty}^{0}\left[\int_S N_{ij}(\hat{\boldsymbol{\xi}})D^{-1}(\hat{\boldsymbol{\xi}})e^{i\xi\hat{\boldsymbol{\xi}}\cdot(\mathbf{x}-\mathbf{y})}\, d\hat{S}(\hat{\boldsymbol{\xi}})\right] d\xi. \tag{2.6.34}$$

Adding these two expressions together gives

$$\begin{aligned} G_{ij}^{\infty}(\mathbf{x}, \mathbf{y}) &= \frac{1}{2(2\pi)^3}\int_{-\infty}^{\infty}\left[\int_S N_{ij}(\hat{\boldsymbol{\xi}})D^{-1}(\hat{\boldsymbol{\xi}})e^{i\xi\hat{\boldsymbol{\xi}}\cdot(\mathbf{x}-\mathbf{y})}\, d\hat{S}(\hat{\boldsymbol{\xi}})\right] d\xi \\ &= \frac{1}{2(2\pi)^3}\int_S\left[\int_{-\infty}^{\infty} e^{i\xi\hat{\boldsymbol{\xi}}\cdot(\mathbf{x}-\mathbf{y})}\, d\xi\right] N_{ij}(\hat{\boldsymbol{\xi}})D^{-1}(\hat{\boldsymbol{\xi}})\, d\hat{S}(\hat{\boldsymbol{\xi}}). \end{aligned} \tag{2.6.35}$$

Making use of the following identity in the above expression,

$$\int_{-\infty}^{\infty} e^{i\xi\hat{\boldsymbol{\xi}}\cdot(\mathbf{x}-\mathbf{y})}\, d\xi = 2\pi\delta[\hat{\boldsymbol{\xi}}\cdot(\mathbf{x}-\mathbf{y})], \tag{2.6.36}$$

we have

$$G_{ij}^{\infty}(\mathbf{x}, \mathbf{y}) = \frac{1}{8\pi^2}\int_S \delta[\hat{\boldsymbol{\xi}}\cdot(\mathbf{x}-\mathbf{y})]N_{ij}(\hat{\boldsymbol{\xi}})D^{-1}(\hat{\boldsymbol{\xi}})\, d\hat{S}(\hat{\boldsymbol{\xi}}). \tag{2.6.37}$$

The integral representations (2.6.18) and (2.6.37) will be used in the following chapters to derive the Eshelby solution.

For isotropic materials,

$$D(\boldsymbol{\xi}) = \mu^2(\lambda + 2\mu)\xi^6, \tag{2.6.38}$$

$$N_{ij}(\boldsymbol{\xi}) = \mu\xi^2\{(\lambda + 2\mu)\delta_{ij}\xi^2 - (\lambda + \mu)\xi_i\xi_j\}. \tag{2.6.39}$$

The integral in (2.6.37) can be evaluated to yield the Green's function for isotropic material:

$$G_{ij}^{\infty}(\mathbf{x}, \mathbf{y}) = \frac{1}{4\pi\mu}\left\{\frac{\delta_{ij}}{|\mathbf{x} - \mathbf{y}|} - \frac{1}{4(1 - \upsilon)}\frac{\partial^2|\mathbf{x} - \mathbf{y}|}{\partial x_i \partial x_j}\right\}$$
$$= \frac{1}{16\pi\mu(1 - \upsilon)|\mathbf{x} - \mathbf{y}|}\left\{(3 - 4\upsilon)\delta_{ij} + \frac{(x_i - y_i)(x_j - y_j)}{|\mathbf{x} - \mathbf{y}|^2}\right\}. \tag{2.6.40}$$

The integral representations discussed above are the general solutions to problems defined in an unbounded domain. When the problem domain is finite, the integral representations will need to include the boundary contributions. In this case, the Green's function approach is often used. Let us consider the boundary value problem stated by (2.5.8)–(2.5.10). To account for the boundary conditions, we define the Green's function by the following equations:

$$L_{ijkl}\frac{\partial^2 G_{km}(\mathbf{x}, \mathbf{y})}{\partial x_l\, \partial x_j} + \delta_{im}\delta(\mathbf{x} - \mathbf{y}) = 0 \quad \text{in } V. \tag{2.6.41}$$

$$G_{ij}(\mathbf{x}, \mathbf{y})|_{\mathbf{x}\in S_u} = 0, \tag{2.6.42}$$

$$L_{ijkl}\frac{\partial G_{km}(\mathbf{x}, \mathbf{y})}{\partial x_l} n_j|_{\mathbf{x}\in S_\sigma} = 0. \tag{2.6.43}$$

Again, by using the reciprocal theorem, it can be easily shown that

$$G_{ij}(\mathbf{x}, \mathbf{y}) = G_{ji}(\mathbf{y}, \mathbf{x}). \tag{2.6.44}$$

Multiplying (2.6.41) by $u_i(\mathbf{x})$ and integrating the results over the volume V lead to

$$\int_V u_i(\mathbf{x})L_{ijkl}\frac{\partial^2 G_{km}(\mathbf{x}, \mathbf{y})}{\partial x_l \partial x_j}\, dV(\mathbf{x}) + u_m(y) = 0. \tag{2.6.45}$$

By using the divergence theorem, we can write the volume integral in (2.6.45) as

$$\int_V u_i(\mathbf{x}) L_{ijkl} \frac{\partial^2 G_{km}(\mathbf{x}, \mathbf{y})}{\partial x_l \partial x_j} \, dV(\mathbf{x})$$

$$= \int_S u_i(\mathbf{x}) L_{ijkl} \frac{\partial G_{km}(\mathbf{x}, \mathbf{y})}{\partial x_l} n_j \, dS(\mathbf{x}) - \int_V u_{i,j}(\mathbf{x}) L_{ijkl} \frac{\partial G_{km}(\mathbf{x}, \mathbf{y})}{\partial x_l} \, dV(\mathbf{x}). \tag{2.6.46}$$

Analogously, multiplying (2.5.8) by $G_{im}(\mathbf{x}, \mathbf{y})$ and integrating the result over the volume V lead to

$$\int_V G_{im}(\mathbf{x}, \mathbf{y}) L_{ijkl} u_{k,lj}(\mathbf{x}) \, dV(\mathbf{x}) + \int_V G_{im}(\mathbf{x}, \mathbf{y}) f_i(\mathbf{x}) \, dV(\mathbf{x}) = 0. \tag{2.6.47}$$

Following the procedures that led to (2.6.46), the first volume integral in the above equation can be written as

$$\int_V G_{im}(\mathbf{x}, \mathbf{y}) L_{ijkl} u_{k,lj}(\mathbf{x}) \, dV(\mathbf{x})$$

$$= \int_S G_{im}(\mathbf{x}, \mathbf{y}) L_{ijkl} u_{k,l}(\mathbf{x}) n_j \, dS(\mathbf{x}) - \int_V \frac{\partial G_{im}(\mathbf{x}, \mathbf{y})}{\partial x_j} L_{ijkl} u_{k,l}(\mathbf{x}) \, dV(\mathbf{x}). \tag{2.6.48}$$

Note that the last volume integral on the right-hand side of (2.6.46) is the same as that of (2.6.48). Therefore, by subtracting (2.6.45) from (2.6.47), we arrive at

$$u_m(\mathbf{y}) = \int_S G_{im}(\mathbf{x}, \mathbf{y}) L_{ijkl} u_{k,l}(\mathbf{x}) n_j \, dS(\mathbf{x}) + \int_V G_{im}(\mathbf{x}, \mathbf{y}) f_i(\mathbf{x}) \, dV(\mathbf{x})$$
$$- \int_S u_i(\mathbf{x}) L_{ijkl} \frac{\partial G_{km}(\mathbf{x}, \mathbf{y})}{\partial x_l} n_j \, dS(\mathbf{x}). \tag{2.6.49}$$

Enforcing the boundary conditions (2.6.42) and (2.6.43) and (2.5.9) and (2.5.10) in the above, we finally arrive at the solution to the boundary value problem (2.5.8)–(2.5.10) in terms of the Green's function $G_{ij}(\mathbf{x}, \mathbf{y})$,

$$u_m(\mathbf{y}) = \int_V G_{im}(\mathbf{x}, \mathbf{y}) f_i(\mathbf{x})\, dV(\mathbf{x}) + \int_{S_\sigma} G_{im}(\mathbf{x}, \mathbf{y}) p_i^0(\mathbf{x})\, dS(\mathbf{x})$$

$$- \int_{S_u} u_i^0(\mathbf{x}) L_{ijkl} \frac{\partial G_{km}(\mathbf{x}, \mathbf{y})}{\partial x_l} n_j\, dS(\mathbf{x}). \qquad (2.6.50)$$

By making use of (2.6.44), the above can be written equivalently as

$$u_i(\mathbf{x}) = \int_V G_{im}(\mathbf{x}, \mathbf{y}) f_m(\mathbf{y})\, dV(\mathbf{y}) + \int_{S_\sigma} G_{im}(\mathbf{x}, \mathbf{y}) p_m^0(\mathbf{y})\, dS(\mathbf{y})$$

$$- \int_{S_u} u_m^0(\mathbf{y}) L_{mjkl} \frac{\partial G_{ik}(\mathbf{x}, \mathbf{y})}{\partial y_l} n_j\, dS(\mathbf{y}). \qquad (2.6.51)$$

PROBLEMS

2.1 If for any function $f(x)$ that is continuous at $x = 0$

$$\int_{-\infty}^{\infty} f(x)\delta(x)\, dx = f(0),$$

then $\delta(x)$ is called the Dirac delta function. Prove the following properties:

$$x\delta(x) = 0, \qquad \delta(ax) = \frac{1}{a}\delta(x), \qquad \delta(-x) = \delta(x).$$

2.2 Show that the Green's function

$$G_{ij}^{\infty}(\mathbf{x}, \mathbf{y}) = \frac{1}{(2\pi)^3} \int_{-\infty}^{\infty} N_{ij}(\boldsymbol{\xi}) D^{-1}(\boldsymbol{\xi}) e^{i\boldsymbol{\xi}\cdot(\mathbf{x}-\mathbf{y})}\, d\boldsymbol{\xi}$$

satisfies the following boundary value problem:

$$L_{ijkl} G_{lm,ik}^{\infty}(\mathbf{x}, \mathbf{y}) + \delta_{jm}\delta(\mathbf{x} - \mathbf{y}) = 0,$$

$$G_{ij}^{\infty}(\mathbf{x}, \mathbf{y}) \to 0 \quad \text{as} \quad |\mathbf{x}| \to \infty,$$

where $\delta(\mathbf{x})$ is the three-dimensional Dirac delta function defined by

$$\int_{V_r} \delta(\mathbf{x})\, dV = 1,$$

where V_r is any spherical volume $x_1^2 + x_2^2 + x_3^2 = r^2$.

2.3 Show (2.6.26) and (2.6.27).

APPENDIX 2.A RELATIONSHIP AMONG ELASTIC CONSTANTS OF ISOTROPIC MATERIALS

	E, v	E, G	K, v	K, G	λ, μ
E	E	E	$3(1-2v)K$	$\dfrac{9K}{1+3K/G}$	$\dfrac{\mu(3+2\mu/\lambda)}{1+\mu/\lambda}$
v	v	$-1+\dfrac{E}{2G}$	v	$\dfrac{1-2G/3K}{2+2G/3K}$	$\dfrac{1}{2(1+\mu/\lambda)}$
G	$\dfrac{E}{2(1+v)}$	G	$\dfrac{3(1-2v)K}{2(1+v)}$	G	μ
K	$\dfrac{E}{3(1-2v)}$	$\dfrac{E}{9-3E/G}$	K	K	$\lambda+\dfrac{2\mu}{3}$
λ	$\dfrac{Ev}{(1+v)(1-2v)}$	$\dfrac{E(1-2G/E)}{3-E/G}$	$\dfrac{3Kv}{1+v}$	$K-\dfrac{2G}{3}$	λ
μ	$\dfrac{E}{2(1+v)}$	G	$\dfrac{3(1-2v)K}{2(1+v)}$	G	μ

APPENDIX 2.B VOIGT ELASTIC CONSTANTS FOR MATERIALS WITH VARIOUS SYMMETRIES

Orthotropic Materials (nine indepenent constants)

$$C = \begin{bmatrix} C_{11} & C_{12} & C_{13} & 0 & 0 & 0 \\ C_{12} & C_{22} & C_{23} & 0 & 0 & 0 \\ C_{13} & C_{23} & C_{33} & 0 & 0 & 0 \\ 0 & 0 & 0 & C_{44} & 0 & 0 \\ 0 & 0 & 0 & 0 & C_{55} & 0 \\ 0 & 0 & 0 & 0 & 0 & C_{66} \end{bmatrix}$$

Transversely Isotropic Materials (five independent constants)

$$\mathbf{C} = \begin{bmatrix} C_{11} & C_{12} & C_{13} & 0 & 0 & 0 \\ C_{12} & C_{11} & C_{13} & 0 & 0 & 0 \\ C_{13} & C_{13} & C_{33} & 0 & 0 & 0 \\ 0 & 0 & 0 & C_{44} & 0 & 0 \\ 0 & 0 & 0 & 0 & C_{44} & 0 \\ 0 & 0 & 0 & 0 & 0 & 0.5(C_{11} - C_{12}) \end{bmatrix}$$

for axis of symmetry in the x_3 direction, and

$$\mathbf{C} = \begin{bmatrix} C_{11} & C_{12} & C_{12} & 0 & 0 & 0 \\ C_{12} & C_{22} & C_{23} & 0 & 0 & 0 \\ C_{12} & C_{23} & C_{22} & 0 & 0 & 0 \\ 0 & 0 & 0 & 0.5(C_{22} - C_{23}) & 0 & 0 \\ 0 & 0 & 0 & 0 & C_{44} & 0 \\ 0 & 0 & 0 & 0 & 0 & C_{44} \end{bmatrix}$$

for axis of symmetry in the x_1 direction.

Cubic Materials (three independent constants)

$$C = \begin{bmatrix} C_{11} & C_{12} & C_{12} & 0 & 0 & 0 \\ C_{12} & C_{11} & C_{12} & 0 & 0 & 0 \\ C_{12} & C_{12} & C_{11} & 0 & 0 & 0 \\ 0 & 0 & 0 & C_{44} & 0 & 0 \\ 0 & 0 & 0 & 0 & C_{44} & 0 \\ 0 & 0 & 0 & 0 & 0 & C_{44} \end{bmatrix}$$

Isotropic Materials (two independent constants)

$$C = \begin{bmatrix} C_{11} & C_{12} & C_{12} & 0 & 0 & 0 \\ C_{12} & C_{11} & C_{12} & 0 & 0 & 0 \\ C_{12} & C_{12} & C_{11} & 0 & 0 & 0 \\ 0 & 0 & 0 & 0.5(C_{11} - C_{12}) & 0 & 0 \\ 0 & 0 & 0 & 0 & 0.5(C_{11} - C_{12}) & 0 \\ 0 & 0 & 0 & 0 & 0 & 0.5(C_{11} - C_{12}) \end{bmatrix}$$

APPENDIX 2.C EXPRESSIONS OF $N_{ij}(\xi)$ AND $D(\xi)$

Transversly Isotropic Materials

$$
\begin{aligned}
D(\xi) &= (\alpha'\eta^2 + \gamma\xi_3^2)\{\alpha\gamma\eta^4 + (\alpha\beta + \gamma^2 - \gamma'^2)\eta^2\xi_3^2 + \beta\gamma\xi_3^4\} \\
&= (\alpha'\eta^2 + \gamma\xi_3^2)\{(\gamma\eta^2 + \beta\xi_3^2)(\alpha\eta^2 + \gamma\xi_3^2) - \gamma'^2\eta^2\xi_3^2\},
\end{aligned}
$$

$$N_{11}(\boldsymbol{\xi}) = (\alpha'\xi_1^2 + \alpha\xi_2^2 + \gamma\xi_3^2)(\gamma\eta^2 + \beta\xi_3^2) - \gamma'^2\xi_2^2\xi_3^2,$$

$$N_{12}(\boldsymbol{\xi}) = \gamma'^2\xi_1\xi_2\xi_3^2 - (\alpha - \alpha')\xi_1\xi_2(\gamma\eta^2 + \beta\xi_3^2),$$

$$N_{13}(\boldsymbol{\xi}) = (\alpha - \alpha')\gamma'\xi_1\xi_2^2\xi_3 - \gamma'\xi_1\xi_3(\alpha'\xi_1^2 + \alpha\xi_2^2 + \gamma\xi_3^2),$$

$$N_{22}(\boldsymbol{\xi}) = (\alpha\xi_1^2 + \alpha'\xi_2^2 + \gamma\xi_3^2)(\gamma\eta^2 + \beta\xi_3^2) - \gamma'^2\xi_1^2\xi_3^2,$$

$$N_{23}(\boldsymbol{\xi}) = (\alpha - \alpha')\gamma'\xi_1^2\xi_2\xi_3 - \gamma'\xi_2\xi_3(\alpha\xi_1^2 + \alpha'\xi_2^2 + \gamma\xi_3^2),$$

$$N_{33}(\boldsymbol{\xi}) = (\alpha\xi_1^2 + \alpha'\xi_2^2 + \gamma\xi_3^2)(\alpha'\xi_1^2 + \alpha\xi_2^2 + \gamma\xi_3^2) - (\alpha - \alpha')^2\xi_1^2\xi_2^2,$$

where

$$\alpha = C_{11} = C_{22}, \qquad \alpha' = C_{66} = \tfrac{1}{2}(C_{11} - C_{12}),$$

$$\beta = C_{33}, \qquad \gamma' - \gamma = C_{13} = C_{23},$$

$$\gamma = C_{44} = C_{55}, \qquad \eta^2 = \xi_1^2 + \xi_2^2.$$

Cubic Materials

$$
\begin{aligned}
D(\boldsymbol{\xi}) = {} & \mu^2(\lambda + 2\mu + \mu')\xi^6 \\
& + \mu\mu'(2\lambda + 2\mu + \mu')\xi^2(\xi_1^2\xi_2^2 + \xi_2^2\xi_3^2 + \xi_3^2\xi_1^2) \\
& + \mu'^2(3\lambda + 3\mu + \mu')\xi_1^2\xi_2^2\xi_3^2,
\end{aligned}
$$

$$N_{11}(\boldsymbol{\xi}) = \mu^2\xi^4 + \beta\xi^2(\xi_2^2 + \xi_3^2) + \gamma\xi_2^2\xi_3^2,$$

$$N_{12}(\boldsymbol{\xi}) = -(\lambda + \mu)\xi_1\xi_2(\mu\xi^2 + \mu'\xi_3^2),$$

and the other components are obtained by the cyclic permutation of 1, 2, 3, where

$$\xi^2 = \xi_i \xi_i,$$

$$\beta = \mu(\lambda + \mu + \mu'),$$

$$\gamma = \mu'(2\lambda + 2\mu + \mu'),$$

$$\lambda = C_{12},$$

$$\mu = C_{44},$$

$$\mu' = C_{11} - C_{12} - 2C_{44}.$$

Isotropic Materials

$$D(\boldsymbol{\xi}) = \mu^2(\lambda + 2\mu)\xi^6,$$

$$N_{ij}(\boldsymbol{\xi}) = \mu\xi^2\{(\lambda + 2\mu)\delta_{ij}\xi^2 - (\lambda + \mu)\xi_i\xi_j\},$$

where $\xi^2 = \xi_k\xi_k$. Also

$$\begin{aligned} L_{jlmn}\xi_l N_{ij}(\boldsymbol{\xi})D^{-1}(\boldsymbol{\xi}) &= (\lambda + 2\mu)^{-1}\xi^{-4}\{\lambda\delta_{mn}\xi_i\xi^2 + (\lambda + 2\mu)\delta_{im}\xi_n\xi^2 \\ &\quad + (\lambda + 2\mu)\delta_{in}\xi_m\xi^2 - 2(\lambda + \mu)\xi_m\xi_n\xi_i\}, \\ L_{ijkl}L_{pqmn}\xi_q\xi_l N_{kp}(\boldsymbol{\xi})D^{-1}(\boldsymbol{\xi}) &= (\lambda + 2\mu)^{-1}\xi^{-4}\{\lambda^2\delta_{ij}\delta_{mn}\xi^4 + 2\lambda\mu\delta_{mn}\xi_i\xi_j\xi^2 \\ &\quad + 2\lambda\mu\delta_{ij}\xi_m\xi_n\xi^2 + \mu(\lambda + 2\mu)(\delta_{im}\xi_j\xi_n + \delta_{jm}\xi_i\xi_n + \delta_{in}\xi_j\xi_m + \delta_{jn}\xi_i\xi_m)\xi^2 \\ &\quad - 4\mu(\lambda + \mu)\xi_i\xi_j\xi_m\xi_n\}. \end{aligned}$$

REFERENCES

Christensen, R. M. (1982). *Theory of Viscoelasticity—An Introduction,* 2nd ed., Academic Press, New York.

Malvern, L. E. (1969). *Introduction to the Mechanics of Continuous Medium,* Prentice-Hall, Englewood Cliffs, NJ.

Ting, T. C. T. (1996). *Anisotropic Elasticity—Theory and Applications,* Oxford University Press, New York.

SUGGESTED READINGS

Fung, Y. C. (1965). *Foundations of Solid Mechanics,* Prentice-Hall, Englewood Cliffs, NJ.

Mura, T. (1987). *Micromechanics of Defects in Solids,* Martinus Nijhoff, Dordrecht.

3

EIGENSTRAINS

In this chapter, we will first define what eigenstrain is and give some examples of it. Then, we will derive the general solutions to the displacements, stresses, and strains when an eigenstrain field is present in a linear elastic solid.

3.1 DEFINITION OF EIGENSTRAINS

Eigenstrain is a generic name introduced in micromechanics to represent inelastic strains such as thermal strains, phase transformation strains, initial strains, plastic strains, misfit strains, and the like. In the literature, other names have been used for eigenstrains by various authors, such as stress-free transformation strains by Eshelby and elastic polarization strains by Kröner. In this book, we will use the term eigenstrain to represent any inelastic strains. For small-strain deformation, we will use the following notation convention:

$$\varepsilon_{ij} = \text{total strain},$$

$$e_{ij} = \text{elastic strain},$$

$$\varepsilon_{ij}^{*} = \text{eigenstrain (inelastic)}.$$

When both elastic strain and eigenstrain coexist in a continuum, the total strain is given by the sum

$$\varepsilon_{ij} = e_{ij} + \varepsilon_{ij}^*. \tag{3.1.1}$$

The above equation implies that both the elastic strain and eigenstrain act together to cause a material point in the continuum to displace from its initial location to its final location in the deformed configuration. The final position of the material particle after the deformation determines the total strain, that is, the total strain is related to the displacement through [see (2.1.12)]

$$\varepsilon_{ij} = \tfrac{1}{2}(u_{i,j} + u_{j,i}), \tag{3.1.2}$$

where u_i is the total displacement field cause by both the elastic strain and the eigenstrain. Although the total strain can be separated into elastic and inelastic parts, it is not always possible to separate the displacement into elastic and inelastic parts. So, we will not introduce the concepts of *elastic displacement* and *eigen displacement.* Therefore, it is unnecessary to use "total" for the displacement. It will be understood that displacement u_i is always the total displacement caused by both strains.

For sufficiently smooth displacement fields, the total strain tensor satisfies Saint Venant's compatibility conditions:

$$R_{ij} = \varepsilon_{ikp}\varepsilon_{jlq}\varepsilon_{pq,kl} = 0, \tag{3.1.3}$$

where ε_{ijk} is the permutation symbol defined by (1.4.15). Note that it is only the total strain, or the sum of the elastic strain and eigenstrain, that must be compatible, that is, satisfying the compatibility condition (3.1.3). The elastic strain and the eigenstrain by themselves alone do not have to be compatible. In fact, we will see in later examples that they often are not compatible by themselves. Another way of saying, for example, that the eigenstrain field is not compatible is that there is no differentiable function u_i^* such that

$$\varepsilon_{ij}^* = \tfrac{1}{2}(u_{i,j}^* + u_{j,i}^*). \tag{3.1.4}$$

Or, equivalently, one can say that $R_{ij}^* = \varepsilon_{ikp}\varepsilon_{jlq}\varepsilon_{pq,kl}^* \neq 0$.

Let us now consider Hooke's law when eigenstrains are present. As discussed in Chapter 2, Hoooke's law describes the stress–strain relationship for linear elastic materials. In other words, it relates the stress to the elastic strain in a linear elastic material. When eigenstrain is present in a linear elastic material, the total strain is given by (3.1.1). Thus the elastic strain is the difference between the total strain and eigenstrain:

$$e_{ij} = \varepsilon_{ij} - \varepsilon_{ij}^*.$$

Consequently, Hooke's law should be written as

$$\sigma_{ij} = L_{ijkl}e_{kl} = L_{ijkl}(\varepsilon_{kl} - \varepsilon_{kl}^*), \tag{3.1.5}$$

or

$$e_{ij} = \varepsilon_{ij} - \varepsilon_{ij}^* = M_{ijkl}\sigma_{kl}. \tag{3.1.6}$$

The interpretation of (3.1.5) is that the stress in a linear elastic material is caused only by elastic strain. If there is no elastic strain, there would be no stress. For this reason, we do not introduce the concept of eigenstress in this book.

As a simple example to understand (3.1.5), let us consider the uniaxial tension of a metal bar. A typical uniaxial stress–strain curve is shown in Figure 3.1. The eigenstrain in this case is the plastic strain ε_p while the elastic strain is ε_e. Clearly, we have the total strain $\varepsilon = \varepsilon_p + \varepsilon_e$. It is clearly seen from this figure that $\sigma = E\varepsilon_e = E(\varepsilon - \varepsilon_p)$. In other words, the stress is the elastic constant (Young's modulus) times the elastic strain, which is the difference between the total strain and the eigenstrain.

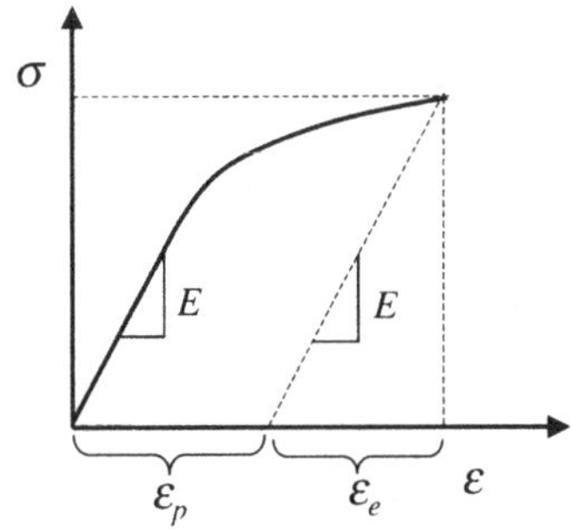

Figure 3.1 Uniaxial stress–strain curve.

3.2 SOME EXAMPLES OF EIGENSTRAINS

First, consider an aluminum ball of radius a at temperature T_0. When the temperature is raised to T, the ball expands. It follows from (2.4.34) that the thermal strain is given by

$$\varepsilon^*_{ij} = \alpha_A \, \Delta T \, \delta_{ij}, \tag{3.2.1}$$

where α_A is the coefficient of thermal expansion for aluminum, δ_{ij} is the Kronecker delta, and $\Delta T = T - T_0$ is the temperature change. Since there is no constraint to the ball, no stress is induced by the temperature rise. Thus, it follows from (3.1.6) that the elastic strain is zero, that is,

$$e_{ij} = \varepsilon_{ij} - \varepsilon^*_{ij} = M_{ijkl}\sigma_{kl} = 0. \tag{3.2.2}$$

Consequently, the eigenstrain in this case is equal to the total strain:

$$\varepsilon_{ij} = e_{ij} + \varepsilon^*_{ij} = \varepsilon^*_{ij} = \alpha_A \, \Delta T \, \delta_{ij}. \tag{3.2.3}$$

We see that thermal strain is an eigenstrain.

Next, let the aluminum ball be embedded in a rigid matrix at temperature T_0. Then, raise the temperature to T. In this case, the total strain of the aluminum ball is zero because of the rigid constraint. This means that

$$\varepsilon_{ij} = e_{ij} + \varepsilon^*_{ij} = e_{ij} + \alpha_A \, \Delta T \, \delta_{ij} = 0, \tag{3.2.4}$$

or

$$e_{ij} = -\varepsilon^*_{ij} = -a_A \, \Delta T \, \delta_{ij}. \tag{3.2.5}$$

It is seen that the elastic strain is no longer zero due to the constraint. Because of the nonzero elastic strain, the stress in the aluminum ball is also nonzero. It is given by Hooke's law:

$$\sigma_{ij} = L_{ijkl}e_{kl} = -\alpha_A L_{nnij} \, \Delta T. \tag{3.2.6}$$

This stress could be very large. For example, for polycrystalline aluminum (isotropic), (3.2.6) can be simplified to

$$\sigma_{ij} = -\frac{E\alpha_A}{1 - 2\nu} \Delta T \, \delta_{ij}, \tag{3.2.7}$$

where for aluminum

$$E = 71 \times 10^9 \text{ Pa}, \qquad \alpha_A = 24 \times 10^{-6}/°\text{C}, \quad \upsilon = 0.3.$$

Therefore, the thermally induced stress is

$$\sigma_{ij} \approx 4.3 \text{ MPa} \frac{\Delta T}{°\text{C}} \delta_{ij}. \tag{3.2.8}$$

So, a temperature change of 10° Celsius would cause 43 MPa stress!

Finally, let us assume that the aluminum ball is embedded in a copper matrix at temperature T_0. Again, let the temperature be raised to T. In this case, the total strain is given by

$$\varepsilon_{ij} = \begin{cases} e_{ij}^A + \alpha_A \, \Delta T \, \delta_{ij} & \text{in the Al ball} \\ e_{ij}^C + \alpha_C \, \Delta T \, \delta_{ij} & \text{in the Cu matrix} \end{cases} \tag{3.2.9}$$

where α_C is the coefficient of thermal expansion of copper. Since the shell is deformable, the total strain is not zero in the aluminum ball, nor is it zero in the copper matrix. The thermally induced stresses cannot be obtained easily in this case; see Problem 3.2. Nevertheless, we know that stresses are induced due to the mismatch between the thermal expansion coefficients.

The above examples show that thermal strain is a type of eigenstrain. Other types of eigenstrains include plastic strain, phase transformation strain, and the like.

3.3 GENERAL SOLUTIONS OF EIGENSTRAIN PROBLEMS

Consider an unbounded elastic body, V, on which an eigenstrain field ε_{ij}^* is prescribed. Without the body force, the equations of equilibrium is given by

$$\sigma_{ij,j} = 0 \quad \text{in } V \tag{3.3.1}$$

Because of the eigenstrains, Hooke's law takes the form of (3.1.5), that is,

$$\sigma_{ij} = L_{ijkl}e_{kl} = L_{ijkl}(\varepsilon_{kl} - \varepsilon^*_{kl}), \tag{3.3.2}$$

where ε_{ij} is the total strain. Since the total strain must be compatible, we have

$$\varepsilon_{ij} = \tfrac{1}{2}(u_{i,j} + u_{j,i}), \tag{3.3.3}$$

where u_i is the total displacement.

Next, making use of (3.3.3) in (3.3.2) and the symmetry of the stiffness tensor, we obtain

$$\sigma_{ij} = L_{ijkl}u_{k,l} - L_{ijkl}\varepsilon^*_{kl}. \tag{3.3.4}$$

Substitution of (3.3.4) into (3.3.1) yields

$$L_{ijkl}u_{k,lj} - L_{ijkl}\varepsilon^*_{kl,j} = 0 \quad \text{in } V. \tag{3.3.5}$$

This is the governing equation for the total displacement field in V. We further assume that the radiation condition is also satisfied, that is,

$$u_i \rightarrow 0 \quad \text{as} \quad |\mathbf{x}| \rightarrow \infty. \tag{3.3.6}$$

This condition implies that the nonzero eigenstrain can only be distributed over a finite domain within V.

Comparing (3.3.5) and (3.3.6) to (2.6.3) and (2.6.4), we conclude that the solution to the eigenstrain problem can be obtained from the elasticity solution if the body force f_i in the elasticity solution is replaced by $-L_{ijkl}\varepsilon^*_{kl,j}$.

To this end, consider the boundary value problem stated by (2.6.3) and (2.6.4). Assuming that the body force is given by the eigenstrain

$$f_i = -L_{ijkl}\varepsilon^*_{kl,j}, \tag{3.3.7}$$

then, through integration by parts, the Fourier transform of f_i can be written as

$$\hat{f}_i(\boldsymbol{\xi}) = \frac{-1}{(2\pi)^3}\int_{-\infty}^{\infty} L_{ijkl}\varepsilon^*_{kl,j}(\mathbf{x})e^{-i\boldsymbol{\xi}\cdot\mathbf{x}}\, d\mathbf{x} = -i\xi_j L_{ijkl}\hat{\varepsilon}^*_{kl}, \tag{3.3.8}$$

where $\hat{\varepsilon}^*_{kl}$ is the Fourier transform of the eigenstrain $\varepsilon^*_{kl}(\mathbf{x})$:

$$\hat{\varepsilon}^*_{kl}(\boldsymbol{\xi}) = \frac{1}{(2\pi)^3}\int_{-\infty}^{\infty} \varepsilon^*_{kl}(\mathbf{x})e^{-i\boldsymbol{\xi}\cdot\mathbf{x}}\, d\mathbf{x}. \tag{3.3.9}$$

Substituting (3.3.8) into (2.6.18), we obtain the solution to the boundary value problem stated by (3.3.5) and (3.3.6):

$$u_i(\mathbf{x}) = -i\int_{-\infty}^{\infty} L_{jlmn}\hat{\varepsilon}^*_{mn}(\boldsymbol{\xi})\xi_l N_{ij}(\boldsymbol{\xi})D^{-1}(\boldsymbol{\xi})e^{i\boldsymbol{\xi}\cdot\mathbf{x}}\, d\boldsymbol{\xi}. \tag{3.3.10}$$

The corresponding strain tensor is

$$\varepsilon_{ij}(\mathbf{x}) = \frac{1}{2}\int_{-\infty}^{\infty} L_{klmn}\hat{\varepsilon}^*_{mn}(\boldsymbol{\xi})\xi_l[\xi_j N_{ik}(\boldsymbol{\xi}) + \xi_i N_{jk}(\boldsymbol{\xi})]D^{-1}(\boldsymbol{\xi})e^{i\boldsymbol{\xi}\cdot\mathbf{x}}\, d\boldsymbol{\xi}. \tag{3.3.11}$$

The stress tensor thus follows from (3.3.4):

$$\sigma_{ij}(\mathbf{x}) = L_{ijkl}\left\{\int_{-\infty}^{\infty} L_{pqmn}\hat{\varepsilon}^*_{mn}(\boldsymbol{\xi})\xi_l\xi_q N_{kp}(\boldsymbol{\xi})D^{-1}(\boldsymbol{\xi})e^{i\boldsymbol{\xi}\cdot\mathbf{x}}\, d\boldsymbol{\xi} - \varepsilon^*_{ij}(\mathbf{x})\right\}. \tag{3.3.12}$$

An alternative approach is to use Green's function. By substituting (3.3.7) into (2.6.21), we obtain

$$u_i(\mathbf{x}) = -\int_{-\infty}^{\infty} L_{mjkl}\varepsilon^*_{kl,j}(\mathbf{y})G^{\infty}_{mi}(\mathbf{x}, \mathbf{y})\, d\mathbf{y}, \tag{3.3.13}$$

where $G^{\infty}_{mi}(\mathbf{x}, \mathbf{y})$ is the infinite domain Green's function given in (2.6.22). Applying divergence theorem to (3.3.13) yields

$$u_i(\mathbf{x}) = \int_{-\infty}^{\infty} L_{mjkl}\varepsilon^*_{kl}(\mathbf{y})\, \frac{\partial G^{\infty}_{mi}(\mathbf{x}, \mathbf{y})}{\partial y_j}\, d\mathbf{y}. \tag{3.3.14}$$

The corresponding strain and stress fields are

$$\varepsilon_{ij}(\mathbf{x}) = \int_{-\infty}^{\infty} L_{klmn}\Gamma_{ijkl}^{\infty}(\mathbf{x} - \mathbf{y})\varepsilon_{mn}^{*}(\mathbf{y})\, d\mathbf{y}, \tag{3.3.15}$$

$$\sigma_{ij}(\mathbf{x}) = L_{ijkl}\left\{\int_{-\infty}^{\infty} L_{pqmn}\varepsilon_{mn}^{*}(\mathbf{y})\Gamma_{klpq}^{\infty}(\mathbf{x} - \mathbf{y})\, dV(\mathbf{y}) - \varepsilon_{kl}^{*}(\mathbf{x})\right\}, \tag{3.3.16}$$

where the fourth-order tensor $\Gamma_{ijkl}^{\infty}(\mathbf{x}, \mathbf{y})$ is given by

$$\Gamma_{ijkl}^{\infty}(\mathbf{x}, \mathbf{y}) = \frac{1}{4}\left[\frac{\partial^2 G_{ki}^{\infty}(\mathbf{x}, \mathbf{y})}{\partial x_j\, \partial y_l} + \frac{\partial^2 G_{kj}^{\infty}(\mathbf{x}, \mathbf{y})}{\partial x_i\, \partial y_l} + \frac{\partial^2 G_{li}^{\infty}(\mathbf{x}, \mathbf{y})}{\partial x_j\, \partial y_k} + \frac{\partial^2 G_{lj}^{\infty}(\mathbf{x}, \mathbf{y})}{\partial x_i\, \partial y_k}\right]. \tag{3.3.17}$$

For future reference, we define

$$P_{ijkl}^{V}(\mathbf{x}) = \int_{V} \Gamma_{ijkl}^{\infty}(\mathbf{x}, \mathbf{y})\, dV(\mathbf{y}), \tag{3.3.18}$$

where the superscript V indicates that the integral is over the volume V.

3.4 EXAMPLES

Straight-Screw Dislocation

Consider the straight-screw dislocation in a cubic crystal shown in Figure 3.2. The eigenstrain corresponding to this screw dislocation is given by

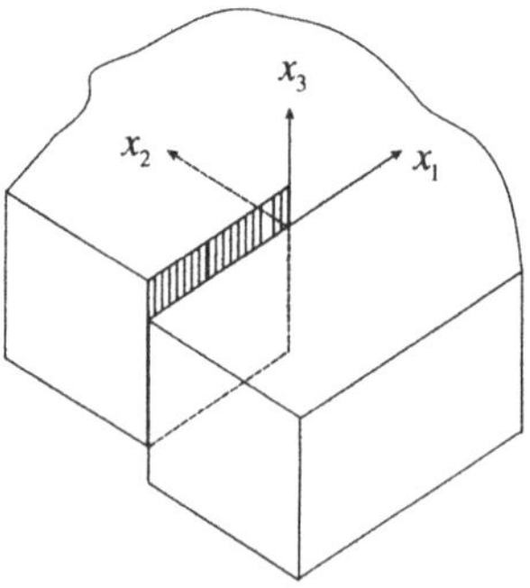

Figure 3.2 Straight-screw dislocation.

$$\varepsilon_{23}^{*} = \tfrac{1}{2} b\delta(x_2)H(-x_1), \tag{3.4.1}$$

where b is the Burgers vector of the dislocation. Thc Fourier transform of ε_{23}^{*} is easily calculated from (3.3.9):

$$\begin{aligned}\bar{\varepsilon}_{23}^{*}(\boldsymbol{\xi}) &= \frac{1}{(2\pi)^3}\int_{-\infty}^{\infty} \frac{1}{2}\, b\delta(x_2)H(-x_1)e^{-i\boldsymbol{\xi}\cdot\mathbf{x}}\, d\mathbf{x} \\ &= \frac{b}{2(2\pi)^3}\int_{-\infty}^{\infty} H(-x_1)e^{-i\xi_1 x_1}\, dx_1 \int_{-\infty}^{\infty} e^{-i\xi_3 x_3}\, dx_3.\end{aligned} \tag{3.4.2}$$

Note that

$$\int_{-\infty}^{\infty} H(-x_1)e^{-i\xi_1 x_1}\, dx_1 = \frac{-1}{i\xi_1}, \tag{3.4.3}$$

$$\int_{-\infty}^{\infty} e^{-i\xi_3 x_3}\, dx_3 = 2\pi\delta(\xi_3). \tag{3.4.4}$$

We have

$$\hat{\varepsilon}_{23}^{*}(\boldsymbol{\xi}) = -\frac{b}{2(2\pi)^2}\frac{\delta(\xi_3)}{i\xi_1}. \tag{3.4.5}$$

Substitution of (3.4.5) into (3.3.10) and use of the properties of delta function yield

$$\begin{aligned}u_i(x) &= \frac{b}{(2\pi)^2}\int_{-\infty}^{\infty} L_{jl23}\frac{\delta(\xi_3)}{\xi_1}\,\xi_l N_{ij}(\boldsymbol{\xi})D^{-1}(\boldsymbol{\xi})e^{i\boldsymbol{\xi}\cdot\mathbf{x}}\, d\boldsymbol{\xi} \\ &= \frac{b}{(2\pi)^2}\int_{-\infty}^{\infty}\int_{-\infty}^{\infty} (L_{jl23}\xi_1^{-1}\xi_l N_{ij}(\boldsymbol{\xi})D^{-1}(\boldsymbol{\xi})e^{i\boldsymbol{\xi}\cdot\mathbf{x}})|_{\xi_3=0}\, d\xi_1\, d\xi_2.\end{aligned} \tag{3.4.6}$$

For materials with cubic symmetry, the integrand in (3.4.6) is simplified to

$$(L_{jl23}\xi_1^{-1}\xi_l N_{1j}(\boldsymbol{\xi})D^{-1}(\boldsymbol{\xi})e^{i\boldsymbol{\xi}\cdot\mathbf{x}})|_{\xi_3=0} = 0, \tag{3.4.7}$$

$$(L_{jl23}\xi_1^{-1}\xi_l N_{2j}(\boldsymbol{\xi})D^{-1}(\boldsymbol{\xi})e^{i\boldsymbol{\xi}\cdot\mathbf{x}})|_{\xi_3=0} = 0, \tag{3.4.8}$$

$$(L_{jl23}\xi_1^{-1}\xi_l N_{3j}(\boldsymbol{\xi})D^{-1}(\boldsymbol{\xi})e^{i\boldsymbol{\xi}\cdot\mathbf{x}})\big|_{\xi_3=0} = \frac{\xi_2}{\xi_1(\xi_1^2 + \xi_2^2)} e^{i(\xi_1 x_1 + \xi_2 x_2)}. \quad (3.4.9)$$

Therefore, we have

$$u_1(\mathbf{x}) = u_2(\mathbf{x}) = 0, \quad (3.4.10)$$

and

$$u_3(\mathbf{x}) = \frac{b}{(2\pi)^2}\int_{-\infty}^{\infty} \frac{\xi_2}{\xi_1(\xi_1^2 + \xi_2^2)} e^{i(\xi_1 x_1 + \xi_2 x_2)}\, d\xi_1\, d\xi_2 = \frac{b}{2\pi}\tan^{-1}\left(\frac{x_2}{x_1}\right). \quad (3.4.11)$$

The only nonzero strain component is the shear strain:

$$\varepsilon_{23} = \frac{1}{2}\frac{\partial u_3}{\partial x_2} = \frac{b}{4\pi}\frac{x_1}{x_1^2 + x_2^2}. \quad (3.4.12)$$

It follows from Hooke's law for cubic materials (2.4.12) that the only nonzero stress component is the shear stress:

$$\sigma_{23} = 2C_{44}\varepsilon_{23} = \frac{C_{44}b}{2\pi}\frac{x_1}{x_1^2 + x_2^2}. \quad (3.4.13)$$

These results are identical to those obtained by Burgers in 1939 for isotropic materials.

Straight-Edge Dislocation

Consider a straight-edge dislocation in a cubic material as shown in Figure 3.3. The eigenstrain corresponding to this dislocation is given by

$$\varepsilon_{12}^* = \tfrac{1}{2}b\delta(x_2)H(-x_1), \quad (3.4.14)$$

where again, b is the Burgers vector of the dislocation. We are to find the corresponding displacements and stresses.

Following the procedures used in deriving (3.4.5), the Fourier transform of ε_{12}^* is easily calculated from (3.3.9):

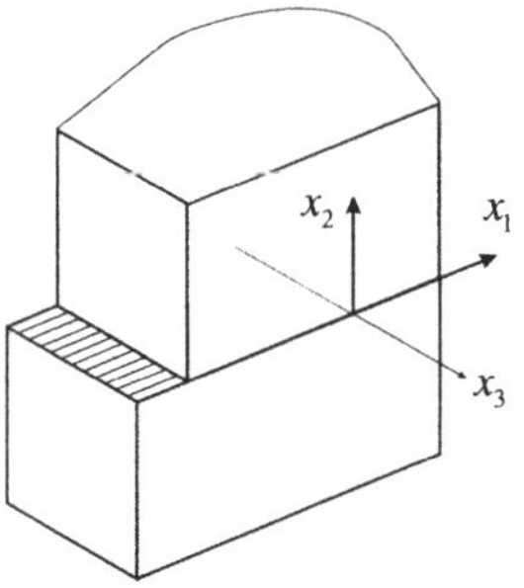

Figure 3.3 Straight-edge dislocation.

$$\bar{\varepsilon}^*_{12}(\boldsymbol{\xi}) = \frac{1}{(2\pi)^3}\int_{-\infty}^{\infty}\frac{1}{2}b\delta(x_2)H(-x_1)e^{i\boldsymbol{\xi}\cdot\mathbf{x}}\,d\mathbf{x} = -\frac{b}{2(2\pi)^2}\frac{\delta(\xi_3)}{i\xi_1}. \quad (3.4.15)$$

Substitution of (3.4.15) into (3.3.10) and use of the properties of the delta function yield

$$u_i(\mathbf{x}) = \frac{b}{(2\pi)^2}\int_{-\infty}^{\infty} L_{jl12}\frac{\delta(\xi_3)}{\xi_1}\,\xi_l N_{ij}(\boldsymbol{\xi})D^{-1}(\boldsymbol{\xi})e^{i\boldsymbol{\xi}\cdot\mathbf{x}}\,d\boldsymbol{\xi}$$

$$= \frac{b}{(2\pi)^2}\int_{-\infty}^{\infty}\int_{-\infty}^{\infty}(L_{jl12}\xi_1^{-1}\xi_l N_{ij}(\boldsymbol{\xi})D^{-1}(\boldsymbol{\xi})e^{i\boldsymbol{\xi}\cdot\mathbf{x}})|_{\xi_3=0}\,d\xi_1\,d\xi_2. \quad (3.4.16)$$

For materials with cubic symmetry, the integrand in (3.4.16) is simplified to

$$(L_{jl12}\xi_1^{-1}\xi_l N_{1j}(\boldsymbol{\xi})D^{-1}(\boldsymbol{\xi})e^{i\boldsymbol{\xi}\cdot\mathbf{x}})|_{\xi_3=0} = \frac{e^{i(\xi_1x_1+\xi_2x_2)}\xi_2(\xi_1^2+\xi_2^2-\alpha\xi_1^2)}{\xi_1[(\xi_1^2+\xi_2^2)^2+\beta\xi_1^2\xi_2^2]}, \quad (3.4.17)$$

$$(L_{jl23}\xi_1^{-1}\xi_l N_{2j}(\boldsymbol{\xi})D^{-1}(\boldsymbol{\xi})e^{i\boldsymbol{\xi}\cdot\mathbf{x}})|_{\xi_3=0} = \frac{e^{i(\xi_1x_1+\xi_2x_2)}(\xi_1^2+\xi_2^2-\alpha\xi_2^2)}{(\xi_1^2+\xi_2^2)^2+\beta\xi_1^2\xi_2^2}, \quad (3.4.18)$$

$$(L_{jl23}\xi_1^{-1}\xi_l N_{3j}(\boldsymbol{\xi})D^{-1}(\boldsymbol{\xi})e^{i\boldsymbol{\xi}\cdot\mathbf{x}})|_{\xi_3=0} = 0, \quad (3.4.19)$$

where

$$\alpha = \frac{2\lambda + 2\mu + \chi}{\lambda + 2\mu + \chi}, \qquad \beta = \frac{\alpha\chi}{\mu}. \tag{3.4.20}$$

Therefore, we have

$$u_3(\mathbf{x}) = 0 \tag{3.4.21}$$

and

$$u_1(\mathbf{x}) = \frac{b}{(2\pi)^2} \int_{-\infty}^{\infty} \int_{-\infty}^{\infty} \frac{e^{i(\xi_1 x_1 + \xi_2 x_2)} \xi_2 (\xi_1^2 + \xi_2^2 - \alpha \xi_1^2)}{\xi_1 [(\xi_1^2 + \xi_2^2)^2 + \beta \xi_1^2 \xi_2^2]} \, d\xi_1 \, d\xi_2, \tag{3.4.22}$$

$$u_2(\mathbf{x}) = \frac{b}{(2\pi)^2} \int_{-\infty}^{\infty} \int_{-\infty}^{\infty} \frac{e^{i(\xi_1 x_1 + \xi_2 x_2)} (\xi_1^2 + \xi_2^2 - \alpha \xi_2^2)}{(\xi_1^2 + \xi_2^2)^2 + \beta \xi_1^2 \xi_2^2} \, d\xi_1 \, d\xi_2. \tag{3.4.23}$$

The nonzero strain components are

$$\varepsilon_{11} = \frac{\partial u_1}{\partial x_1} = \frac{ib}{(2\pi)^2} \int_{-\infty}^{\infty} \int_{-\infty}^{\infty} \frac{e^{i(\xi_1 x_1 + \xi_2 x_2)} \xi_2 (\xi_1^2 + \xi_2^2 - \alpha \xi_1^2)}{[(\xi_1^2 + \xi_2^2)^2 + \beta \xi_1^2 \xi_2^2]} \, d\xi_1 \, d\xi_2, \tag{3.4.24}$$

$$\varepsilon_{22} = \frac{\partial u_2}{\partial x_2} = \frac{ib}{(2\pi)^2} \int_{-\infty}^{\infty} \int_{-\infty}^{\infty} \frac{e^{i(\xi_1 x_1 + \xi_2 x_2)} \xi_2 (\xi_1^2 + \xi_2^2 - \alpha \xi_2^2)}{(\xi_1^2 + \xi_2^2)^2 + \beta \xi_1^2 \xi_2^2]} \, d\xi_1 \, d\xi_2, \tag{3.4.25}$$

$$\begin{aligned} \varepsilon_{12} &= \frac{1}{2} \left(\frac{\partial u_1}{\partial x_2} + \frac{\partial u_2}{\partial x_1} \right) \\ &= \frac{ib}{2(2\pi)^2} \int_{-\infty}^{\infty} \int_{-\infty}^{\infty} \frac{e^{i(\xi_1 x_1 + \xi_2 x_2)} [(\xi_1^2 + \xi_2^2)^2 - 2\alpha \xi_1^2 \xi_2^2])}{\xi_1 [(\xi_1^2 + \xi_2^2)^2 + \beta \xi_1^2 \xi_2^2]} \, d\xi_1 \, d\xi_2. \end{aligned} \tag{3.4.26}$$

In terms of the integrals defined in Appendix 3.A, these strains can be written as

$$\varepsilon_{11} = \frac{ib}{(2\pi)^2}[(1-\alpha)I_1 + I_2], \tag{3.4.27}$$

$$\varepsilon_{22} = \frac{ib}{(2\pi)^2} = [I_1 + (1-\alpha)I_2], \tag{3.4.28}$$

$$\varepsilon_{12} = \frac{ib}{2(2\pi)^2} = [I_3 - (2\alpha + \beta)I_1], \tag{3.4.29}$$

where I_1, I_2, and I_3 are given in Appendix 3.A.

For isotropic materials, $\chi = 0$, $\lambda = 2\mu v/(1-2v)$. Thus,

$$\alpha = \frac{1}{1-v}, \qquad \beta = 0. \tag{3.4.30}$$

The nonzero displacement components in this case are simplified to

$$u_1(\mathbf{x}) = \frac{b}{(2\pi)^2}\int_{-\infty}^{\infty}\int_{-\infty}^{\infty} e^{i(\xi_1 x_1 + \xi_2 x_2)}\left[\frac{\xi_2}{\xi_1(\xi_1^2 + \xi_2^2)} - \frac{\alpha\xi_2\xi_1}{(\xi_1^2 + \xi_2^2)^2}\right] d\xi_1\, d\xi_2, \tag{3.4.31}$$

$$u_2(\mathbf{x}) = \frac{b}{(2\pi)^2}\int_{-\infty}^{\infty}\int_{-\infty}^{\infty} e^{i(\xi_1 x_1 + \xi_2 x_2)}\left[\frac{1}{\xi_1^2 + \xi_2^2} - \frac{\alpha\xi_2^2}{(\xi_1^2 + \xi_2^2)^2}\right] d\xi_1\, d\xi_2. \tag{3.4.32}$$

Making use of the integrals given in Appendix 3.B, we arrive at the displacement components for isotropic materials:

$$u_1(\mathbf{x}) = \frac{b}{4\pi}\left[2\tan^{-1}\left(\frac{x_2}{x_1}\right) + \frac{1}{1-v}\frac{x_1 x_2}{x_1^2 + x_2^2}\right], \tag{3.4.33}$$

$$u_2(\mathbf{x}) = \frac{b}{4\pi(1-v)}\left[\frac{x_2^2}{x_1^2 + x_2^2} - \frac{1-2v}{2}\log(x_1^2 + x_2^2)\right]. \tag{3.4.34}$$

The corresponding strain components are obtained by taking the derivatives of the displacements:

$$\varepsilon_{11} = \frac{\partial u_1}{\partial x_1} = \frac{-bx_2[x_1^2(3 - 2v) + x_2^2(1 - 2v)]}{4\pi(1 - v)(x_1^2 + x_2^2)^2}, \quad (3.4.35)$$

$$\varepsilon_{22} = \frac{\partial u_2}{\partial x_2} = \frac{-bx_2[x_1^2(1 + 2v) - x_2^2(1 - 2v)]}{4\pi(1 - v)(x_1^2 + x_2^2)^2}, \quad (3.4.36)$$

$$\varepsilon_{12} = \frac{1}{2}\left(\frac{\partial u_1}{\partial x_2} + \frac{\partial u_2}{\partial x_1}\right) = \frac{bx_1(x_1^2 - x_2^2)}{4\pi(1 - v)(x_1^2 + x_2^2)^2}. \quad (3.4.37)$$

The corresponding stresses are obtained by using Hooke's law for isotropic materials:

$$\sigma_{11} = \frac{-b\mu x_2(3x_1^2 + x_2^2)}{2\pi(1 - v)(x_1^2 + x_2^2)^2}, \quad (3.4.38)$$

$$\sigma_{22} = \frac{b\mu x_2(x_1^2 - x_2^2)}{2\pi(1 - v)(x_1^2 + x_2^2)^2}, \quad (3.4.39)$$

$$\sigma_{12} = \frac{b\mu x_1(x_1^2 - x_2^2)}{2\pi(1 - v)(x_1^2 - x_2^2)^2}. \quad (3.4.40)$$

These are the classic solutions for a straight-edge dislocation.

Dislocation of General Nature

Consider a dislocation loop L inside a crystalline material as shown in Figure 3.4. Let the area of the slip plane be denoted by S. To describe the slip caused by the dislocation, one side of the slip plane is labeled S^+ and the other side is labeled S^-. The slip is a result of S^+ moving relative to S^- by a Burgers vector $\mathbf{b}$. The unit normal vector of S^+

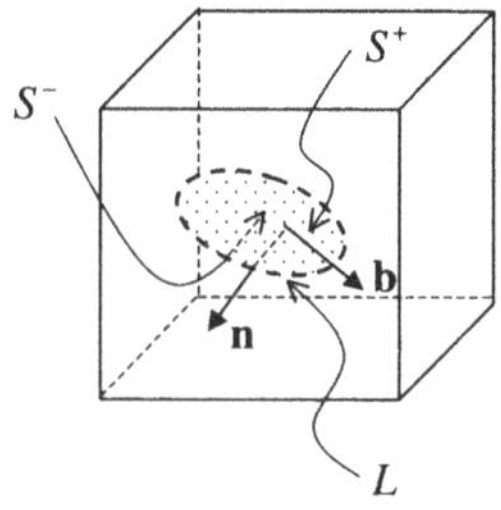

Figure 3.4 Dislocation of general nature.

pointing to S^- is denoted by $\mathbf{n}$. The eigenstrain corresponding to such a dislocation loop can then be written as

$$\varepsilon_{ij}^*(\mathbf{x}) = -\tfrac{1}{2}(b_i n_j + b_j n_i)\bar{\delta}(S - \mathbf{x}), \tag{3.4.41}$$

where $\bar{\delta}(S - \mathbf{x})$ is called the surface Dirac delta function defined by

$$\bar{\delta}(S - \mathbf{x}) = \int_S \delta(\mathbf{x} - \mathbf{z})\, dS(\mathbf{z}). \tag{3.4.42}$$

It can be easily shown that for any continuous function defined in the volume V that contains the surface S, the surface Dirac delta function satisfies the following:

$$\int_V f(\mathbf{x})\bar{\delta}(S - \mathbf{x})\, dV(\mathbf{x}) = \int_S f(\mathbf{x})\, dS(\mathbf{x}). \tag{3.4.43}$$

Substituting (3.4.41) into (3.3.14) and making use of (3.4.43), we obtain the displacement field due to the dislocation loop L,

$$\begin{aligned} u_i(\mathbf{x}) &= -\frac{1}{2}\int_{-\infty}^{\infty} L_{mjkl}(b_k n_l + b_l n_k)\bar{\delta}(S - \mathbf{y})\, \frac{\partial G_{mi}^{\infty}(\mathbf{x}, \mathbf{y})}{\partial y_j}\, d\mathbf{y} \\ &= -\int_S L_{mjkl} b_k n_l\, \frac{\partial G_{mi}^{\infty}(\mathbf{x}, \mathbf{y})}{\partial y_j}\, dS(\mathbf{y}). \end{aligned} \tag{3.4.44}$$

Upon using the properties of Green's function (2.6.27), the above equation can be written as

$$u_i(\mathbf{x}) = \int_S L_{mjkl} b_k n_l\, \frac{\partial G_{mi}^{\infty}(\mathbf{x}, \mathbf{y})}{\partial x_j}\, dS(\mathbf{y}). \tag{3.4.45}$$

This is the well-known Volterra formula.

The displacement gradient thus follows from the Volterra formula:

$$u_{i,j}(\mathbf{x}) = \int_S L_{mnkl} b_k n_l\, \frac{\partial G_{mi}^{\infty}(\mathbf{x}, \mathbf{y})}{\partial x_n\, \partial x_j}\, dS(\mathbf{y}). \tag{3.4.46}$$

The stress caused by the dislocation loop L is thus given by

$$\sigma_{ij} = L_{ijkl}(\varepsilon_{kl} - \varepsilon^*_{kl}) = L_{ijkl}u_{k,l} - L_{ijkl}\varepsilon^*_{kl}. \tag{3.4.47}$$

Note that

$$L_{ijkl}\varepsilon^*_{kl} = -\tfrac{1}{2}L_{ijkl}(b_k n_l + b_l n_k)\bar{\delta}(S - \mathbf{x}) = -L_{ijkl}b_k n_l \bar{\delta}(S - \mathbf{x}). \tag{3.4.48}$$

Introducing (3.4.48) into (3.4.47) yields

$$\sigma_{ij} = L_{ijkl}[u_{k,l} + b_k n_l \bar{\delta}(S - \mathbf{x})] = L_{ijkl}\beta_{kl}, \tag{3.4.49}$$

where

$$\beta_{ij} = \int_S L_{mnkl} b_k n_l \frac{\partial^2 G^\infty_{mi}(\mathbf{x}, \mathbf{y})}{\partial x_n \, \partial x_j} \, dS(\mathbf{y}) + b_i n_j \bar{\delta}(S - \mathbf{x}) \tag{3.4.50}$$

is called the elastic distortion. We next show that the surface integral in (3.4.50) can be converted to a line integral along the dislocation loop. To this end, consider the following line integral along the dislocation loop:

$$\int_L \varepsilon_{jnh} L_{pqmn} \frac{\partial G^\infty_{ip}(\mathbf{x}, \mathbf{y})}{\partial x_q} b_m t_h \, dL(\mathbf{y}), \tag{3.4.51}$$

where t_h is the tangential unit vector along the dislocation loop. Making use of the Stokes' theorem (2.3.3),

$$\int_L \varepsilon_{jnh} L_{pqmn} \frac{\partial G^\infty_{ip}(\mathbf{x}, \mathbf{y})}{\partial x_q} b_m t_h \, dL(\mathbf{y})$$
$$= -\int_S \varepsilon_{klh}\varepsilon_{jnh} L_{pqmn} \frac{\partial G^\infty_{ip}(\mathbf{x}, \mathbf{y})}{\partial x_q \, \partial x_l} b_m n_k \, dS(\mathbf{y}), \tag{3.4.52}$$

where the negative sign is due to the properties of Green's function (2.6.27). Using the $\varepsilon - \delta$ relationship (1.4.16), the right-hand side of (3.4.52) can be written as

$$
\begin{aligned}
-\int_S & \varepsilon_{klh}\varepsilon_{jnh}L_{pqmn} \frac{\partial G_{ip}^{\infty}(\mathbf{x}, \mathbf{y})}{\partial x_q \, \partial x_l} b_m n_k \, dS(\mathbf{y}) \\
&= -\int_S \left(L_{pqmn} \frac{\partial G_{ip}^{\infty}(\mathbf{x}, \mathbf{y})}{\partial x_q \, \partial x_n} b_m n_j - L_{pqmn} \frac{\partial G_{ip}^{\infty}(\mathbf{x}, \mathbf{y})}{\partial x_q \, \partial x_j} b_m n_n \right) dS(\mathbf{y}).
\end{aligned}
\tag{3.4.53}
$$

Since

$$
L_{pqmn} \frac{\partial G_{ip}^{\infty}(\mathbf{x}, \mathbf{y})}{\partial x_q \, \partial x_n} + \delta_{mi}\delta(\mathbf{x} - \mathbf{y}) = 0, \tag{3.4.54}
$$

we can simplify (3.4.53) to

$$
\begin{aligned}
-\int_S & \varepsilon_{klh}\varepsilon_{jnh}L_{pqmn} \frac{\partial G_{ip}^{\infty}(\mathbf{x}, \mathbf{y})}{\partial x_q \, \partial x_l} b_m n_k \, dS(\mathbf{y}) \\
&= b_i n_j \bar{\delta}(S - \mathbf{x}) + \int_S L_{pqmn} \frac{\partial G_{ip}^{\infty}(\mathbf{x}, \mathbf{y})}{\partial x_q \, \partial x_j} b_m n_n \, dS(\mathbf{y}).
\end{aligned}
\tag{3.4.55}
$$

Comparison of (3.4.55) and (3.4.50) leads to

$$
\beta_{ij} = -\int_S \varepsilon_{klh}\varepsilon_{jnh}L_{pqmn} \frac{\partial G_{ip}^{\infty}(\mathbf{x}, \mathbf{y})}{\partial x_q \, \partial x_l} b_m n_k \, dS(\mathbf{y}). \tag{3.4.56}
$$

It then follows from (3.4.52) that

$$
\begin{aligned}
\beta_{ij} &= \int_S L_{mnkl} b_k n_l \frac{\partial^2 G_{mi}^{\infty}(\mathbf{x}, \mathbf{y})}{\partial x_n \, \partial x_j} \, dS(\mathbf{y}) + b_i n_j \bar{\delta}(S - \mathbf{x}) \\
&= \int_L \varepsilon_{jnh} L_{pqmn} \frac{\partial G_{ip}^{\infty}(\mathbf{x}, \mathbf{y})}{\partial x_q} b_m t_h \, dL(\mathbf{y}).
\end{aligned}
\tag{3.4.57}
$$

This is known as the Mura formula (Mura, 1987). The stress field thus follows from (3.4.49):

$$\sigma_{ij} = L_{ijkl} \int_L \varepsilon_{lnh} L_{pqmn} \frac{\partial G^{\infty}_{kp}(\mathbf{x}, \mathbf{y})}{\partial x_q} b_m t_h \, dL(\mathbf{y}). \tag{3.4.58}$$

The fact that the stress field is represented as a line integral along the dislocation loop indicates that the stress depends only on the loop, not the slip plane area S.

PROBLEMS

3.1 Consider a composite bar (one dimensional) whose ends are fixed on rigid walls as shown in Figure 3.5. When the temperature is T_0, the interface between Al and Cu is at the middle $x = L/2$. Please find the position of the interface when the temperature is raised to $T > T_0$. Assume the Young's moduli and the thermal expansion coefficients of Al and Cu are, respectively, E_a, E_c, α_a, and α_c.

3.2 Consider a spherical aluminum ball embedded in a copper matrix of infinite extent at temperature T. When the temperature rises by ΔT, find the stress fields in both the ball and the matrix.

3.3 Consider a Cu water pipe in your house. Assuming the pipe has an outer diameter of 2.54 cm with wall thickness of 1 mm. Is it safe to expose this pipe to -20°C with water in it? If the required factor of safety is 3, what is the lowest temperature this water pipe can stand? The properties of Cu and ice are given below.

	Young's Modulus	Poisson's Ratio	CTE	Tensile Strength
Cu	125 GPa	0.33	17×10^{-6}/°C	70 MPa
Ice	9.5 GPa	0.33	-50×10^{-6}/°C	N/A

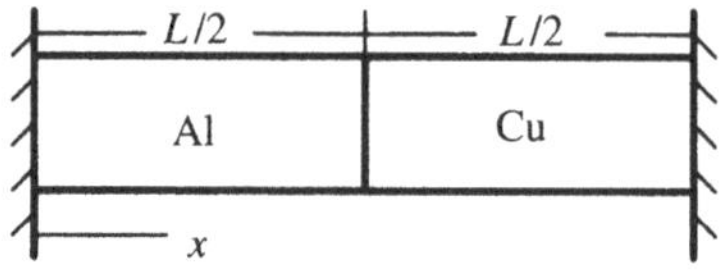

Figure 3.5

APPENDIX 3.A

We introduce the following integrals:

$$
\begin{aligned}
I_1 &= \int_{-\infty}^{\infty}\int_{-\infty}^{\infty} \frac{e^{i(\xi_1 x_1+\xi_2 x_2)}\xi_2\xi_1^2}{[(\xi_1^2+\xi_2^2)^2+\beta\xi_1^2\xi_2^2]}\,d\xi_1\,d\xi_2 \\
&= \int_{-\infty}^{\infty}\int_{-\infty}^{\infty} \frac{e^{i(\xi_1 x_1+\xi_2 x_2)}\xi_1\xi_2^2}{[(\xi_1^2+\xi_2^2)^2+\beta\xi_1^2\xi_2^2]}\,d\xi_1\,d\xi_2,
\end{aligned} \tag{3.A.1}
$$

$$
\begin{aligned}
I_2 &= \int_{-\infty}^{\infty}\int_{-\infty}^{\infty} \frac{e^{i(\xi_1 x_1+\xi_2 x_2)}\xi_2^3}{(\xi_1^2+\xi_2^2)^2+\beta\xi_1^2\xi_2^2}\,d\xi_1\,d\xi_2 \\
&= \int_{-\infty}^{\infty}\int_{-\infty}^{\infty} \frac{e^{i(\xi_1 x_1+\xi_2 x_2)}\xi_1^3}{(\xi_1^2+\xi_2^2)^2+\beta\xi_1^2\xi_2^2}\,d\xi_1\,d\xi_2,
\end{aligned} \tag{3.A.2}
$$

$$
I_3 = \int_{-\infty}^{\infty}\int_{-\infty}^{\infty} \frac{e^{i(\xi_1 x_1+\xi_2 x_2)}}{\xi_1}\,d\xi_1\,d\xi_2. \tag{3.A.3}
$$

Let us consider (3.A.3) first. Making use of the well-known integrals,

$$
\int_{-\infty}^{\infty} e^{i\xi_2 x_2}\,d\xi_2 = 2\pi\delta(x_2), \qquad \int_{-\infty}^{\infty} \frac{e^{i\xi_1 x_1}}{\xi_1}\,d\xi_1 = i\pi\,\mathrm{sgn}(x_1), \tag{3.A.4}
$$

we can reduce (3.A.3) to

$$
I_3 = \int_{-\infty}^{\infty} e^{i\xi_2 x_2}\,d\xi_2 \int_{-\infty}^{\infty} \frac{e^{i\xi_1 x_1}}{\xi_1}\,d\xi_1 = 2i\pi^2\delta(x_2)\mathrm{sgn}(x_1). \tag{3.A.5}
$$

To evaluate the other integrals, we introduce the polar coordinate systems

$$
x_1 = r\cos\theta, \qquad x_2 = r\sin\theta, \qquad \xi_1 = \rho\cos\varphi, \qquad \xi_2 = \rho\sin\varphi.
$$

The above integrals can then be written as

$$I_1 = \int_0^{2\pi} \int_0^{\infty} \frac{e^{ir\rho\cos(\theta-\varphi)} \sin\varphi \cos^2\varphi \, d\rho}{[1 + \beta\cos^2\varphi \sin^2\varphi]} \, d\varphi, \tag{3.A.6}$$

$$I_2 = \int_0^{2\pi} \int_0^{\infty} \frac{e^{ir\rho\cos(\theta-\varphi)} \sin^3\varphi \, d\rho}{[1 + \beta\cos^2\varphi \sin^2\varphi]} \, d\varphi. \tag{3.A.7}$$

Next, consider the integral

$$I = \int_0^{\infty} e^{ir\rho\cos(\theta-\varphi)} \, d\rho. \tag{3.A.8}$$

Making use of the expansion

$$e^{ir\rho\cos(\theta-\varphi)} = J_0(r\rho) + 2\sum_{k=1}^{\infty} i^k J_k(r\rho)\cos[k(\theta-\varphi)], \tag{3.A.9}$$

we have

$$I = \int_0^{\infty} \left(J_0(r\rho) + 2\sum_{k=1}^{\infty} i^k J_k(r\rho)\cos[k(\theta-\varphi)] \right) d\rho. \tag{3.A.10}$$

where $J_k(\bullet)$ is the Bessel function of order k. Since

$$\int_0^{\infty} J_k(r\rho) \, d\rho = \frac{1}{r}, \qquad \text{for any } k \geq 0 \text{ and } r > 0, \tag{3.A.11}$$

we obtain

$$\begin{aligned} I &= \frac{1}{r}\left(1 + 2\sum_{k=1}^{\infty} i^k \cos[k(\theta-\varphi)]\right) \\ &= \frac{1}{r}\left(1 + 2\sum_{k=1}^{\infty} i^k [\cos(k\theta)\cos(k\varphi) + \sin(k\theta)\sin(k\varphi)]\right). \end{aligned} \tag{3.A.12}$$

Substituting (3.A.12) into (3.A.6)–(3.A.7) yields

$$I_1 = \frac{2}{r} \sum_{k=0}^{\infty} i^k [a_{1k} \cos(k\theta) + b_{1k} \sin(k\theta)], \tag{3.A.13}$$

$$I_2 = \frac{2}{r} \sum_{k=0}^{\infty} i^k [a_{2k} \cos(k\theta) + b_{2k} \sin(k\theta)], \tag{3.A.14}$$

where

$$a_{10} = \frac{1}{2} \int_0^{2\pi} \frac{\sin \varphi \cos^2 \varphi \, d\varphi}{[1 + \beta \cos^2 \varphi \sin^2 \varphi]}, \tag{3.A.15}$$

$$a_{1k} = \int_0^{2\pi} \frac{\cos(k\varphi) \sin \varphi \cos^2 \varphi}{[1 + \beta \cos^2 \varphi \sin^2 \varphi]} \, d\varphi, \tag{3.A.16}$$

$$b_{1k} = \int_0^{2\pi} \frac{\sin(k\theta) \sin \varphi \cos^2 \varphi}{[1 + \beta \cos^2 \varphi \sin^2 \varphi]} \, d\varphi, \tag{3.A.17}$$

$$a_{20} = \frac{1}{2} \int_0^{2\pi} \frac{\sin^3 \varphi \, d\varphi}{[1 + \beta \cos^2 \varphi \sin^2 \varphi]}, \tag{3.A.18}$$

$$a_{2k} = \int_0^{2\pi} \frac{\cos(k\varphi) \sin^3 \varphi \, d\varphi}{[1 + \beta \cos^2 \varphi \sin^2 \varphi]}, \tag{3.A.19}$$

$$b_{2k} = \int_0^{2\pi} \frac{\sin(k\theta) \sin^3 \varphi \, d\varphi}{[1 + \beta \cos^2 \varphi \sin^2 \varphi]}. \tag{3.A.20}$$

For isotropic materials, $\beta = 0$. The nonzero coefficients from (3.A.15) to (3.A.20) are

$$b_{11} = b_{13} = \frac{\pi}{4}, \qquad b_{21} = \frac{3\pi}{4}, \qquad b_{23} = -\frac{\pi}{4}.$$

Thus,

$$I_1 = \frac{i\pi}{2r} [\sin \theta - \sin(3\theta)] = \frac{i\pi}{r} [2 \sin^3 \theta - \sin \theta],$$

$$I_2 = \frac{i\pi}{2r} [3 \sin \theta + \sin(3\theta)] = \frac{i\pi}{r} [3 \sin \theta - 2 \sin^3 \theta].$$

APPENDIX 3.B

The following integrals are given in Mura (1987, pp. 17):

$$\int_{-\infty}^{\infty}\int_{-\infty}^{\infty}\frac{\xi_2 e^{i(\xi_1 x_1+\xi_2 x_2)}}{\xi_1(\xi_1^2+\xi_2^2)}\,d\xi_1\,d\xi_2 = 2\pi\tan^{-1}\left(\frac{x_2}{x_1}\right), \tag{3.B.1}$$

$$\int_{-\infty}^{\infty}\int_{-\infty}^{\infty}\frac{\xi_2\xi_1 e^{i(\xi_1 x_1+\xi_2 x_2)}}{(\xi_1^2+\xi_2^2)^2}\,d\xi_1\,d\xi_2 = \frac{-\pi x_1 x_2}{x_1^2+x_2^2}, \tag{3.B.2}$$

$$\int_{-\infty}^{\infty}\int_{-\infty}^{\infty}\frac{e^{i(\xi_1 x_1+\xi_2 x_2)}}{\xi_1^2+\xi_2^2}\,d\xi_1\,d\xi_2 = -\pi\log(x_1^2+x_2^2), \tag{3.B.3}$$

$$\int_{-\infty}^{\infty}\int_{-\infty}^{\infty}\frac{\xi_2^2 e^{i(\xi_1 x_1+\xi_2 x_2)}}{(\xi_1^2+\xi_2^2)^2}\,d\xi_1\,d\xi_2 = -\frac{\pi}{2}\log(x_1^2+x_2^2) - \frac{\pi x_2^2}{x_1^2+x_2^2}. \tag{3.B.4}$$

REFERENCES

Burgers, J. M. (1939). "Some considerations on the fields of stress connected with dislocations in a regular crystal lattice, I, II," *Proc. Kon. Nederl. Akad. Wetensch.,* Vo. 42, pp. 293–324 and pp. 378–399.

Mura, T. (1987). *Micromechanics of Defects in Solids,* Martinus Nijhoff, Boston, MA, Chapter 1.

SUGGESTED READING

Nemat-Nasser, S. and M. Hori. (1993). *Micromechanics: Overall Properties of Heterogeneous Materials,* North-Holland, New York.

4

INCLUSIONS AND INHOMOGENEITIES

In this chapter, we first introduce the concepts of inclusions and inhomogeneities. We will then derive the stress and displacement fields due to the presence of inclusions and inhomogeneities. Finally, we will derive explicitly solutions to ellipsoidal inclusions and inhomogeneities.

4.1 DEFINITIONS OF INCLUSIONS AND INHOMOGENEITIES

Although the terms inclusion and inhomogeneity are used interchangeably in the literature, we will use them differently throughout this book. The readers are advised that the distinction between these two terms made in this book is mainly for the convenience of discussion.

An *inclusion* is defined as a subdomain Ω in a domain D, where eigenstrain $\varepsilon^*_{ij}(\mathbf{x})$ is given in Ω and is zero in $D - \Omega$. The material properties in Ω and in $D - \Omega$ are the *same*. The domain outside Ω, that is, $D - \Omega$, is called the matrix; see Figure 4.1. For linear elastic material, this can be stated as

$$\begin{aligned} \Omega \text{ is an inclusion} \Leftrightarrow \varepsilon^*_{ij}(\mathbf{x}) &\neq 0 \quad \text{for } \mathbf{x} \in \Omega \in D \\ \varepsilon^*_{ij}(\mathbf{x}) &= 0 \quad \text{for } \mathbf{x} \in D - \Omega \\ L_{ijkl} &\text{ is uniform throughout } D \end{aligned}$$

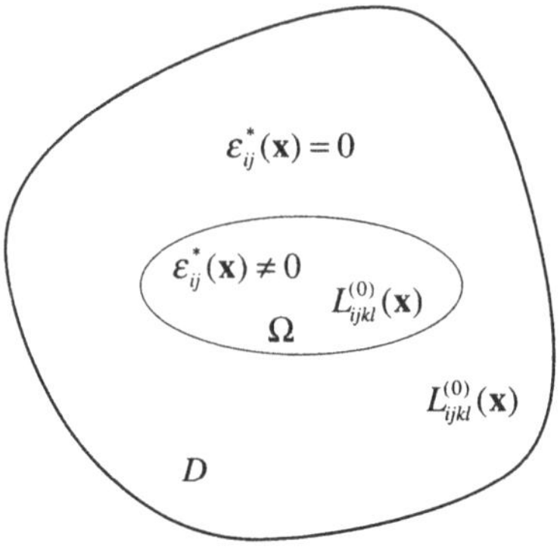

Figure 4.1 Inclusion.

An *inhomogeneity* is defined as a subdomain Ω in domain D, where the material properties in Ω and in $D - \Omega$ are *different*. The domain outside Ω, that is, $D - \Omega$, is called the matrix; see Figure 4.2. For linear elastic material, this can be stated as

$$\Omega \text{ is an inhomogeneity} \Leftrightarrow L_{ijkl}(\mathbf{x}) = \begin{cases} L^{(1)}_{ijkl} & \text{in } \Omega \in D \\ L^{(0)}_{ijkl} & \text{in } D - \Omega \end{cases}$$
$$\text{no eigenstrain throughout } D$$

From the above definition, we see that inclusion is nothing but a distribution of eigenstrains in an otherwise homogeneous material. The presence of eigenstrain may cause stresses in the material. On the other hand, inhomogeneity is a foreign material embedded in an otherwise homogeneous matrix material. If the inhomogeneity fits into the sur-

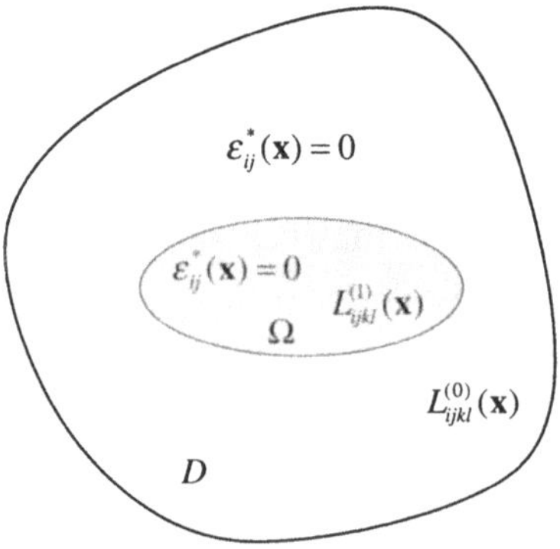

Figure 4.2 Inhomogeneity.

rounding matrix perfectly, there will not be stress produced either in the matrix or in the inhomogeneity.

4.2 INTERFACE CONDITIONS

Consider an inclusion Ω in an elastic domain D of elastic stiffness tensor L_{ijkl}. The interface between Ω and its surrounding material will be denoted by S. We wish to investigate the continuity of field quantities (displacements, strain, stress) across this interface. If the material across the interface S is continuous (e.g., no gap and slip), we call it a perfect interface. For perfect interface, the displacement and the interfacial traction across the interface must be continuous, that is,

$$\Delta u_i \equiv u_i(S^+) - u_i(S^-) = 0, \tag{4.2.1}$$

$$\Delta \sigma_{ij} n_j \equiv [\sigma_{ij}(S^+) - \sigma_{ij}(S^-)] n_j = 0. \tag{4.2.2}$$

where n_j is the outward unit normal to the interface S, and $u_i(S^+)$ and $\sigma_{ij}(S^+)$ are the values of $u_i(\mathbf{x})$ and $\sigma_{ij}(\mathbf{x})$ evaluated at the positive side of S, while $u_i(S^-)$ and $\sigma_{ij}(S^-)$ are the values of $u_i(\mathbf{x})$ and $\sigma_{ij}(\mathbf{x})$ evaluated at the negative side of S. The positive side of S is the side to which n_j points; see Figure 4.3.

It then follows from (4.2.1) that the displacement gradient may have a jump across the interface given by

$$\Delta u_{i,j} \equiv u_{i,j}(S^+) - u_{i,j}(S^-) = \lambda_i n_j, \tag{4.2.3}$$

where λ_i is the magnitude of the jump that will be determined shortly. Note that the right-hand side of (4.2.3) is orthogonal to the tangent vector of S. Thus, Eq. (4.2.3) simply says that, due to the constraint of (4.2.1), the tangential derivatives of the displacements along S must be continuous.

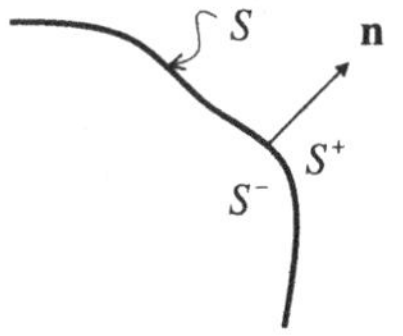

Figure 4.3 Interface with positive and negative sides.

Next, making use of Hooke's law (3.1.5) in (4.2.2) and the fact that the stiffness tensor is the same inside and outside the inclusion yields

$$L_{ijkl}(\Delta u_{k,l} - \Delta\varepsilon^*_{kl})n_j = 0, \tag{4.2.4}$$

where

$$\Delta\varepsilon^*_{kl} \equiv \varepsilon^*_{kl}(S^+) - \varepsilon^*_{kl}(S^-) = -\varepsilon^*_{kl}(S^-). \tag{4.2.5}$$

Since

$$\varepsilon^*_{kl}(S^+) = 0, \tag{4.2.6}$$

substitution of (4.2.3) into (4.2.4) gives

$$L_{ijkl}n_l n_j \lambda_k = -L_{ijkl}\varepsilon^*_{kl}(S^-)n_j. \tag{4.2.7}$$

This is a system of algebraic equations for λ_k. Comparing it with (2.6.13), one can easily write out the solution

$$\lambda_i = -L_{jkmn}\varepsilon^*_{mn}(S^-)n_k N_{ij}(\mathbf{n})/D(\mathbf{n}), \tag{4.2.8}$$

where

$$N_{ij}(\mathbf{n}) = \tfrac{1}{2}\varepsilon_{ikl}\varepsilon_{jmn}K_{km}(\mathbf{n})K_{ln}(\mathbf{n}), \tag{4.2.9}$$

$$D(\mathbf{n}) = \varepsilon_{mnl}K_{m1}(\mathbf{n})K_{n2}(\mathbf{n})K_{l3}(\mathbf{n}), \tag{4.2.10}$$

$$K_{ik}(\mathbf{n}) = L_{ijkl}n_i n_j, \tag{4.2.11}$$

$$\varepsilon_{ijk} = \text{permutation tensor.}$$

Notice that N_{ij} is the cofactor of K_{ij} and D is the determinant of K_{ij}. Therefore, N_{ij}/D is the inverse of K_{ij}. Expressions of N_{ij} and D for isotropic, cubic, and transversely isotropic materials are given in Appendix 4.A.

Substituting (4.2.8) into (4.2.3) we arrived at

$$\Delta u_{i,j} = -L_{lkmn}\varepsilon^*_{mn}(S^-)n_k n_j N_{il}(\mathbf{n})/D(\mathbf{n}). \tag{4.2.12}$$

It thus follows that

$$\Delta\varepsilon_{ij} = -\tfrac{1}{2}L_{lkmn}\varepsilon^*_{mn}(S^-)n_k\{n_jN_{il}(\mathbf{n}) + n_iN_{jl}(\mathbf{n})\}/D(\mathbf{n}), \quad (4.2.13)$$

$$\Delta\sigma_{ij} \equiv \sigma_{ij}(S^+) - \sigma_{ij}(S^-) = L_{ijkl}(\Delta u_{k,l} - \Delta\varepsilon^*_{kl})$$
$$= L_{ijkl}\{-L_{lkmn}\varepsilon^*_{mn}(S^-)n_kn_jN_{il}(\mathbf{n})/D(\mathbf{n}) + \varepsilon^*_{kl}(S^{-1})\}. \quad (4.2.14)$$

Note that for a given inclusion the right-hand sides of (4.2.13) and (4.2.14) involve only known quantities. Therefore, if the strain and stress are known within the inclusion, the corresponding strain and stress just outside the inclusion can be computed from these equations. This proves the following uniqueness theorem.

Uniqueness Theorem for Inclusion–Matrix Interface If the stress or strain tensor is known locally at one side of the interface between an inclusion and the surrounding matrix, then their jumps and consequent values at the other side of the interface are explicitly determinable in terms of the matrix moduli, the eigenstrain in the inclusion, and the interface normal.

Let us now consider the continuity conditions at the interface between an inhomogeneity and its surrounding matrix. Let the stiffness tensor of inhomogeneity be $L^{(1)}_{ijkl}$ and that of the matrix be L_{ijkl}. Further, we assume that the interface is a perfect one, that is, the continuity conditions (4.2.1) and (4.2.2) are satisfied. Consequently, Eq. (4.2.3) also holds.

Making use of Hooke's law, we have

$$\sigma_{ij}(S^+) = L_{ijkl}u_{k,l}(S^+), \qquad \sigma_{ij}(S^-) = L^{(1)}_{ijkl}u_{k,l}(S^-). \quad (4.2.15)$$

Substituting the above into (4.2.2) yields

$$L_{ijkl}u_{k,l}(S^+)n_j = L^{(1)}_{ijkl}u_{k,l}(S^-)n_j. \quad (4.2.16)$$

Eliminating $u_{k,l}(S^+)$ in the above equation by using (4.2.3), we arrive at the following equation for λ_i:

$$L_{ijkl}\lambda_kn_ln_j = \Delta L_{ijkl}u_{k,l}(S^-)n_j, \quad (4.2.17)$$

where

$$\Delta L_{ijkl} = L_{ijkl}^{(1)} - L_{ijkl}. \tag{4.2.18}$$

Similar to (4.2.7), Eq. (4.2.17) can be solved to yield

$$\lambda_i = \Delta L_{jkmn} u_{m,n}(S^-) n_k N_{ij}(\mathbf{n})/D(\mathbf{n}) = \Delta L_{jkmn} \varepsilon_{mn}(S^-) n_k N_{ij}(\mathbf{n})/D(\mathbf{n}). \tag{4.2.19}$$

Making use of the above in (4.2.3) yields

$$\Delta u_{i,j} = \Delta L_{lkmn} \varepsilon_{mn}(S^-) n_l n_j N_{ik}(\mathbf{n})/D(\mathbf{n}). \tag{4.2.20}$$

Thus, we have

$$\Delta \varepsilon_{ij} = \tfrac{1}{2} \Delta L_{lkmn} \varepsilon_{mn}(S^{-1}) n_l \{ n_j N_{ik}(\mathbf{n}) + n_i N_{jk}(\mathbf{n}) \} / D(\mathbf{n}). \tag{4.2.21}$$

Further, combination of (4.2.20) and the first of (4.2.15) gives

$$\sigma_{ij}(S^+) = L_{ijkl} \{ \varepsilon_{kl}(S^-) + \Delta L_{pqmn} \varepsilon_{mn}(S^-) n_l n_q N_{kp}(\mathbf{n})/D(\mathbf{n}) \}$$

or

$$\Delta \sigma_{ij} = \Delta L_{pqmn} \varepsilon_{mn}(S^-) [L_{ijkl} n_l n_q N_{kp}(\mathbf{n})/D(\mathbf{n}) - \delta_{pq}]. \tag{4.2.22}$$

Again, we notice from (4.2.21) and (4.2.22) that for a given inhomogeneity, if the strain or the stress field on one inside the interface is known, the corresponding strain and stress immediately on the other side of interface can be computed. Therefore, similar to the inclusion case, we can state a uniqueness theorem for the inhomogeneity.

Uniqueness Theorem for Inhomogeneity–Matrix Interface If the stress or strain tensor is known locally at one side of the interface between an inhomogeneity and its surrounding matrix, then their jumps and consequent values at the other side of the interface are explicitly determinable in terms of the moduli of the inhomogeneity and the matrix and the interface normal.

4.3 ELLIPSOIDAL INCLUSION WITH UNIFORM EIGENSTRAINS (ESHELBY SOLUTION)

Consider an infinite domain D containing an ellipsoidal inclusion Ω with uniform eigenstrain ε_{ij}^*. Let the stiffness tensor of the material be L_{ijkl}, and the ellipsoidal inclusion (Figure 4.4) be described by

$$\Omega = \{x_1, x_2, x_3; (x_1/a_1)^2 + (x_2/a_2)^2 + (x_3/a_3)^2 \leq 1\}, \quad (4.3.1)$$

where a_1, a_2, and a_3 are the semiaxes of the ellipsoid.

Since

$$\varepsilon_{ij}^*(\mathbf{x}) = \begin{cases} \varepsilon_{ij}^* & \text{for } \mathbf{x} \subset \Omega \\ 0 & \text{for } \mathbf{x} \not\subset \Omega, \end{cases} \quad (4.3.2)$$

it then follows from Green's function formulation of the general solution to eigenstrain problems (3.3.14), (3.3.15), and (3.3.16) that

$$u_i(\mathbf{x}) = L_{mjkl}\varepsilon_{kl}^* \int_\Omega \frac{\partial G_{mi}^\infty(\mathbf{x} - \mathbf{y})}{\partial y_j}\, dV(\mathbf{y}), \quad (4.3.3)$$

$$\varepsilon_{ij}(\mathbf{x}) = L_{klmn}\varepsilon_{mn}^* P_{ijkl}^\Omega(\mathbf{x}), \quad (4.3.4)$$

$$\sigma_{ij}(\mathbf{x}) = L_{ijkl}(L_{pqmn}\varepsilon_{mn}^* P_{klpq}^\Omega(\mathbf{x}) - \varepsilon_{kl}^*), \quad (4.3.5)$$

where

$$\Gamma_{ijkl}^\infty(\mathbf{x}, \mathbf{y}) = \frac{1}{4}\left[\frac{\partial^2 G_{ki}^\infty(\mathbf{x}, \mathbf{y})}{\partial x_j\, \partial y_l} + \frac{\partial^2 G_{kj}^\infty(\mathbf{x}, \mathbf{y})}{\partial x_i\, \partial y_l} + \frac{\partial^2 G_{li}^\infty(\mathbf{x}, \mathbf{y})}{\partial x_j\, \partial y_k} + \frac{\partial^2 G_{lj}^\infty(\mathbf{x}, \mathbf{y})}{\partial x_i\, \partial y_k}\right] \quad (4.3.6)$$

was introduced by (3.3.17), and

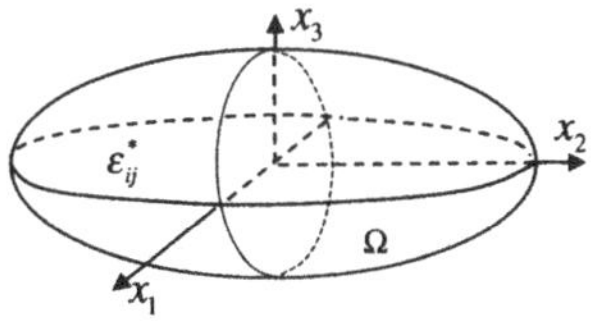

Figure 4.4 Ellipsoidal inclusion Ω.

$$P^{\Omega}_{ijkl}(\mathbf{x}) = \int_{\Omega} \Gamma^{\infty}_{ijkl}(\mathbf{x}, \mathbf{y})\, dV(\mathbf{y}). \tag{4.3.7}$$

For convenience, we introduce

$$S_{ijkl}(\mathbf{x}) = L_{mnkl} P^{\Omega}_{ijmn}(\mathbf{x}). \tag{4.3.8}$$

Then (4.3.4) and (4.3.5) can be rewritten as

$$\varepsilon_{ij}(\mathbf{x}) = S_{ijkl}(\mathbf{x}) \varepsilon^*_{kl}, \tag{4.3.9}$$

$$\sigma_{ij}(\mathbf{x}) = L_{ijkl}\{S_{klmn}(\mathbf{x})\varepsilon^*_{mn} - \varepsilon^*_{kl}\} = L_{ijkl}\{S_{klmn}(\mathbf{x}) - I_{klmn}\}\varepsilon^*_{mn}, \tag{4.3.10}$$

where I_{ijkl} is the fourth-order identity tensor given by

$$I_{ijkl} \equiv \tfrac{1}{2}(\delta_{ik}\delta_{jl} + \delta_{il}\delta_{jk}). \tag{4.3.11}$$

It is noted that (4.3.3)–(4.3.10) are valid for both $\mathbf{x}$ inside and outside the inclusion Ω. We now show that $P^{\Omega}_{ijkl}(\mathbf{x})$ is a constant fourth-order tensor for $\mathbf{x}$ inside Ω, that is,

$$P^{\Omega}_{ijkl}(\mathbf{x}) = P_{ijkl} = \frac{a_1 a_2 a_3}{4\pi} \int_{\hat{S}} H_{ijkl}(\boldsymbol{\xi}) a^{-3} D^{-1}(\boldsymbol{\xi})\, d\hat{S}(\boldsymbol{\xi}), \quad \mathbf{x} \subset \Omega \tag{4.3.12}$$

where P_{ijkl} is referred to as the Hill polarization tensor, and

$$H_{ijkl}(\boldsymbol{\xi}) = N_{ik}(\boldsymbol{\xi})\xi_j\xi_l + N_{jk}(\boldsymbol{\xi})\xi_i\xi_l + N_{il}(\boldsymbol{\xi})\xi_j\xi_k + N_{jl}(\boldsymbol{\xi})\xi_i\xi_k, \tag{4.3.13}$$

$$a = \sqrt{(a_1\xi_1)^2 + (a_2\xi_2)^2 + (a_3\xi_3)^2}. \tag{4.3.14}$$

In (4.3.12), the integration is over the surface of the unit sphere $\hat{S}$ as shown in Figure 4.5.

To this end, let us consider the integral

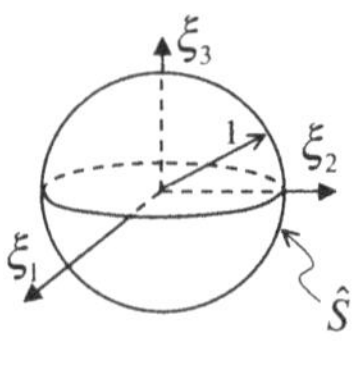

Figure 4.5 Unit sphere $\hat{S}$.

$$\overline{G}_{ijkl}(\mathbf{x}) = \int_{\Omega} \frac{\partial^2 G^{\infty}_{ki}(\mathbf{x}, \boldsymbol{\xi})}{\partial x_j \, \partial \xi_l} \, d\boldsymbol{\xi} = -\frac{\partial^2}{\partial x_j \, \partial x_l} \int_{\Omega} G_{ik}(\mathbf{x}, \boldsymbol{\xi}) \, d\boldsymbol{\xi}. \quad (4.3.15)$$

Replacing Green's function in the above integral by its expression (2.6.37), we have

$$\overline{G}_{ijkl}(\mathbf{x}) = -\frac{1}{8\pi^2} \frac{\partial^2}{\partial x_j \, \partial x_l} \int_{\Omega} \left\{ \int_{S} \delta[\hat{\boldsymbol{\xi}} \cdot (\mathbf{x} - \boldsymbol{\xi})] N_{ik}(\hat{\boldsymbol{\xi}}) D^{-1}(\hat{\boldsymbol{\xi}}) \, d\hat{S}(\hat{\boldsymbol{\xi}}) \right\} d\boldsymbol{\xi}. \quad (4.3.16)$$

Exchanging the order of integration,

$$\overline{G}_{ijkl}(\mathbf{x}) = -\frac{1}{8\pi^2} \frac{\partial^2}{\partial x_j \, \partial x_l} \int_{S} \left\{ \int_{\Omega} \delta[\hat{\boldsymbol{\xi}} \cdot (\mathbf{x} - \boldsymbol{\xi})] \, d\boldsymbol{\xi} \right\} N_{ik}(\hat{\boldsymbol{\xi}}) D^{-1}(\hat{\boldsymbol{\xi}}) \, d\hat{S}(\hat{\boldsymbol{\xi}}). \quad (4.3.17)$$

Making use of the result in Appendix 4.A, we have

$$\begin{aligned} \overline{G}_{ijkl}(\mathbf{x}) &= -\frac{a_1 a_2 a_3}{8\pi} \frac{\partial^2}{\partial x_j \, \partial x_l} \int_{S} \frac{1}{a^3} [a^2 - (\hat{\boldsymbol{\xi}} \cdot \mathbf{x})^2] N_{ik}(\hat{\boldsymbol{\xi}}) D^{-1}(\hat{\boldsymbol{\xi}}) \, d\hat{S}(\hat{\boldsymbol{\xi}}) \\ &= \frac{a_1 a_2 a_3}{4\pi} \int_{S} \hat{\xi}_j \hat{\xi}_l a^{-3} N_{ik}(\hat{\boldsymbol{\xi}}) D^{-1}(\hat{\boldsymbol{\xi}}) \, d\hat{S}(\hat{\boldsymbol{\xi}}). \end{aligned} \quad (4.3.18)$$

Substituting (4.3.18) into (4.3.7) yields (4.3.12). Consequently, it follows from (4.3.8) that for $\mathbf{x}$ inside the inclusion Ω,

$$S_{ijkl} = L_{mnkl} P_{ijmn} = l_{mnkl} \frac{a_1 a_2 a_3}{4\pi} \int_{S} H_{ijmn}(\boldsymbol{\xi}) a^{-3} D^{-1}(\boldsymbol{\xi}) \, d\hat{S}(\boldsymbol{\xi}). \quad (4.3.19)$$

This proves (4.3.12).

Consequently, it follows from (4.3.9) that the total strain (and the stress as well) is uniform inside the ellipsoidal inclusion when the eigenstrain is uniform, that is,

$$\varepsilon_{ij}(\mathbf{x}) = S_{ijkl} \varepsilon^*_{kl} \quad \text{for } \mathbf{x} \subset \Omega \quad (4.3.20)$$

where the fourth-order tensor S_{ijkl} is commonly referred to as the Eshelby inclusion tensor, and (4.3.20) is called the Eshelby ellipsoidal inclusion solution. It can be seen from (4.3.12) that

$$S_{ijkl} = S_{jikl} = S_{ijlk}. \tag{4.3.21}$$

However, the Eshelby tensor does not possess the diagonal symmetry, that is, in general, $S_{ijkl} \neq S_{klij}$.

Having obtained the total strain within the inclusion, the stress field within the inclusion can be computed from Hooke's law (3.1.5):

$$\sigma_{ij} = L_{ijkl}(\varepsilon_{kl} - \varepsilon_{kl}^*) = L_{ijkl}\varepsilon_{kl} + \tau_{ij}^*, \tag{4.3.22}$$

where

$$\tau_{ij}^* = -L_{ijkl}\varepsilon_{kl}^* \tag{4.3.23}$$

is called the stress polarization. It is the stress in the inclusion caused by the eigenstrain ε_{ij}^* when the inclusion is not allowed to deform at all (i.e., the total strain is zero, $\varepsilon_{kl} = 0$). This can be easily understood if we recalled the example in Chapter 3, where an aluminum ball is heated while the ball is constrained so the total strain is zero. The corresponding stress generated in the ball, according to (3.2.6), is

$$\sigma_{ij} = \tau_{ij}^* = -\alpha_A L_{nnij}\, \Delta T$$

because the eigenstrain in this case is $\alpha_A\, \Delta T$.

It then follows from (4.3.19) that the Eshelby solution (4.3.20) can be written as

$$\varepsilon_{ij} = S_{ijkl}\varepsilon_{kl}^* = P_{ijmn}L_{mnkl}\varepsilon_{kl}^* = -P_{ijmn}\tau_{mn}^*, \tag{4.3.24}$$

where

$$P_{ijmn} = S_{ijkl}M_{klmn}$$

is called the stress polarization. Note that this is not the stress on the inclusion. The stress on the inclusion (4.3.22) is given by

$$\sigma_{ij} = L_{ijkl}(S_{klmn}\varepsilon_{mn}^* - \varepsilon_{kl}^*) = -Q_{ijkl}\varepsilon_{kl}^*,$$

where the fourth-order tensor

$$Q_{ijkl} = L_{ijmn}(I_{mnkl} - S_{mnkl}) \tag{4.3.25}$$

can be viewed as the dual of the stress polarization tensor P_{ijkl}.

One can also show that

$$\mathbf{L}(\mathbf{I} - \mathbf{S}) = (\mathbf{I} - \mathbf{S}^{\mathrm{T}})\mathbf{L} \tag{4.3.26}$$

is positive definite. To this end, consider the total strain energy of the ellipsoidal inclusion problem:

$$\begin{aligned} U &= \frac{1}{2}\int_D \sigma_{ij}(\varepsilon_{ij} - \varepsilon^*_{ij})\, dv = \frac{1}{2}\int_D \sigma_{ij}u_{i,j}\, dv - \frac{1}{2}\int_\Omega \sigma_{ij}\varepsilon^*_{ij}\, dv \\ &= \frac{1}{2}\int_D [(\sigma_{ij}u_i)_{,j} - \sigma_{ij,j}u_i]\, dv - \frac{1}{2}\int_\Omega L_{ijkl}(\varepsilon_{kl} - \varepsilon^*_{kl})\varepsilon^*_{ij}\, dv \\ &= \frac{1}{2}\int_{S_\infty} \sigma_{ij}u_i n_j\, ds - \frac{1}{2}\int_D \sigma_{ij,j}u_i\, dv - \frac{1}{2}\int_\Omega L_{ijkl}(\varepsilon_{kl} - \varepsilon^*_{kl})\varepsilon^*_{ij}\, dv, \end{aligned} \tag{4.3.27}$$

where S_∞ is the boundary of D. Since D is an infinite domain and the stress vanishes at infinity, the surface integral must be zero. By virtue of the equilibrium equation, the second term on the right-hand side of (4.3.27) is also zero. Therefore, we have

$$\begin{aligned} U &= \frac{1}{2}\int_\Omega (\boldsymbol{\varepsilon}^* - \boldsymbol{\varepsilon})\mathbf{L}\boldsymbol{\varepsilon}^*\, dv = \frac{1}{2}\int_\Omega (\boldsymbol{\varepsilon}^* - \mathbf{S}\boldsymbol{\varepsilon}^*)\mathbf{L}\boldsymbol{\varepsilon}^*\, dv \\ &= \frac{1}{2}\int_\Omega \boldsymbol{\varepsilon}^*(\mathbf{I} - \mathbf{S}^{\mathrm{T}})\mathbf{L}\boldsymbol{\varepsilon}^*\, dv = \frac{1}{2}\int_\Omega \boldsymbol{\varepsilon}^*\mathbf{L}(\mathbf{I} - \mathbf{S})\boldsymbol{\varepsilon}^*\, dv. \end{aligned} \tag{4.3.28}$$

Since the strain energy must be positive and becomes zero only if the eigenstrain is zero, we must have

$$\int_\Omega \boldsymbol{\varepsilon}^*(\mathbf{I} - \mathbf{S}^{\mathrm{T}})\mathbf{L}\boldsymbol{\varepsilon}^*\, dv = \int_\Omega \boldsymbol{\varepsilon}^*\mathbf{L}(\mathbf{I} - \mathbf{S})\boldsymbol{\varepsilon}^*\, dv \geq 0. \tag{4.3.29}$$

Note that S is independent of the size of Ω and (4.3.29) should hold for any size of Ω, we can conclude from (4.3.29) that

$$\boldsymbol{\varepsilon}^*(\mathbf{I} - \mathbf{S}^{\mathrm{T}})\mathbf{L}\boldsymbol{\varepsilon}^* = \boldsymbol{\varepsilon}^*\mathbf{L}(\mathbf{I} - \mathbf{S})\boldsymbol{\varepsilon}^* \geq 0, \tag{4.3.30}$$

and the equal sign is realized only when $\boldsymbol{\varepsilon}^* = \mathbf{0}$. This proves that $\mathbf{L}(\mathbf{I} - \mathbf{S}) = (\mathbf{I} - \mathbf{S}^{\mathrm{T}})\mathbf{L}$ is positive definite. Two corollaries of (4.3.30) are

$$\mathbf{LS} = \mathbf{S}^{\mathrm{T}}\mathbf{L}, \tag{4.3.31}$$

and $(\mathbf{I} - \mathbf{S})$ is nonsingular.

Next, we show that $\mathbf{S}$ is also nonsingular. We begin by computing the strain energy stored in the region outside the inclusion:

$$\begin{aligned} U_m &= \frac{1}{2}\int_{D-\Omega} \sigma_{ij}\varepsilon_{ij}\, dv = \frac{1}{2}\int_{D-\Omega} \sigma_{ij}u_{i,j}\, dv \\ &= \frac{1}{2}\int_{D-\Omega} [(\sigma_{ij}u_i)_{,j} - \sigma_{ij,j}u_i]\, dv. \end{aligned} \tag{4.3.32}$$

Again, the second term on the right hand of (4.3.32) vanishes because of equilibrium. The first term can be converted to a surface integral by using the divergence theorem:

$$U_m = \frac{1}{2}\int_{S_\infty} \sigma_{ij}u_i n_j\, ds - \frac{1}{2}\int_{S} \sigma_{ij}u_i n_j\, ds, \tag{4.3.33}$$

where the negative sign of the second term comes about because n_j is the outward normal vector of Ω. The first term vanishes because the stresses go to zero at infinity. Thus, we have

$$U_m = -\frac{1}{2}\int_{S} \sigma_{ij}u_i n_j\, ds = -\frac{1}{2}\int_{\Omega} \sigma_{ij}u_{i,j}\, dv = -\frac{1}{2}\int_{\Omega} \sigma_{ij}\varepsilon_{ij}\, dv. \tag{4.3.34}$$

If $\mathbf{S}$ is singular, there must exist an eigenstrain field

$$\hat{\varepsilon}^*_{kl} \neq 0 \tag{4.3.35}$$

such that

$$\varepsilon_{ij} = S_{ijkl}\hat{\varepsilon}^*_{kl} = 0 \quad \text{on } \Omega. \tag{4.3.36}$$

In what follows, we will show that (4.3.36) contradicts (4.3.35), that is, if (4.3.36) is satisfied, then (4.3.35) cannot be true, thus proving that $\mathbf{S}$ cannot be singular.

To this end, first compute the corresponding stress on Ω when (4.3.36) holds:

$$\sigma_{ij} = L_{ijkl}(\varepsilon_{kl} - \hat{\varepsilon}^*_{kl}) = -L_{ijkl}\hat{\varepsilon}^*_{kl}. \qquad (4.3.37)$$

Substitution of (4.3.36) into (4.3.34) leads to $U_m = 0$. It then follows from (4.3.32) that $\sigma_{ij} = \varepsilon_{ij} = 0$ in the region outside the inclusion, or $\sigma_{ij}(S^+) = 0$.

On the other hand, the traction continuity condition on the inclusion/matrix interface yields

$$0 = \sigma_{ij}(S^+)n_j = \sigma_{ij}(S^-)n_j = -L_{ijkl}\hat{\varepsilon}^*_{kl}n_j, \qquad (4.3.38)$$

where (4.3.37) has been used in deriving the last equation of (4.3.38). Note that $L_{ijkl}\hat{\varepsilon}^*_{kl}$ is a constant second-order tensor while n_j varies along the boundary. For (4.3.38) to hold for the entire boundary, we must have $L_{ijkl}\hat{\varepsilon}^*_{kl} = 0$. Since L_{ijkl} is nonsingular, we end up with $\hat{\varepsilon}^*_{kl} = 0$. This contradicts (4.3.35). Therefore, **S** cannot be singular.

We also note that S_{ijkl} is independent of the eigenstrain. But, it does depend on the matrix material. For general anisotropic materials, numerical methods are typically required to carry out the integral in S_{ijkl}. For isotropic materials, the integral can be written in terms of elliptical integrals and explicit expression in terms of elemental functions can be obtained for certain special cases. For example, the Eshelby tensor S_{ijkl} for a spherical inclusion in an isotropic material is a fourth-order isotropic tensor:

$$S_{ijkl} = \gamma\delta_{ij}\delta_{kl} + \delta(\delta_{ik}\delta_{jl} + \sigma_{il}\delta_{jk} - \tfrac{2}{3}\delta_{ij}\delta_{kl}), \qquad (4.3.39)$$

where

$$\gamma = \frac{K}{3K + 4\mu} = \frac{1 + v}{9(1 - v)}, \qquad \delta = \frac{3(K + 2\mu)}{5(3K + 4\mu)} = \frac{4 - 5v}{15(1 - v)}. \qquad (4.3.40)$$

Using the symbolic notation introduced in (1.3.3), we can simply write

$$\mathbf{S} = (3\gamma, 2\delta). \qquad (4.3.41)$$

The expressions of S_{ijkl} for inclusions of special shapes are given in Appendix 4.B.

Before closing this section, we mention that (4.3.20) is no longer valid for a material point outside the inclusion Ω. To obtain the strain and stress fields outside the inclusion, one needs to evaluate the integral in (4.3.7) for $\mathbf{x}$ outside the ellipsoid. An alternative approach is to use the continuity conditions at the interface between an inclusion and its surrounding matrix. Since the stresses and strains are known inside the inclusion, the stresses and strains, based on the uniqueness theorem, can be obtained just outside the inclusion. This allows us to compute the strain and stress field anywhere outside the inclusion. The stresses and strains outside the inclusion will not be needed in this book. So, we will not present them. However, we do note that for $\mathbf{x}$ outside of Ω_r, $\mathbf{S}_r(\mathbf{x})$ decades as $\|\mathbf{x}\|^{-3}$ as $\|\mathbf{x}\| \to \infty$. Readers are referred to the Suggested Readings at end of this chapter for further details.

4.4 ELLIPSOIDAL INHOMOGENEITIES

Consider an elastic body D with elastic modulus $\mathbf{L}_0$ containing an ellipsoidal inhomogeneity Ω with elastic modulus tensor $\mathbf{L}_1$. Let D be subjected to surface traction $\mathbf{p}^0 = \boldsymbol{\sigma}^0 \cdot \mathbf{n}$ on the boundary of D, as shown in Figure 4.6. We are to find the stress fields in D.

Obviously, when $\mathbf{L}_1 = \mathbf{L}_0$, the material is homogeneous. Consequently, the total stress field becomes uniform throughout D (see Figure 4.7), and is given by

$$\boldsymbol{\sigma}^t = \boldsymbol{\sigma}^0 \tag{4.4.1}$$

It is therefore conceivable that, through the principle of linear superposition, when $\mathbf{L}_1$ is different from $\mathbf{L}_0$, the total stress field in D can be written as

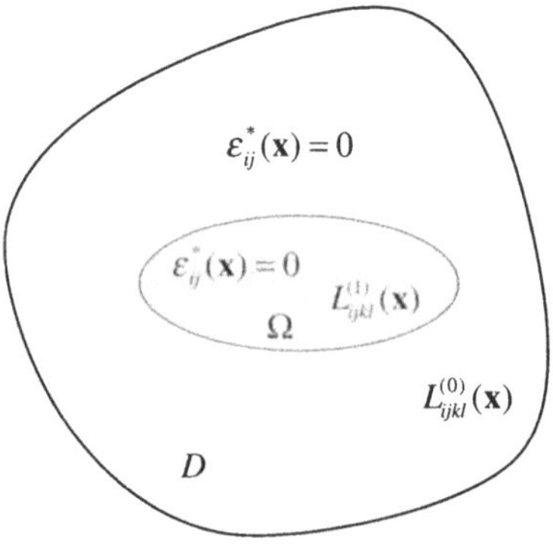

Figure 4.6 Ellipsoidal inhomogeneity.

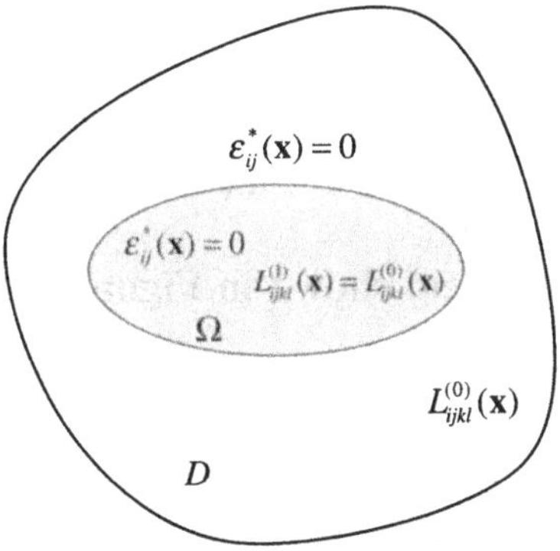

Figure 4.7 Uniform stress field in homogeneous body.

$$\boldsymbol{\sigma}^t = \boldsymbol{\sigma}^0 + \boldsymbol{\sigma}, \tag{4.4.2}$$

where $\boldsymbol{\sigma}$ is the stress perturbation due to the presence of the inhomogeneity Ω. Obviously, the perturbed stress field $\boldsymbol{\sigma}$ should satisfy the equilibrium equations and the homogeneous boundary condition

$$\boldsymbol{\sigma} \cdot \mathbf{n} = \mathbf{0} \quad \text{on } S. \tag{4.4.3}$$

If $\boldsymbol{\varepsilon}^0$ is used to denote the strain field corresponding to $\boldsymbol{\sigma}^0$, that is,

$$\boldsymbol{\sigma}^0 = \mathbf{L}_0\boldsymbol{\varepsilon}^0, \tag{4.4.4}$$

then the total strain on the inclusion can be written as

$$\boldsymbol{\varepsilon}^t = \boldsymbol{\varepsilon}^0 + \boldsymbol{\varepsilon}, \tag{4.4.5}$$

where $\boldsymbol{\varepsilon}$ is the perturbed strain field due to the presence of the inhomogeneity. Application of Hooke's law in D and $D - \Omega$ yields, respectively,

$$\boldsymbol{\sigma}^t = \boldsymbol{\sigma}^0 + \boldsymbol{\sigma} = \mathbf{L}_1(\boldsymbol{\varepsilon}^0 + \boldsymbol{\varepsilon}) \quad \text{in } \Omega, \tag{4.4.6}$$

$$\boldsymbol{\sigma}^t = \boldsymbol{\sigma}^0 + \boldsymbol{\sigma} = \mathbf{L}_0(\boldsymbol{\varepsilon}^0 + \boldsymbol{\varepsilon}) \quad \text{in } D - \Omega, \tag{4.4.7}$$

Our task is to solve for $\boldsymbol{\varepsilon}$ and $\boldsymbol{\sigma}$ by the so-called equivalent inclusion method.

To this end, let us consider a homogeneous body D with elastic modulus tensor $\mathbf{L}_0$ everywhere, containing an inclusion Ω with eigenstrain $\boldsymbol{\varepsilon}^*$ (the value of which is to be determined); see Figure 4.8. The eigenstrain $\boldsymbol{\varepsilon}^*$ is introduced here to simulate the inhomogeneity. In other words, we are trying to create an inclusion problem which, by properly adjusting the value of $\boldsymbol{\varepsilon}^*$, has the same stress field as that of the inhomogeneity problem.

For the inclusion problem shown in Figure 4.8, Hooke's law gives

$$\boldsymbol{\sigma}^t = \boldsymbol{\sigma}^0 + \boldsymbol{\sigma} = \mathbf{L}_0(\boldsymbol{\varepsilon}^0 + \boldsymbol{\varepsilon} - \boldsymbol{\varepsilon}^*) \quad \text{in } \Omega, \tag{4.4.8}$$

$$\boldsymbol{\sigma}^t = \boldsymbol{\sigma}^0 + \boldsymbol{\sigma} = \mathbf{L}_0(\boldsymbol{\varepsilon}^0 + \boldsymbol{\varepsilon}) \quad \text{in } D - \Omega, \tag{4.4.9}$$

where $\boldsymbol{\sigma}^0$ and $\boldsymbol{\varepsilon}^0$ are the fields in D due to the applied traction $\mathbf{p}^0 = \boldsymbol{\sigma}^0 \cdot \mathbf{n}$ when the inclusion is absent, and $\boldsymbol{\sigma}$ and $\boldsymbol{\varepsilon}$ are the fields due to eigenstrain $\boldsymbol{\varepsilon}^*$ in the inclusion Ω. From the Eshelby solution (4.3.20),

$$\boldsymbol{\varepsilon} = \mathbf{S}\boldsymbol{\varepsilon}^*, \tag{4.4.10}$$

we can rewrite (4.4.8) as

$$\boldsymbol{\sigma}^t = \boldsymbol{\sigma}^0 + \boldsymbol{\sigma} = \mathbf{L}_0(\boldsymbol{\varepsilon}^0 + \mathbf{S}\boldsymbol{\varepsilon}^* - \boldsymbol{\varepsilon}^*) \quad \text{in } \Omega. \tag{4.4.11}$$

Now, recall that we created this inclusion problem to simulate the inhomogeneity problem. We want to adjust the eigenstrain $\boldsymbol{\varepsilon}^*$ so that the stress field in the inclusion Ω given by (4.4.11) is the same as the stress field in the inhomogeneity Ω given by (4.4.6), namely,

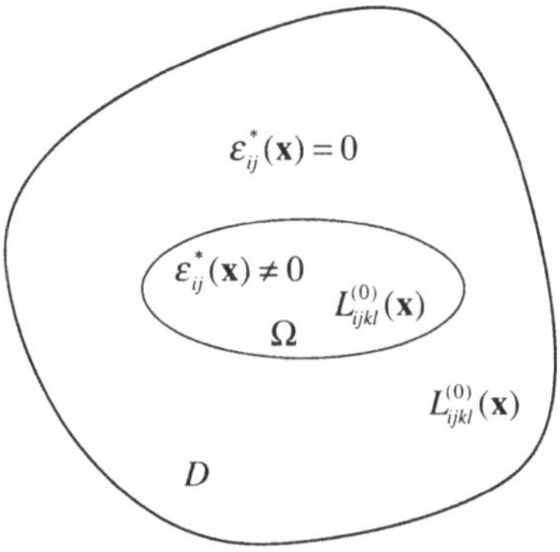

Figure 4.8 Ellipsoidal inclusion.

$$\mathbf{L}_0(\boldsymbol{\varepsilon}^0 + \mathbf{S}\boldsymbol{\varepsilon}^* - \boldsymbol{\varepsilon}^*) = \mathbf{L}_1(\boldsymbol{\varepsilon}^0 + \boldsymbol{\varepsilon}) \quad \text{in } \Omega. \tag{4.4.12}$$

Making use of the Eshelby solution (4.3.20) again yields

$$\mathbf{L}_0(\boldsymbol{\varepsilon}^0 + \mathbf{S}\boldsymbol{\varepsilon}^* - \boldsymbol{\varepsilon}^*) = \mathbf{L}_1(\boldsymbol{\varepsilon}^0 + \mathbf{S}\boldsymbol{\varepsilon}^*) \quad \text{in } \Omega. \tag{4.4.13}$$

This is called the equivalent inclusion equation. From this equation, the eigenstrain needed to simulate the inhomogeneity can be solved:

$$\begin{aligned}\boldsymbol{\varepsilon}^* &= -[(\mathbf{L}_1 - \mathbf{L}_0)\mathbf{S} + \mathbf{L}_0]^{-1}(\mathbf{L}_1 - \mathbf{L}_0)\boldsymbol{\varepsilon}^0 \\ &= -((\mathbf{L}_1 - \mathbf{L}_0)^{-1}[(\mathbf{L}_1 - \mathbf{L}_0)\mathbf{S} + \mathbf{L}_0])^{-1}\boldsymbol{\varepsilon}^0 \\ &= -[\mathbf{S} + (\mathbf{L}_1 - \mathbf{L}_0)^{-1}\mathbf{L}_0)]^{-1}\boldsymbol{\varepsilon}^0. \end{aligned} \tag{4.4.14}$$

The total strain on the inclusion thus follows from (4.4.5):

$$\boldsymbol{\varepsilon}^t = \boldsymbol{\varepsilon}^0 + \boldsymbol{\varepsilon} = \boldsymbol{\varepsilon}^0 + \mathbf{S}\boldsymbol{\varepsilon}^* = \mathbf{T}\boldsymbol{\varepsilon}^0, \tag{4.4.15}$$

where

$$\begin{aligned}\mathbf{T} &= \mathbf{I} - \mathbf{S}[\mathbf{S} + (\mathbf{L}_1 - \mathbf{L}_0)^{-1}\mathbf{L}_0]^{-1} \\ &= [\mathbf{S} + (\mathbf{L}_1 - \mathbf{L}_0)^{-1}\mathbf{L}_0) - \mathbf{S}][\mathbf{S} + (\mathbf{L}_1 - \mathbf{L}_0)^{-1}\mathbf{L}_0]^{-1} \\ &= (\mathbf{L}_1 - \mathbf{L}_0)^{-1}\mathbf{L}_0[\mathbf{S} + (\mathbf{L}_1 - \mathbf{L}_0)^{-1}\mathbf{L}_0]^{-1} \\ &= ([\mathbf{S} + (\mathbf{L}_1 - \mathbf{L}_0)^{-1}\mathbf{L}_0]\mathbf{L}_0^{-1}(\mathbf{L}_1 - \mathbf{L}_0))^{-1} \\ &= [\mathbf{S}\mathbf{L}_0^{-1}(\mathbf{L}_1 - \mathbf{L}_0) + \mathbf{I}]^{-1} \end{aligned} \tag{4.4.16}$$

or

$$\mathbf{T} = [\mathbf{I} + \mathbf{S}\mathbf{L}_0^{-1}(\mathbf{L}_1 - \mathbf{L}_0)]^{-1}. \tag{4.4.17}$$

Since the stresses in the inclusion are the same as those in the inhomogeneity, the total stress in the inhomogeneity is, therefore, given by

$$\boldsymbol{\sigma}^t = \boldsymbol{\sigma}^0 + \boldsymbol{\sigma} = \mathbf{L}_0(\boldsymbol{\varepsilon}^0 + \mathbf{S}\boldsymbol{\varepsilon}^* - \boldsymbol{\varepsilon}^*), \tag{4.4.18}$$

where $\boldsymbol{\varepsilon}^*$ is given by (4.4.14). This completes the solution of the inhomogeneity problem.

To close this section, we need to mention that when the eigenstrain is given by (4.4.14), the stresses for the inclusion problem and the inhomogeneity problem are identical not only inside Ω but also in $D - \Omega$. This can be shown by using the uniqueness theorem for inclusions and inhomogeneities.

4.5 INHOMOGENEOUS INHOMOGENEITIES

Sometimes an inhomogeneity may possesses its own eigenstrains. Such inhomogeneities are called inhomogeneous inhomogeneities. In other words, a subdomain occupied by foreign material may involve a distribution of eigenstrains as well. Examples of such inhomogeneous inhomogeneities include the formation of martensite blades in quenched carbon steels and precipitations in alloys.

To solve for the inhomogeneous inhomogeneity problem, let us consider the inhomogeneity shown in Figure 4.1. Let D be subjected to surface traction $\mathbf{p}^0 = \boldsymbol{\sigma}^0 \cdot \mathbf{n}$ on the boundary. Furthermore, we assume that a uniform eigenstrain $\boldsymbol{\varepsilon}^p$ is prescribed on the inhomogeneities Ω. We are to find the stress field in D.

If $\boldsymbol{\varepsilon}^0$ is used to denote the strain field corresponding to $\boldsymbol{\sigma}^0$, that is,

$$\boldsymbol{\sigma}^0 = \mathbf{L}_0 \boldsymbol{\varepsilon}^0, \tag{4.5.1}$$

then application of Hooke's law in D and $D - \Omega$ yields, respectively,

$$\boldsymbol{\sigma}^t = \boldsymbol{\sigma}^0 + \boldsymbol{\sigma} = \mathbf{L}_1(\boldsymbol{\varepsilon}^0 + \boldsymbol{\varepsilon} - \boldsymbol{\varepsilon}^p) \quad \text{in } \Omega, \tag{4.5.2}$$

$$\boldsymbol{\sigma}^t = \boldsymbol{\sigma}^0 + \boldsymbol{\sigma} = \mathbf{L}_0(\boldsymbol{\varepsilon}^0 + \boldsymbol{\varepsilon}) \quad \text{in } D - \Omega, \tag{4.5.3}$$

where $\boldsymbol{\sigma}$ and $\boldsymbol{\varepsilon}$ are the perturbed stress and strain fields, respectively, caused by the sum of two factors, the inhomogeneity and the eigenstrain $\boldsymbol{\varepsilon}^p$ in the inhomogeneity. Our task is to solve for $\boldsymbol{\varepsilon}$ and $\boldsymbol{\sigma}$.

Imagine that the inhomogeneous inhomogeneity is being simulated by an inclusion in a homogeneous matrix with elastic stiffness tensor $\mathbf{L}$, where the total eigenstrain on the inclusion is $\boldsymbol{\varepsilon}^p + \boldsymbol{\varepsilon}^*$. Then, (4.5.2) and (4.5.3) can be rewritten as

$$\boldsymbol{\sigma}^t = \boldsymbol{\sigma}^0 + \boldsymbol{\sigma} = \mathbf{L}_0(\boldsymbol{\varepsilon}^0 + \boldsymbol{\varepsilon} - \boldsymbol{\varepsilon}^p - \boldsymbol{\varepsilon}^*) \quad \text{in } \Omega, \tag{4.5.4}$$

$$\boldsymbol{\sigma}^t = \boldsymbol{\sigma}^0 + \boldsymbol{\sigma} = \mathbf{L}_0(\boldsymbol{\varepsilon}^0 + \boldsymbol{\varepsilon}) \quad \text{in } D - \Omega. \tag{4.5.5}$$

Consequently, the equivalent inclusion equation becomes

$$\mathbf{L}_0(\boldsymbol{\varepsilon}^0 + \boldsymbol{\varepsilon} - \boldsymbol{\varepsilon}^p - \boldsymbol{\varepsilon}^*) = \mathbf{L}_1(\boldsymbol{\varepsilon}^0 + \boldsymbol{\varepsilon} - \boldsymbol{\varepsilon}^p) \quad \text{in } \Omega. \tag{4.5.6}$$

The Eshelby solution (4.3.20) in this case takes the form of

$$\boldsymbol{\varepsilon} = \mathbf{S}\boldsymbol{\varepsilon}^{**} = \mathbf{S}(\boldsymbol{\varepsilon}^p + \boldsymbol{\varepsilon}^*), \tag{4.5.7}$$

where

$$\boldsymbol{\varepsilon}^{**} = \boldsymbol{\varepsilon}^p + \boldsymbol{\varepsilon}^*. \tag{4.5.8}$$

Introducing (4.5.7) into (4.5.6) yields

$$\mathbf{L}_0(\boldsymbol{\varepsilon}^0 + \mathbf{S}\boldsymbol{\varepsilon}^{**} - \boldsymbol{\varepsilon}^{**}) = \mathbf{L}_1(\boldsymbol{\varepsilon}^0 + \mathbf{S}\boldsymbol{\varepsilon}^{**} - \boldsymbol{\varepsilon}^p) \quad \text{in } \Omega. \tag{4.5.9}$$

This can be solved to obtain

$$\boldsymbol{\varepsilon}^{**} = [\mathbf{L}_0(\mathbf{S} - \mathbf{I}) - \mathbf{L}_1\mathbf{S}]^{-1}(\Delta\mathbf{L}\ \mathbf{M}_0\boldsymbol{\sigma}^0 - \mathbf{L}_1\boldsymbol{\varepsilon}^p), \tag{4.5.10}$$

where

$$\Delta\mathbf{L} = \mathbf{L}_1 - \mathbf{L}_0. \tag{4.5.11}$$

Note that

$$\begin{aligned}\mathbf{S}[\mathbf{L}_0(\mathbf{S} - \mathbf{I}) - \mathbf{L}_1\mathbf{S}]^{-1} &= \mathbf{S}[\mathbf{L}_0(\mathbf{S} - \mathbf{I}) - \mathbf{L}_1\mathbf{S}]^{-1} \\ &= -(\mathbf{L}_0\mathbf{S}^{-1} - \mathbf{L}_0 + \mathbf{L}_1)^{-1} = -(\mathbf{H} + \mathbf{L}_1)^{-1},\end{aligned} \tag{4.5.12}$$

where

$$\mathbf{H} = \mathbf{L}_0\mathbf{S}^{-1} - \mathbf{L}_0 = \mathbf{L}_0(\mathbf{S}^{-1} - \mathbf{I}) = \mathbf{P}^{-1} - \mathbf{L}_0 \tag{4.5.13}$$

is known as Hill's constraint tensor. Therefore, (4.5.10) can be written alternatively,

$$\boldsymbol{\varepsilon}^{**} = -\mathbf{S}^{-1}(\mathbf{H} + \mathbf{L}_1)^{-1}(\Delta\mathbf{L}\ \mathbf{M}_0\boldsymbol{\sigma}^0 - \mathbf{L}_1\boldsymbol{\varepsilon}^p). \tag{4.5.14}$$

Once the eigenstrain $\boldsymbol{\varepsilon}^{**}$ is known, the total strain on the inhomogeneous inhomogeneity can be computed as

$$\boldsymbol{\varepsilon}^t = \boldsymbol{\varepsilon}^0 + \boldsymbol{\varepsilon} = \boldsymbol{\varepsilon}^0 + \mathbf{S}\boldsymbol{\varepsilon}^{**}. \tag{4.5.15}$$

Therefore, substituting (4.5.14) into (4.5.15), we can write the total strain in the inhomogeneous inhomogeneity as

$$\boldsymbol{\varepsilon}^t = \boldsymbol{\varepsilon}^0 + \boldsymbol{\varepsilon} = \boldsymbol{\varepsilon}^0 - (\mathbf{H} + \mathbf{L}_1)^{-1}(\Delta\mathbf{L}\ \mathbf{M}_0\boldsymbol{\sigma}^0 + \boldsymbol{\tau}^p), \tag{4.5.16}$$

where the stress polarization is defined by

$$\boldsymbol{\tau}^p = -\mathbf{L}_1\boldsymbol{\varepsilon}^p. \tag{4.5.17}$$

Similarly, the total stress on the inhomogeneous inhomogeneity is given by (4.5.4)

$$\begin{aligned}\boldsymbol{\sigma}^t &= \mathbf{L}_0(\boldsymbol{\varepsilon}^0 + \mathbf{S}\boldsymbol{\varepsilon}^{**} - \boldsymbol{\varepsilon}^{**}) \\ &= \boldsymbol{\sigma}^0 - \mathbf{L}_0(\mathbf{I} - \mathbf{S}^{-1})(\mathbf{H} + \mathbf{L}_1)^{-1}(\Delta\mathbf{L}\ \mathbf{M}_0\boldsymbol{\sigma}^0 + \boldsymbol{\tau}^p) \quad \text{in } \Omega \\ &= \boldsymbol{\sigma}^0 + \mathbf{H}(\mathbf{H} + \mathbf{L}_1)^{-1}(\Delta\mathbf{L}\ \mathbf{M}_0\boldsymbol{\sigma}^0 + \boldsymbol{\tau}^p).\end{aligned} \tag{4.5.18}$$

Making use of (4.5.16) in (4.5.18) gives an alternative form:

$$\boldsymbol{\sigma}^t = \boldsymbol{\sigma}^0 - \mathbf{H}(\boldsymbol{\varepsilon}^t - \boldsymbol{\varepsilon}^0) \quad \text{in } \Omega. \tag{4.5.19}$$

In the absence of the applied loading, (4.5.16) and (4.5.18) become, respectively,

$$\boldsymbol{\varepsilon}^t = -(\mathbf{H} + \mathbf{L}_1)^{-1}\boldsymbol{\tau}^p, \tag{4.5.20}$$

$$\boldsymbol{\sigma}^t = \mathbf{H}(\mathbf{H} + \mathbf{L}_1)^{-1}\boldsymbol{\tau}^p = -\mathbf{H}\boldsymbol{\varepsilon}^t \quad \text{in } \Omega. \tag{4.5.21}$$

PROBLEMS

4.1 A spherical cavity of radius a is present in an otherwise uniform, isotropic, linearly elastic solid of infinite extent with Poisson's ratio ν. A uniaxial tension $\sigma_{ij}^0 = \sigma_0 \delta_{i3} \delta_{j3}$ is applied at infinity. (See Fig. 4.9.)

1. Use the equivalent inclusion method to find the stress components $\sigma_{ij}(x_1, x_2, x_3)$ at points (0, 0, a) and (0, a, 0).
2. Define the stress concentration factors by

$$\eta_t = \frac{\sigma_{33}(0, a, 0)}{\sigma_0}, \qquad \eta_c = \frac{\sigma_{11}(0, 0, a)}{\sigma_0}$$

and plot the stress concentration factors versus the Poisson's ratio ν.

4.2 Prove that the stress polarization tensor P_{ijkl} symmetric and positive-definite. Further,

$$\tau_{ij}^* M_{ijkl} \tau_{kl}^* \geq \tau_{ij}^* P_{ijkl} \tau_{kl}^* \geq 0$$

for any arbitrary second-order tensor τ_{ij}^*.

4.3 Prove the identity

$$-(\mathbf{L}_1 - \mathbf{L}_0)^{-1}\mathbf{L}_0 = \mathbf{M}_1(\mathbf{M}_1 - \mathbf{M}_0)^{-1}.$$

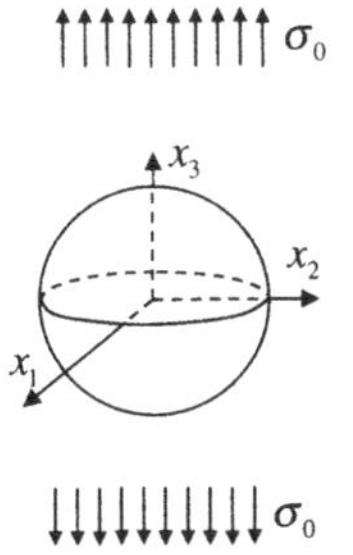

Figure 4.9

APPENDIX 4A

Consider the integral

$$I = \int_{\Omega} \delta[\hat{\boldsymbol{\xi}} \cdot (\mathbf{x} - \boldsymbol{\xi})] \, d\boldsymbol{\xi}, \tag{4.A.1}$$

where the integration doman Ω is an ellipsoid defined by (4.3.1). To carry out this integral, we introduce the following substitutions:

$$\bar{\boldsymbol{\xi}} = \begin{bmatrix} \xi_1/a_1 \\ \xi_2/a_2 \\ \xi_3/a_3 \end{bmatrix}, \qquad \bar{\mathbf{x}} = \begin{bmatrix} x_1/a_1 \\ x_2/a_2 \\ x_3/a_3 \end{bmatrix}, \qquad \boldsymbol{\eta} = \frac{1}{a} \begin{bmatrix} a_1\hat{\xi}_1 \\ a_2\hat{\xi}_2 \\ a_3\hat{\xi}_3 \end{bmatrix}, \tag{4.A.2}$$

where

$$a = \sqrt{(a_1\hat{\xi}_1)^2 + (a_2\hat{\xi}_2)^2 + (a_3\hat{\xi}_3)^2}. \tag{4.A.3}$$

Then, the volume integral over the ellipsoid can be converted to a volume integral overa unit sphere:

$$I = \int_{\Omega} \delta[\hat{\boldsymbol{\xi}} \cdot (\mathbf{x} - \boldsymbol{\xi})] \, d\boldsymbol{\xi} = \int_{\bar{\xi}_i \bar{\xi}_i \le 1} \delta[a(\boldsymbol{\eta} \cdot \bar{\mathbf{x}} - \boldsymbol{\eta} \cdot \bar{\boldsymbol{\xi}})] a_1 a_2 a_3 \, d\bar{\boldsymbol{\xi}}. \tag{4.A.4}$$

In the case $\boldsymbol{\eta} \cdot \bar{\mathbf{x}} < 1$, the above integral can be written as

$$\begin{aligned} I &= \frac{a_1 a_2 a_3}{a} \int_0^{2\pi} \left\{ \int_{-1}^{1} \left[\int_0^{\sqrt{1-z^2}} \delta(\boldsymbol{\eta} \cdot \bar{\mathbf{x}} - z) \, dr \right] dz \right\} d\varphi \\ &= \frac{a_1 a_2 a_3}{2a} \int_0^{2\pi} \left\{ \int_{-1}^{1} (1 - z^2) \delta(\boldsymbol{\eta} \cdot \bar{\mathbf{x}} - z) \, dz \right\} d\varphi \\ &= \frac{a_1 a_2 a_3}{2a} \int_0^{2\pi} [1 - (\boldsymbol{\eta} \cdot \bar{\mathbf{x}})^2] \, d\varphi \\ &= \frac{\pi a_1 a_2 a_3}{a} [1 - (\boldsymbol{\eta} \cdot \bar{\mathbf{x}})^2]. \end{aligned} \tag{4.A.5}$$

See Figure 4.10.

Returning to the original variables, we have

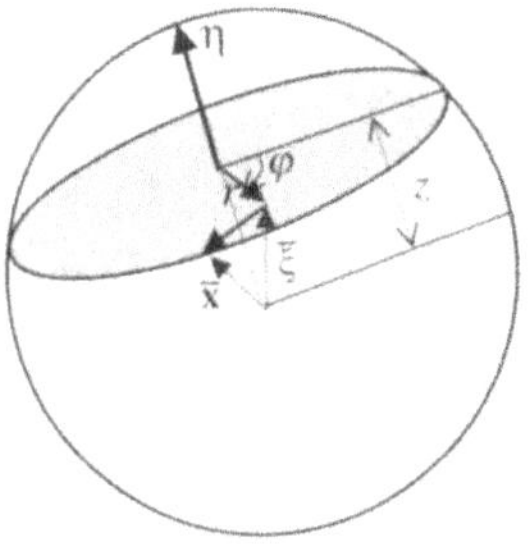

Figure 4.10 Domain of integration.

$$I = \frac{\pi a_1 a_2 a_3}{a}\,[1 - (\boldsymbol{\eta}\cdot\bar{\mathbf{x}})^2] = \frac{\pi a_1 a_2 a_3}{a^3}\,[a^2 - (\hat{\boldsymbol{\xi}}\cdot\mathbf{x})^2]. \quad (4.A.6)$$

For a point $\mathbf{x}$ inside the ellipsoid, the corresponding $\bar{\mathbf{x}}$ is inside a unit sphere, that is, $\|\bar{\mathbf{x}}\| < 1$. Since $\boldsymbol{\eta}$ is a unit vector, we have $\boldsymbol{\eta}\cdot\bar{\mathbf{x}} < 1$ for any point $\mathbf{x}$ inside the ellipsoid Ω. Therefore, the above result is valid for any point $\mathbf{x}$ inside the ellipsoid Ω.

APPENDIX 4B ESHELBY TENSOR FOR ISOTROPIC MATERIALS WITH POISSON'S RATIO ν

Ellipsoid ($a_1 > a_2 > a_3$)

$$S_{1111} = \frac{3}{8\pi(1-\upsilon)}\,a_1^2 I_{11} + \frac{1-2\upsilon}{8\pi(1-\upsilon)}\,I_1,$$

$$S_{1122} = \frac{1}{8\pi(1-\upsilon)}\,a_2^2 I_{12} - \frac{1-2\upsilon}{8\pi(1-\upsilon)}\,I_1,$$

$$S_{1133} = \frac{1}{8\pi(1-\upsilon)}\,a_3^2 I_{13} - \frac{1-2\upsilon}{8\pi(1-\upsilon)}\,I_1,$$

$$S_{1212} = \frac{a_1^2 + a_2^2}{16\pi(1-\upsilon)}\,I_{12} + \frac{1-2\upsilon}{16\pi(1-\upsilon)}\,(I_1 + I_2).$$

All other nonzero components are obtained by the cyclic permutation of the above equations. The components that cannot be obtained by the cyclic permutation are zero; for instance, $S_{1112} = S_{1223} = S_{1232} = 0$.

In the above equations, the constants I_1, I_2, I_{11}, I_{12}, and I_{13} are given by the following:

$$I_1 = \frac{4\pi a_1 a_2 a_3}{(a_1^2 - a_2^2)(a_1^2 - a_3^2)^{1/2}} \{F(\theta, k) - E(\theta, k)\},$$

$$I_3 = \frac{4\pi a_1 a_2 a_3}{(a_2^2 - a_3^2)(a_1^2 - a_3^2)^{1/2}} \left\{\frac{a_2(a_1^2 - a_3^2)^{1/2}}{a_1 a_3} - E(\theta, k)\right\},$$

where the elliptical integrals are defined by

$$F(\theta, k) = \int_0^\theta \frac{dw}{(1 - k^2 \sin^2 w)^{1/2}},$$

$$E(\theta, k) = \int_0^\theta (1 - k^2 \sin^2 w)^{1/2}\, dw,$$

$$\theta = \sin^{-1}(1 - a_3^2/a_1^2)^{1/2}, \qquad k = \{(a_1^2 - a_2^2)/(a_1^2 - a_3^2)\}^{1/2}.$$

Furthermore

$$I_1 + I_2 + I_3 = 4\pi,$$

$$3I_{11} + I_{12} + I_{13} = 4\pi/a_1^2,$$

$$3a_1^2 I_{11} + a_2^2 I_{12} + a_3^2 I_{13} = 3I_1,$$

$$I_{12} = (I_2 - I_1)/(a_1^2 - a_2^2).$$

These equations and their cyclic counterparts give sufficient relations to express I_i and I_{ij} in terms of I_1 and I_3.

Sphere ($a_1 = a_2 = a_3 = a$)

$$I_1 = I_2 = I_3 = 4\pi/3,$$

$$I_{11} = I_{22} = I_{33} = I_{12} = I_{23} = I_{31} = 4\pi/5a^2,$$

$$S_{1111} = S_{2222} = S_{3333} = \frac{7 - 5\upsilon}{15(1 - \upsilon)},$$

$$S_{1122} = S_{2233} = S_{3311} = S_{1133} = S_{2211} = S_{3322} = \frac{5v - 1}{15(1 - v)},$$

$$S_{1212} = S_{2323} = S_{3131} = \frac{4 - 5v}{15(1 - v)}.$$

Elliptic Cylinder ($a_3 \to \infty$)

$$I_1 = 4\pi a_2/(a_1 + a_2), \qquad I_2 = 4\pi a_1/(a_1 + a_2), \quad I_3 = 0$$

$$I_{12} = 4\pi/(a_1 + a_2)^2, \qquad 3I_{11} = 4\pi/a_1^2 - I_{12},$$

$$3I_{22} = 4\pi/a_2^2 - I_{12}, \qquad I_{13} = I_{23} = I_{33} = 0,$$

$$a_3^2 I_{13} = I_1, \qquad a_3^2 I_{23} = I_2, \qquad a_3^2 I_{33} = 0.$$

$$S_{1111} = \frac{1}{2(1 - v)} \left\{ \frac{a_2^2 + 2a_1 a_2}{(a_1 + a_2)^2} + (1 - 2v) \frac{a_2}{a_1 + a_2} \right\},$$

$$S_{2222} = \frac{1}{2(1 - v)} \left\{ \frac{a_1^2 + 2a_1 a_2}{(a_1 + a_2)^2} + (1 - 2v) \frac{a_1}{a_1 + a_2} \right\}, \qquad S_{3333} = 0,$$

$$S_{1122} = \frac{1}{2(1 - v)} \left\{ \frac{a_2^2}{(a_1 + a_2)^2} - (1 - 2v) \frac{a_2}{a_1 + a_2} \right\},$$

$$S_{2233} = \frac{1}{2(1 - v)} \frac{2v a_1}{a_1 + a_2}, \qquad S_{3311} = 0,$$

$$S_{1133} = \frac{1}{2(1 - v)} \frac{2v a_2}{a_1 + a_2},$$

$$S_{2211} = \frac{1}{2(1 - v)} \left\{ \frac{a_1^2}{(a_1 + a_2)^2} - (1 - 2v) \frac{a_1}{a_1 + a_2} \right\},$$

$$S_{3322} = 0, \qquad S_{1212} = \frac{1}{2(1 - v)} \left\{ \frac{a_1^2 + a_2^2}{2(a_1 + a_2)^2} + \frac{1 - 2v}{2} \right\},$$

$$S_{2323} = \frac{a_1}{2(a_1 + a_2)}, \qquad S_{3131} = \frac{a_2}{2(a_1 + a_2)}.$$

Penny Shape ($a_1 = a_2 >> a_3$)

$$I_1 = I_2 = \pi^2 a_3/a_1, \qquad I_3 = 4\pi - 2\pi^2 a_3/a_1,$$

$$I_{12} = I_{21} = 3\pi^2 a_3/4a_1^3,$$

$$I_{13} = I_{23} = I_{31} = I_{32} = 3(\tfrac{4}{3}\pi - \pi^2 a_3/a_1)/a_1^2,$$

$$I_{11} = I_{22} = 3\pi^2 a_3/4a_1^3, \qquad I_{33} = \tfrac{4}{3}\pi/a_3^2;$$

$$S_{1111} = S_{2222} = \frac{13 - 8v}{32(1 - v)}\pi\frac{a_3}{a_1}, \qquad S_{3333} = 1 - \frac{1 - 2v}{1 - v}\frac{\pi}{4}\frac{a_3}{a_1},$$

$$S_{1122} = S_{2211} = \frac{8v - 1}{32(1 - v)}\pi\frac{a_3}{a_1}, \qquad S_{2233} = \frac{2v - 1}{8(1 - v)}\pi\frac{a_3}{a_1},$$

$$S_{3311} = S_{3322} = \frac{v}{1 - v}\left(1 - \frac{4v + 1}{8v}\pi\frac{a_3}{a_1}\right),$$

$$S_{1212} = \frac{7 - 8v}{32(1 - v)}\pi\frac{a_3}{a_1}, \qquad S_{1313} = S_{2323} = \frac{1}{2}\left(1 + \frac{v - 2}{1 - v}\frac{\pi}{4}\frac{a_3}{a_1}\right),$$

$$S_{kk11} = S_{kk22} = \frac{1 - 2v}{1 - v}\frac{\pi}{4}\frac{a_3}{a_1} + \frac{v}{1 - v}, \qquad S_{kk33} = 1 - \frac{1 - 2v}{1 - v}\frac{\pi}{2}\frac{a_3}{a_1}.$$

When $a_3 = 0$,

$$I_1 = I_2 = 0, \quad I_3 = 4\pi,$$

$$I_{12} = 0, \qquad I_{23} = 4\pi/a_2^2, \qquad I_{31} = 4\pi/a_1^2,$$

$$I_{11} = I_{22} = 0, \qquad a_3^2 I_{33} = 4\pi/3,$$

$$S_{2323} = S_{3131} = \tfrac{1}{2},$$

$$S_{3311} = S_{3322} = v/(1 - v),$$

$$S_{3333} = 1, \qquad \text{and all other } S_{ijkl} = 0.$$

Flat Ellipsoid ($a_1 > a_2 >> a_3$)

$$I_1 = 4\pi a_2 a_3\{F(k) - E(k)\}/(a_1^2 - a_2^2),$$

$$I_2 = 4\pi a_3 E(k)/a_2 - 4\pi a_2 a_3\{F(k) - E(k)\}/(a_1^2 - a_2^2),$$

$$I_3 = 4\pi - 4\pi a_3 E(k)/a_2,$$

$$I_{12} = [4\pi a_3 E(k)/a_2 - 8\pi a_2 a_3\{F(k) - E(k)\}/(a_1^2 - a_2^2)]/(a_1^2 - a_2^2),$$

$$I_{23} = [4\pi - 8\pi a_3 E(k)/a_2 + 4\pi a_2 a_3\{F(k) - E(k)\}/(a_1^2 - a_2^2)]/a_2^2,$$

$$I_{31} = [4\pi - 4\pi a_2 a_3\{F(k) - E(k)\}/(a_1^2 - a_2^2) - 4\pi a_3 E(k)/a_2]/a_1^2,$$

$$I_{33} = 4\pi/3a_3^2,$$

where $F(k)$ and $E(k)$ are the complete elliptic integrals of the first and the second kind, respectively,

$$E(k) = \int_0^{\pi/2} (1 - k^2 \sin^2 \phi)^{1/2}\, d\phi,$$

$$F(k) = \int_0^{\pi/2} (1 - k^2 \sin^2 \phi)^{-1/2}\, d\phi,$$

$$k^2 = (a_1^2 - a_2^2)/a_1^2.$$

Oblate Spheroid ($a_1 = a_2 > a_3$)

$$I_1 = I_2 = \frac{2\pi a_1^2 a_3}{(a_1^2 - a_3^2)^{3/2}} \left\{ \cos^{-1}\frac{a_3}{a_1} - \frac{a_3}{a_1}\left(1 - \frac{a_3^2}{a_1^2}\right)^{1/2} \right\},$$

$$I_3 = 4\pi - 2I_1,$$

$$I_{11} = I_{22} = I_{12},$$

$$I_{12} = \pi/a_1^2 - \tfrac{1}{4} I_{13} = \pi/a_1^2 - \frac{I_1 - I_3}{4(a_3^2 - a_1^2)},$$

$$I_{13} = I_{23} = (I_1 - I_3)/(a_3^2 - a_1^2), \qquad 3I_{33} = 4\pi/a_3^2 - 2I_{13}.$$

Prolate Spheroid ($a_1 > a_2 = a_3$)

$$I_2 = I_3 = \frac{2\pi a_1 a_3^2}{(a_1^2 - a_3^2)^{3/2}} \left\{ \frac{a_1}{a_3} \left(\frac{a_1^2}{a_3^2} - 1 \right)^{1/2} - \cosh^{-1} \frac{a_1}{a_3} \right\},$$

$$I_1 = 4\pi - 2I_2, \qquad I_{12} = (I_2 - I_1)/(a_1^2 - a_2^2),$$

$$3I_{11} = 4\pi/a_1^2 - 2I_{12},$$

$$I_{22} = I_{33} = I_{23}, \quad 3I_{22} = 4\pi/a_2^2 - I_{23} - (I_2 - I_1)/(a_1^2 - a_2^2),$$

$$I_{23} = \pi/a_2^2 - (I_2 - I_1)/4(a_1^2 - a_2^2).$$

SUGGESTED READING

Mura, T. (1987). *Micromechanics of Defects in Solids*, Martinus Nijhoff, Boston, Chapters 3–4.

Nemat-Nasser, S. and M. Hori. (1993). *Micromechanics: Overall Properties of Heterogeneous Materials,* North-Holland, New York, Chapter 7.

5

DEFINITIONS OF EFFECTIVE MODULI OF HETEROGENEOUS MATERIALS

Most engineering materials are heterogeneous in nature. They generally consist of different constituents or phases, which are distinguishable at specific scales. Each constituent may show different physical properties (e.g., elastic moduli, thermal expansion, yield strength, electrical conductivity, thermal conductivity, etc.) and/or material orientations. However, in many engineering applications, a structure component may contain numerous such constituents such that it is impractical or even impossible to account for each and every one of them for engineering design and analysis. What is really relevant is the overall or effective property of the material from which the component is made. In this chapter, we will lay out the basic definitions of overall or effective properties of engineering materials that are heterogeneous at the length scale of interest.

Generally speaking, heterogeneous materials can be classified into two categories, periodic and nonperiodic. In this text, we will focus primarily on heterogeneous materials with nonperiodic microstructures. So, unless indicated otherwise, the term heterogeneous material refers to heterogeneous materials with nonperiodic microstructures. Examples of such materials may include most of the fiber or particulate reinforced composites and polycrystalline materials.

5.1 HETEROGENEITY AND LENGTH SCALES

Generally speaking, engineering materials are heterogeneous, that is, they have microstructure and they contain heterogeneities. However, what constitutes a heterogeneity depends on the length scales used in the observation. Certain constituents or phases in a given material are identifiable only at or below a specific length scale. At such a length scale, each constituent may be homogeneous, but when observed at a smaller length scale, the constituent itself may become heterogeneous. A polycrystal copper, for example, may seem to be homogeneous with the naked eye ($\sim$10 μm). But, under optical microscope ($\sim$1 μm), the polycrystal copper is composed of many grains—small single copper crystals with different orientations. At this length scale, although each grain can be treated as a homogeneous material, the polycrystal copper as an assembly of grains cannot be treated as a homogeneous material any more because the microstructure attributes, such as grain size, orientation, and grain boundaries, dictate how the material behaves at this length. When we go down further along the length scales, we know that each grain also has its own microstructure. The behavior and properties of materials at each length scale are controlled by the observable microstructure at the corresponding length scale. Therefore, whether a material is heterogeneous or not depends on the length scale used in the observation. In other words, when studying the properties of a real material, which is always heterogeneous, we need to define the length scale at which the properties of interest are directly relevant. Microstructure features not observable at this length can be neglected. In fact, for practical purposes, it is only certain averaged effects of the microstructure that are of interest. So, in the study of heterogeneous materials, we often speak of overall properties. By that we mean the properties averaged over a certain volume of the heterogeneous material. For such overall properties to be meaningful, the average taken over any arbitrary volume element comparable with the relevant length scale must be the same with the heterogeneous material sample under consideration. Heterogeneous materials that meet this requirement are said to be macroscopically homogeneous. We will give a more quantitative definition of macroscopic homogeneity later in this chapter.

For now, let us consider, for example, a fiber-reinforced composite laminate. It is certainly impractical and unnecessary to deal with each individual fiber when the overall bending rigidity of the laminate is of primary interest. On the other hand, the bending rigidity of the laminate is very closely related to the lay up of the various layers within the laminate. To derive the overall rigidity of the composite, the laminate must be treated as a stack of plies each being different in composition and/or orientation, albeit each ply can be treated as a homogeneous layer.

In the examples given above, we have implicitly implied that there is a separation of length scales D and d on which the macroscopic properties (e.g., bending rigidity of the composite laminate) and microscopic properties (elasticity of the plies) are defined, respectively. The microscale length d corresponds to the smallest constituent whose physical properties, orientation, and shape are judged to have direct first-order effects on the macroscopic overall physical properties of the heterogeneous material at the length scale D. The choice of d is generally adapted to the problem under consideration. An appropriate choice should be guided by systematic "multiscale" experimental observations. Generally speaking, an optimum choice would be the one that includes a good balance between the definitions of the microscale that have a first-order effect on the overall properties, and the simplicity of the resulting model. The macroscale length D should be large enough so that the microscale fluctuations (perturbations) of the stress and strain fields due to the local variation of microstructure at the scale d influence the overall effective property only through their averages. In fact, in the framework of micromechanics, the stress and strain fields are split into contributions corresponding to different length scales. It is assumed that these length scales are sufficiently different in terms of their order of magnitude, so that for each pair of them, the fluctuations of stress and strain fields (micro or local quantities) at the smaller length scale influence the overall (or macroscopic) behavior at the larger length scale only through their averages, and, conversely, fluctuations of stress and strain fields as well as the compositional gradients (macro or global quantities) are not significant at the smaller length scale. Therefore, at this scale, these macrofields are locally uniform and can be described as uniformly applied stresses and strains. To meet these conditions, the typical dimension of the microscale constituents should be orders of magnitude smaller than the macroscale element so that, $d/D << 1$. In the composite laminate example mentioned above, if the thickness of each single ply is on the order of $d \sim 0.1$ mm, the thickness of the laminate should be at least on the order of $D \sim 1$ mm so that the macroscopic effective bending rigidity can be defined meaningfully.

Note that the identification of d and D is dependent of the length scale of interest. In the above example, since our interest is in the overall bending rigidity of the laminate, we have taken the overall thickness of the laminate to be our macroscopic length parameter and the thickness of each ply to be our microscopic parameter. This way, it implies that each ply is a homogeneous material with no microstructure. However, if we are interested in the behavior of each individual ply, that is, our interest is in the length scale of the ply thickness, we must then identify the ply thickness as our macroscopic length parameter. In this case, the microscopic length parameter d would be the diameter of the individual fibers within the ply. Each ply as a whole can no longer be viewed as a homogeneous material. Instead, it becomes a composite consisting of polymer resin and the fibers. Each constituent can be considered a homogeneous material.

5.2 REPRESENTATIVE VOLUME ELEMENT

In light of the above discussions, we introduce the concept of microhomogeneity. The microstructure of heterogeneous materials is, at any given length scale, very complex, and, to a certain extent, random. To describe the precise topographic features of the microstructure variation is usually impractical and very often impossible. In fact, for practical purposes, it is only certain averaged effects of the microstructure that are of interest. So, in the study of heterogeneous materials, we often speak of overall properties. By that we mean the properties averaged over a certain volume of the heterogeneous material. For such overall properties to be meaningful, the average taken over any arbitrary volume element comparable with the relevant length scale must be the same with the heterogeneous material sample under consideration. Heterogeneous materials that meet this requirement are said to be macroscopically heterogeneous. In light of the discussions in the preceding paragraph, if in a heterogeneous material the microscopic length parameter d and the macroscopic length parameter D can be identified for a length scale of interest such that $d/D \ll 1$, then the heterogeneous material is microscopically homogeneous at the length scale D. A volume element with characteristic dimension of D is called a representative volume element (RVE) because the overall properties on any RVE would be the same. In other words, the overall properties of each RVE represent the overall properties of the heterogeneous material.

5.3 RANDOM MEDIA

The foregoing discussions assumed that the microstructure of the heterogeneous material is known. In reality, the microstructures of engineering materials are very complex and very often random in nature. It is impractical and even impossible to have knowledge on every detail of the microstructure configuration at any given length scale, except in cases such as those displaying periodicity. Typically, only certain "average" features of the microstructure will be known, and such information is likely to be statistical in character. In this section, we will discuss how such statistical information can be incorporated in the definition of RVE.

Consider a batch of particulate-reinforced composite blocks made under the same processing conditions. Let $\mathfrak{R}$ denote the batch and N be the total number of blocks in $\mathfrak{R}$. Assume that all the blocks are the same size D, which is much larger than the particle diameter d, and the particle content (e.g., volume fraction) c is the same for all blocks. However, due to manufacture variability, the distribution of the particles in each block is inevitably different one from the other. Consequently, if we choose a small volume at the same location $\mathbf{x}$ on each block, the particle content $c(\mathbf{x}, \alpha)$ in this small volume may not be the same among all the blocks, and thus, the average

$$\langle c(\mathbf{x}) \rangle = \frac{1}{N} \sum_{\alpha=1}^{N} c(\mathbf{x}, \alpha), \tag{5.3.1}$$

may depend on N. However, it is conceivable that if N is very large, the average value computed from (5.3.1) would become relatively independent of N. In the limit of $N \to \infty$, the summation can be replaced by an integral over the sample space $\mathfrak{R}$, and the average becomes independent of N:

$$\langle c(\mathbf{x}) \rangle = \lim_{N \to \infty} \frac{1}{N} \sum_{\alpha=1}^{N} c(\mathbf{x}, \alpha) = \int_{\mathfrak{R}} c(\mathbf{x}, \alpha)\, d\alpha. \tag{5.3.2}$$

Such an averaging scheme is called the ensemble average.

Now, let us consider a different average by picking out a block randomly from the batch; call it the αth block. First we divide this block into M subdomains of volume V_i, where $\Sigma_{i=1}^{M} V_i = V$ with V being the total volume of the block. Then, we calculate the particle volume fraction $c(\mathbf{x}, \alpha)$ on each of the subdomains, where $\mathbf{x}_i$ is the coordinate of a representative point, for example, the centroid of V_i. Again, due to manufacture variability, the particle volume fraction on each of the subdomains may not be the same among all the subdomains, and, thus, the average

$$\overline{c(\alpha)} = \frac{1}{V} \sum_{i=1}^{M} c(\mathbf{x}_i) V_i \tag{5.3.3}$$

may depend on M. However, it is conceivable that if M is very large, the average value computed from (5.3.3) would become relatively independent of M. In the limit of $M \to \infty$, the summation can be replaced by an integral over the entire volume V and the average becomes independent of M:

$$\overline{c(\alpha)} = \lim_{M\to\infty} \frac{1}{V} \sum_{i=1}^{M} c(\mathbf{x}_i) V_i = \frac{1}{V} \int_V c(\mathbf{x})\, dv. \tag{5.3.4}$$

We will call this average scheme the volume (or spatial) average.

Note that the ensemble average is carried out over a group of blocks made under the same process, while the volume average is performed over an individual block out of the group. The nature of manufacture variability is such that the local particle volume fraction $c(\mathbf{x}, \alpha)$ at $\mathbf{x}$ can be viewed as a random function of the sample α, as well as a random function of the position $\mathbf{x}$; each represents a different random process. The ensemble average gives the mean value among all the samples in the sample space, while the volume average tells us the mean value over a particular sample. Imagine that the particle distribution in each and every block were exactly the same; it is plausible that the two average schemes should give the same value, that is,

$$\langle c(\mathbf{x}) \rangle = \overline{c(\alpha)}, \tag{5.3.5}$$

or

$$\int_{\Re} c(\mathbf{x}, \alpha)\, d\alpha = \frac{1}{V} \int_V c(\mathbf{x}, \alpha)\, dV(\mathbf{x}). \tag{5.3.6}$$

This statement, that is, ensemble average equals volume average, is known as the ergodic hypothesis, a concept used frequently in stochastic analysis, where time average is used in lieu of the volume average defined in (5.3.4).

Let us now generalize the ergodic hypothesis to the description of microstructure features in heterogeneous materials. Consider, generally, a group of heterogeneous materials. Let us name the group $\Re$ and each member of the group will be characterized by the label α. The value of α is taken as defining the member (one of the materials in the group) completely. Further, assume that, associated with α, there is a probability distribution function $p(\alpha)$ defined over $\Re$, which satisfies

$$\int_{\Re} p(\alpha)\, d\alpha = 1. \tag{5.3.7}$$

Next, assume that each member of the group has a total volume V and consists of N different types of inhomogeneities or phases. The r th phase in sample α is assumed to occupy the volume $V_r(\alpha)$, $r = 1, 2, 3, \ldots, N$. To describe the material's microstructure or the spatial distribution of inhomogeneities, it is convenient to introduce the characteristic function that takes the value of 1 if $\mathbf{x}$ lies within the volume $V_r(\alpha)$ and zero otherwise, that is,

$$f_r(\mathbf{x}, \alpha) = \begin{cases} 1 & \mathbf{x} \in V_r(\alpha) \\ 0 & \mathbf{x} \notin V_r(\alpha) \end{cases}. \tag{5.3.8}$$

Clearly, the characteristic function $f_r(\mathbf{x}, \alpha)$ satisfies the following consistency condition:

$$\sum_{r=1}^{N} f_r(\mathbf{x}, \alpha) = 1. \tag{5.3.9}$$

Thus, the ensemble average of $f_r(\mathbf{x}, \alpha)$ defines the probability $P_r(\mathbf{x})$ of finding phase r at location $\mathbf{x}$:

$$P_r(\mathbf{x}) = \langle f_r(\mathbf{x}, \alpha) \rangle \equiv \int_{\Re} f_r(\mathbf{x}, \alpha) p(\alpha)\, d\alpha. \tag{5.3.10}$$

The function $P_r(\mathbf{x})$ is also referred to as the one-point correlation function for the characteristic function $f_r(\mathbf{x}, \alpha)$. Likewise, the probability $P_{rs}(\mathbf{x}, \mathbf{x}')$ of finding simultaneously phase r at $\mathbf{x}$ and phase s at $\mathbf{x}'$ is

$$P_{rs}(\mathbf{x}, \mathbf{x}') = \langle f_r(\mathbf{x}, \alpha) f_s(\mathbf{x}', \alpha) \rangle \equiv \int_{V} f_r(\mathbf{x}, \alpha) f_s(\mathbf{x}', \alpha) p(\alpha)\, d\alpha. \tag{5.3.11}$$

The function $P_{rs}(\mathbf{x}, \mathbf{x}')$ is also referred to as the two-point correlation function for the characteristic function $f_r(\mathbf{x}, \alpha)$. Following this approach, probabilities involving, for example, k points, or k-point correlation function, can be defined similarly.

Let us now develop an understanding of the geometrical interpretations of the correlation functions. Consider an epoxy matrix containing spherical silica particles, a two-phase composite used commonly in the electronics industry. If we take a large group of the material samples, each has the same particle content and a volume V large enough to contain many particles, we can, in principle, compute the one-point correlations $P_r(\mathbf{x})$ ($r = 1, 2$) by knowing the distribution density $p(\alpha)$ and the characteristic function $f_r(\mathbf{x}, \alpha)$. Let $r = 1$ be the matrix phase and $r = 2$ be the particle phase. If we found that $P_2(\mathbf{x})$ is a constant (i.e., independent of $\mathbf{x}$), then we can say that the particles are uniformly distributed through each sample. This is because $P_2(\mathbf{x}) = \text{const.}$ means that the chance of finding a particle is the same everywhere in V. Note that this does not necessarily mean that the particles are everywhere in V. It just gives us a sense of how "uniformly" the particles are distributed. Next, we take $\mathbf{x}$ to be the center of a particle and $\mathbf{x}'$ to be the center of another particle, and compute $P_{22}(\mathbf{x}, \mathbf{x}')$. Let us assume that it satisfies $P_{22}(\mathbf{x}, \mathbf{x}') = P_{22}(\mathbf{x} - \mathbf{x}')$, that is, the two-point correlation function $P_{22}(\mathbf{x}, \mathbf{x}')$ depends only on the distance between the two points, regardless where these two points are located. What that means is that if one takes a measuring stick shorter than the smallest dimension of V, and places the stick inside V, the chance that there is a particle at each end of the stick is the same regardless of where the stick is placed. The other two-point correlation functions $P_{11}(\mathbf{x}, \mathbf{x}')$, $P_{12}(\mathbf{x}, \mathbf{x}')$, and $P_{21}(\mathbf{x}, \mathbf{x}')$ can be interpreted similarly. Clearly the two-point correlation functions give us another measure of the "uniformity" of the heterogeneous materials. Said differently, if a material is statistically uniform, then its correlation functions $P_r(\mathbf{x})$, $P_{rs}(\mathbf{x}, \mathbf{x}')$, and so forth are insensitive to translation, that is, $P_r(\mathbf{x})$ reduces to a constant, $P_{rs}(\mathbf{x}, \mathbf{x}')$ becomes a function of $(\mathbf{x} - \mathbf{x}')$ only, and so on.

More generally, in micromechanics, a heterogeneous medium is called statistically homogeneous (or uniform) of grade k if the correlation functions up to the k point are all translation invariant. Since it is very unlikely that enough information is available to computer the k-point correlation function for k greater than 2, in this book we will typically call a statistically homogenous of grade 2 simply statistically homogeneous.

In what follows, we show that if the characteristic function of a heterogeneous material satisfies the ergodic hypothesis, the material is statistically homogeneous. To this end, consider the volume average of the characteristic function $f_r(\mathbf{x}, \alpha)$ for sample α:

$$\overline{f_r(\mathbf{x}, \alpha)} = \frac{1}{V}\int_V f_r(\mathbf{x}, \alpha)\, dV(\mathbf{x}) = \frac{V_r}{V} \equiv c_r, \tag{5.3.12}$$

where c_r is the volume fraction of the rth phase in the heterogeneous material. The ergodic hypothesis thus means

$$P_r(\mathbf{x}) = \langle f_r(\mathbf{x}, \alpha)\rangle \equiv \overline{f_r(\mathbf{x}, \alpha)} = c_r. \tag{5.3.13}$$

Next, consider the volume average of the two-point correlation:

$$\overline{f_r(\mathbf{x}, \alpha) f_s(\mathbf{x}', \alpha)} = \frac{1}{V}\int_V f_r(\mathbf{x}, \alpha) f_s(\mathbf{x}', \alpha)\, dV(\mathbf{x}). \tag{5.3.14}$$

Let

$$\mathbf{x}'' = \mathbf{x} - \mathbf{x}'. \tag{5.3.15}$$

Then, we have

$$\frac{1}{V}\int_V f_r(\mathbf{x}, \alpha) f_s(\mathbf{x}', \alpha)\, dV(\mathbf{x}) = \frac{1}{V}\int_{V_r} f_s(\mathbf{x} - \mathbf{x}'', \alpha)\, dV(\mathbf{x}) = g_{rs}(\mathbf{x}''), \tag{5.3.16}$$

or

$$\overline{f_r(\mathbf{x}, \alpha) f_s(\mathbf{x}', \alpha)} = g_{rs}(\mathbf{x}'') = g_{rs}(\mathbf{x} - \mathbf{x}'). \tag{5.3.17}$$

Therefore, according to the ergodic hypothesis,

$$P_{rs}(\mathbf{x}, \mathbf{x}') = \langle f_r(\mathbf{x}, \alpha) f_s(\mathbf{x}', \alpha)\rangle = \overline{f_r(\mathbf{x}, \alpha) f_s(\mathbf{x}', \alpha)} = g_{rs}(\mathbf{x} - \mathbf{x}').$$

This completes the proof.

We can summarize the foregoing discussions as follows:

1. The microstructure or the distribution of heterogeneities in a heterogeneous material can be characterized quantitatively by the correlation functions $P_r(\mathbf{x})$, $P_{rs}(\mathbf{x}, \mathbf{x}')$, and so forth, which are the ensemble average of the characteristic function $f_r(\mathbf{x}, \alpha)$.

2. The heterogeneous material is considered statistically homogeneous if the correlation functions are translation invariant, that is, $P_r(\mathbf{x}) = c_r$, $P_{rs}(\mathbf{x}, \mathbf{x}') = g_{rs}(\mathbf{x} - \mathbf{x}')$, and so forth.
3. If the correlation functions are ergodic, that is, the ensemble average equals the volume average, then they are translation invariant. Therefore, if the correlation functions of a heterogeneous material are ergodic, the material is considered statistically homogeneous.

Having considered the random media and introduced the concepts of ergodicity and statistical homogeneity, we can now give the representative volume element (RVE) a more precise definition. To define the length scale corresponding to the properties of interest, we need to select an RVE of the heterogeneous material in question. The size of the RVE must be such that it includes a very large number of inhomogeneities and in the meanwhile be statistically homogeneous and representative of the local continuum properties, so that appropriate averaging schemes over these domains give rise to the same mechanical properties, corresponding to the overall or effective mechanical properties.

Finally, to close this section, we mention that in the rest of this book, we will be mainly concerned with the statistically homogeneous media. Therefore, it will be assumed implicitly, unless indicted otherwise, that the ensemble average and the volume average are the same, that is,

$$\langle h(\mathbf{x}) \rangle = \int_{\Re} h(\mathbf{x}, \alpha)\, d\alpha = \frac{1}{V} \int_V h(\mathbf{x}, \alpha)\, dV(\mathbf{x}) = \overline{h(\alpha)}, \qquad (5.3.18)$$

where $h(\mathbf{x}, \alpha)$ can be either a scalar or a tensor function.

5.4 MACROSCOPIC AVERAGES

In this section, attention is focused on developing averaging theorems devoted to heterogeneous materials with arbitrary constituents independent of their thermomechanical behavior.

The average stresses over a domain D is defined as

$$\bar{\sigma}_{ij} = \frac{1}{D} \int_D \sigma_{ij}\, dV, \quad \text{or} \quad \bar{\boldsymbol{\sigma}} = \frac{1}{D} \int_D \boldsymbol{\sigma}\, dV. \qquad (5.4.1)$$

Similarly, the average strain over in a domain D is defined as

$$\bar{\varepsilon}_{ij} = \frac{1}{D}\int_D \varepsilon_{ij}\, dV, \quad \text{or} \quad \bar{\boldsymbol{\varepsilon}} = \frac{1}{D}\int_D \boldsymbol{\varepsilon}\, dV. \tag{5.4.2}$$

In these definitions, the domain D does not need to be homogeneous, nor does it need to be a single domain. Note that, as mentioned earlier, the volume average and ensemble average are the same for statistically homogeneous materials.

Average-Stress Theorem Let σ_{ij}^0 be a given constant stress tensor and S be the entire boundary of a domain D with outward normal n_j. If

$$\sigma_{ij} n_j|_S = p_j^0 = \sigma_{ij}^0 n_j \tag{5.4.3}$$

is prescribed on S, then the average stress on D is given by

$$\bar{\sigma}_{ij} = \sigma_{ij}^0. \tag{5.4.4}$$

To prove this theorem, consider the definition of average stress:

$$\bar{\sigma}_{ij} = \frac{1}{D}\int_D \sigma_{ij}\, dV = \frac{1}{D}\int_D \sigma_{ik}\delta_{jk}\, dV. \tag{5.4.5}$$

Note that $x_{j,k} = \delta_{jk}$, so

$$\bar{\sigma}_{ij} = \frac{1}{D}\int_D \sigma_{ik} x_{j,k}\, dV = \frac{1}{D}\int_D [(\sigma_{ik} x_j)_{,k} - \sigma_{ik,k}]\, dV. \tag{5.4.6}$$

When there is no body force as assumed here, the second term in the above integral vanishes. Making use of the divergence theorem and then the traction boundary condition (5.4.3), we have

$$\bar{\sigma}_{ij} = \frac{1}{D}\int_S \sigma_{ik} x_j n_k\, dS = \sigma_{ik}^0 \frac{1}{D}\int_S x_j n_k\, dS. \tag{5.4.7}$$

Making use of the divergence theorem again leads to

$$\bar{\sigma}_{ij} = \sigma_{ik}^0 \frac{1}{D}\int_D x_{j,k}\, dV = \sigma_{ik}^0 \frac{1}{D}\int_D \delta_{jk}\, dV = \sigma_{ij}^0. \tag{5.4.8}$$

This completes the proof of the average stress theorem.

What the average stress theorem tells us is that when a body is subjected to the traction boundary condition (5.4.3) with σ_{ij}^0 being a constant stress tensor, the stress averaged over the entire body is the same as σ_{ij}^0, regardless the complexity of the stress field within the domain. Therefore, in the rest of this book, traction boundary conditions will be written as

$$\sigma_{ij} n_j|_S = \overline{\sigma}_{ij} n_j, \tag{5.4.9}$$

with the understanding that $\overline{\sigma}_{ij}$ is the average stress tensor in the body enclosed by S when (5.4.9) is applied.

Average-Strain Theorem Let ε_{ij}^0 be a given constant strain tensor and S be the entire boundary of a domain D with outward normal n_j. If

$$u_i|_S = \varepsilon_{ij}^0 x_j \tag{5.4.10}$$

is given on S, then the average strain on D is given by

$$\overline{\varepsilon}_{ij} = \varepsilon_{ij}^0. \tag{5.4.11}$$

To prove this theorem, consider the definition of average strain:

$$\begin{aligned} \overline{\varepsilon}_{ij} &= \frac{1}{D} \int_D \varepsilon_{ij} \, dV = \frac{1}{2D} \int_D (u_{i,j} + u_{j,i}) \, dV \\ &= \frac{1}{2D} \int_D u_{i,j} \, dV + \frac{1}{2D} \int_D u_{j,i} \, dV. \end{aligned} \tag{5.4.12}$$

Making use of the divergence theorem yields

$$\overline{\varepsilon}_{ij} = \frac{1}{2D} \int_S u_i n_j \, dS + \frac{1}{2D} \int_S u_j n_i \, dS. \tag{5.4.13}$$

Substitute the boundary condition (5.4.10) into the above integrals:

$$\bar{\varepsilon}_{ij} = \frac{1}{2D}\int_S \varepsilon^0_{ik}x_k n_j \, dS + \frac{1}{2D}\int_S \varepsilon^0_{jk}x_k n_i \, dS.$$

Making use of the divergence theorem again,

$$\bar{\varepsilon}_{ij} = \frac{1}{2D}\int_D \varepsilon^0_{ik}x_{k,j} \, dV + \frac{1}{2D}\int_D \varepsilon^0_{jk}x_{k,i} \, dV = \frac{1}{2}\left(\varepsilon^0_{ij} + \varepsilon^0_{ji}\right) = \varepsilon^0_{ij}. \tag{5.4.14}$$

This completes the proof of the average strain theorem.

According to average strain theorem, if the displacements along the entire boundary of a given body is prescribed by (5.4.10) with ε^0_{ij} being a constant strain tensor, then the strain field averaged over the entire body is ε^0_{ij}, regardless the complexity of strain field within the body. Thus, in the rest of this book, displacement boundary conditions will be written as

$$u_i|_S = \bar{\varepsilon}_{ij}x_j, \tag{5.4.15}$$

with the understanding that $\bar{\varepsilon}_{ij}$ is the average strain tensor in the body enclosed by S when (5.4.15) is applied.

5.5 HILL'S LEMMA

Again, consider an RVE with volume V and boundary S. For any stress and strain fields σ_{ij} and ε_{ij} at a given point in the RVE under prescribed boundary traction or boundary displacement condition, one has the following result:

$$\overline{\sigma_{ij}\varepsilon_{ij}} - \bar{\sigma}_{ij}\bar{\varepsilon}_{ij} = \frac{1}{D}\int_S (u_i - x_j\bar{\varepsilon}_{ij})(\sigma_{ik}n_k - \bar{\sigma}_{ik}n_k)\, dS, \tag{5.5.1}$$

where the overbar stands for the volume average; for example,

$$\overline{\sigma_{ij}\varepsilon_{ij}} = \frac{1}{V}\int_V \sigma_{ij}\varepsilon_{ij}\, dv. \tag{5.5.2}$$

To prove Hill's lemma, let us first expand the integrand of the surface integral on the right-hand side of (5.5.1):

$$\int_S (u_i - x_j\bar{\varepsilon}_{ij})(\sigma_{ik}n_k - \bar{\sigma}_{ik}n_k)\, dS$$
$$= \int_S (u_i\sigma_{ik}n_k - u_i n_k\bar{\sigma}_{ik} - \sigma_{ik}n_k x_j\bar{\varepsilon}_{ij} + x_j n_k\bar{\varepsilon}_{ij}\bar{\sigma}_{ik})\, dS. \quad (5.5.3)$$

The following surface integrals can be easily evaluated:

$$\int_S u_i\sigma_{ik}n_k\, dS = \int_V u_{i,j}\sigma_{ij}\, dV = \int_V \varepsilon_{ij}\sigma_{ij}\, dV = D\overline{\sigma_{ij}\varepsilon_{ij}},$$

$$\int_S u_i n_k\bar{\sigma}_{ik}\, dS = \bar{\sigma}_{ik}\int_V u_{i,k}\, dV = D\bar{\sigma}_{ik}\bar{\varepsilon}_{ik},$$

$$\int_S \sigma_{ik}n_k x_j\, dS = \int_V \sigma_{ik}\delta_{jk}\, dV = D\bar{\sigma}_{ij},$$

$$\int_S x_j n_k\, dS = \int_V \delta_{jk}\, dV = D\delta_{jk}.$$

Substituting the above into (5.5.3) yields

$$\frac{1}{D}\int_S (u_i - x_j\bar{\varepsilon}_{ij})(\sigma_{ik}n_k - \bar{\sigma}_{ik}n_k)\, dS = \overline{\sigma_{ij}\varepsilon_{ij}} - \bar{\varepsilon}_{ik}\bar{\sigma}_{ik} - \bar{\sigma}_{ij}\bar{\varepsilon}_{ij} + \bar{\varepsilon}_{ik}\bar{\sigma}_{ik}$$
$$= \overline{\sigma_{ij}\varepsilon_{ij}} - \bar{\varepsilon}_{ij}\bar{\sigma}_{ij}. \quad (5.5.4)$$

This proves Hill's lemma (5.5.1).

A corollary of Hill's lemma can be stated as follows:

$$\overline{\sigma_{ij}\varepsilon_{ij}} = \bar{\varepsilon}_{ij}\bar{\sigma}_{ij}, \quad (5.5.5)$$

if $\sigma_{ij}n_j|_S = \bar{\sigma}_{ij}n_j$, or $u_i|_S = \bar{\varepsilon}_{ij}x_j$. In other words, for statically admissible stress fields ($\sigma_{ij}n_j|_S = \bar{\sigma}_{ij}n_j$), or kinematically admissible displacement fields ($u_i|_S = \bar{\varepsilon}_{ij}x_j$), the volume average of the product $\overline{\sigma_{ij}\varepsilon_{ij}}$ is the same as the product of the volume averages $\bar{\varepsilon}_{ij}$ and $\bar{\sigma}_{ij}$. Equation (5.5.5) is also known as Hill's macrohomogeneity condition or Mandel–Hill condition. Note that $\sigma_{ij}\varepsilon_{ij}$ is twice the value of strain energy density, (5.5.5) implies that the volume-averaged strain energy density of a heterogeneous material can be obtained from the volume averages of the stresses and strains. Accordingly, homogenization can be interpreted as finding a homogeneous comparison material that is energetically equivalent to a given microstructured material. This idea will be used to define the effective properties of heterogeneous media next.

5.6 DEFINITIONS OF EFFECTIVE MODULUS OF HETEROGENEOUS MEDIA

In this section, we introduce the concept of effective modulus of a statistically homogeneous heterogeneous medium. It is assumed that such a medium can be represented by an RVE consisting of N distinct phases dispersed throughout the RVE as inhomogeneities (or heterogeneities) of dimension much less than the size of the RVE. For the moment, perfect bonding will be assumed across the interfaces between the different phases. Nonperfect interfaces will be considered in Chapter 10.

Some common engineering materials are special cases of the heterogeneous material described in the preceding paragraph. For example, a polycrystalline material can be viewed as a heterogeneous material with infinite ($N = \infty$) number of distinct phases. Fiber- (particulate-) reinforced composites contain two distinct phases ($N = 2$), namely fibers (particles) and the matrix. Heterogeneous materials with a distinct matrix phase are called composite materials.

For convenience, body force is not considered in the remaining part of the book unless indicated otherwise.

Consider a heterogeneous material D bounded by surface S. Assume that the heterogeneous material consists of randomly oriented and shaped inhomogeneities embedded in a matrix with stiffness tensor $\mathbf{L}_0$. Let the stiffness tenors of the inhomogeneities be $\mathbf{L}_1, \mathbf{L}_2, \mathbf{L}_3, \ldots, \mathbf{L}_N$, as shown in Figure 5.1.

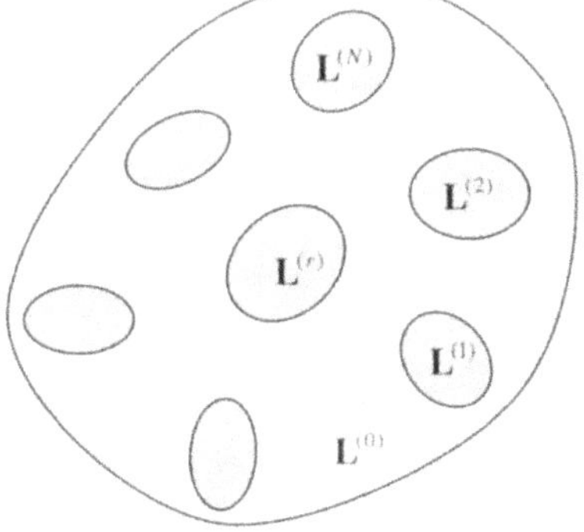

Figure 5.1 Multiphase heterogeneous material.

A straightforward definition of the effective modulus (stiffness) tensor of the heterogeneous material is given by the following relationship:

$$\overline{\boldsymbol{\sigma}} = \overline{\mathbf{L}}\overline{\boldsymbol{\varepsilon}}, \tag{5.6.1}$$

where $\overline{\boldsymbol{\sigma}}$ and $\overline{\boldsymbol{\varepsilon}}$ are, respectively, the average stress and strain tensors, and the fourth tensor $\overline{\mathbf{L}}$ defined by (5.6.1) is call the *effective modulus* or *effective stiffness* tensor of the heterogeneous material.

Similarly, the effective compliance tensor of the heterogeneous material is defined by

$$\overline{\boldsymbol{\varepsilon}} = \overline{\mathbf{M}}\overline{\boldsymbol{\sigma}}, \tag{5.6.2}$$

where the fourth-order tensor $\overline{\mathbf{M}}$ is called the effective compliance tensor of the heterogeneous material. Obviously, it follows from the definitions that

$$\overline{\mathbf{L}}\overline{\mathbf{M}} = \overline{\mathbf{M}}\overline{\mathbf{L}} = \mathbf{I}, \quad \text{or} \quad \overline{\mathbf{L}}^{-1} = \overline{\mathbf{M}}. \tag{5.6.3}$$

Although such definitions stated above are straightforward, they are not very convenient to use.

An alternative definition of the effective modulus tensor can be given based on the strain energy concept. To this end, let us consider two materials of the same geometry. One is a heterogeneous material, the other is a homogeneous one with stiffness tensor $\overline{\mathbf{L}}$, as shown in Figure 5.2. First, assume displacements are prescribed on the boundary of both materials, namely

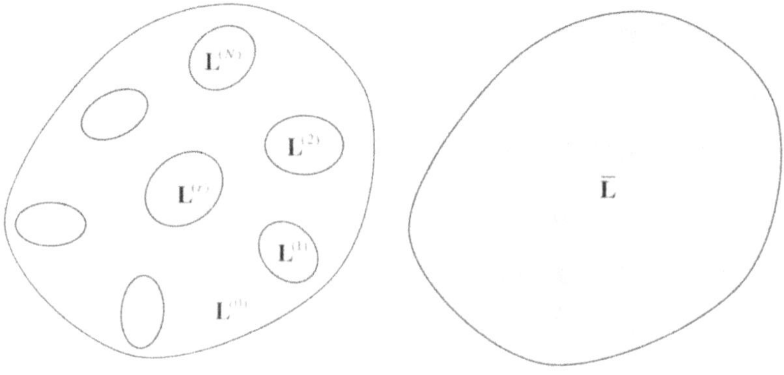

(*a*) (*b*)

Figure 5.2 (*a*) Heterogeneous material and (*b*) its comparison material.

$$u_i|_S = \bar{\varepsilon}_{ij} x_j, \tag{5.6.4}$$

where $\bar{\varepsilon}_{ij}$, as discussed in the previous section, would be the average strain in both materials.

The stiffness tensor $\bar{\mathbf{L}}$ of the homogeneous material is said to be the effective stiffness tensor of the heterogeneous material shown in Figure 5.2(*a*) if and only if

$$U_c = U_h, \tag{5.6.5}$$

where U_c and U_h are the strain energy stored in the heterogeneous material and in the homogeneous material, respectively.

It can be easily shown that the two definitions of the stiffness tenor given above are equivalent. To this end, we calculate first the strain energy stored in the composite:

$$U_c = \frac{1}{2} \int_D \sigma_{ij} \varepsilon_{ij} \, dv = D \overline{\sigma_{ij} \varepsilon_{ij}}. \tag{5.6.6}$$

Making use of (5.6.4) in conjunction with Hill's lemma (5.5.5), we arrive at

$$U_c = \frac{D}{2} \bar{\varepsilon}_{ij} \bar{\sigma}_{ij}. \tag{5.6.7}$$

In matrix notation, we have

$$U_c = \frac{D}{2} \bar{\boldsymbol{\varepsilon}} \bar{\boldsymbol{\sigma}}. \tag{5.6.8}$$

For the homogeneous material, it is obvious the strain and stress fields are uniform when (5.6.4) is applied. They are $\varepsilon_{ij} = \bar{\varepsilon}_{ij}$ and $\sigma_{ij} = \bar{L}_{ijkl} \bar{\varepsilon}_{kl}$. So, the total strain energy of the homogeneous material is given by

$$U_h = \frac{D}{2} \varepsilon_{ij} \sigma_{ij} = \frac{D}{2} \varepsilon_{ij} L_{ijkl} \sigma_{kl}, \tag{5.6.9}$$

or in matrix notation

$$U_h = \frac{D}{2}\,\bar{\boldsymbol{\varepsilon}}\bar{\mathbf{L}}\bar{\boldsymbol{\varepsilon}}. \tag{5.6.10}$$

Comparing (5.6.8) with (5.6.10), we conclude that $U_c = U_h$ for any arbitrary $\bar{\boldsymbol{\varepsilon}}$, if and only if $\bar{\boldsymbol{\sigma}} = \bar{\mathbf{L}}\bar{\boldsymbol{\varepsilon}}$. Consequently, we have proved that the two foregoing definitions of stiffness tenors for heterogeneous materials are equivalent.

An alternative definition of the effective compliance tensor also can be given based on the strain energy concept. For example, consider the two materials shown in Figure 5.2. Assume that the homogeneous material has a compliance tensor $\bar{\mathbf{M}}$. This time let the traction $\sigma_{ij}n_j|_S = \bar{\sigma}_{ij}n_j$ be prescribed on the entire boundary of both materials. Then, the compliance tensor $\bar{\mathbf{M}}$ is said to be the effective compliance tensor of the composite shown in Figure 5.2(*a*) if and only if

$$U_c = U_h, \tag{5.6.11}$$

where again U_c and U_h are the strain energy stored in the composite and in the homogeneous material, respectively.

Proof of the equivalency between this definition and the one given by (5.6.2) is straightforward; see Problem 5.1.

Although the definition based on strain energy does not seem as straightforwards as the direct definition, it is nevertheless convenient for the determination of the effective properties, as will be seen in later chapters.

5.7 CONCENTRATION TENSORS AND EFFECTIVE PROPERTIES

Consider the composite material shown in Figure 5.2(*a*). Let the displacement boundary condition be given by

$$u_i|_S = \bar{\varepsilon}_{ij}x_j. \tag{5.7.1}$$

Then, it follows from (5.4.11) that

$$\bar{\boldsymbol{\varepsilon}} = \sum_{r=0}^{N} c_r\bar{\boldsymbol{\varepsilon}}_r, \tag{5.7.2}$$

where $\bar{\varepsilon}_r$ is the average strain on the rth inhomogeneity and c_r is the volume fraction of the rth inhomogeneity. The case of $r = 0$ is for the matrix. Note that $\bar{\varepsilon}$ is the average strain of the entire composite material and $\bar{\varepsilon}_r$ is the average strain of the rth inhomogeneity. Thus, a "strain concentration tensor" $\mathbf{A}_r$ ($r > 0$) can be defined for the rth inhomogeneity through

$$\bar{\boldsymbol{\varepsilon}}_r = \mathbf{A}_r\bar{\boldsymbol{\varepsilon}}. \tag{5.7.3}$$

Clearly, the "magnitude" of $\mathbf{A}_r$ represents the strain concentration on the inhomogeneity. Substituting (5.7.3) into (5.7.2) yields

$$c_0\bar{\boldsymbol{\varepsilon}}_0 = \bar{\boldsymbol{\varepsilon}} - \sum_{r=1}^{N} c_r\bar{\boldsymbol{\varepsilon}}_r = \bar{\boldsymbol{\varepsilon}} - \sum_{r=1}^{N} c_r\mathbf{A}_r\bar{\boldsymbol{\varepsilon}}. \tag{5.7.4}$$

Next, consider the average stress in the composite

$$\bar{\boldsymbol{\sigma}} = \sum_{r=0}^{N} c_r\bar{\boldsymbol{\sigma}}_r = c_0\mathbf{L}_0\bar{\boldsymbol{\varepsilon}}_0 + \sum_{r=1}^{N} c_r\mathbf{L}_r\bar{\boldsymbol{\varepsilon}}_r, \tag{5.7.5}$$

where $\bar{\boldsymbol{\sigma}}_r$ is the average stress tensor in the rth inhomogeneity, and Hooke's law $\bar{\boldsymbol{\sigma}}_r = \mathbf{L}_r\bar{\boldsymbol{\varepsilon}}_r$ for each inhomogeneity has been used. Making use of (5.7.4) in (5.7.5) gives the following:

$$\bar{\boldsymbol{\sigma}} = \left[\mathbf{L}_0 + \sum_{r=1}^{N} c_r(\mathbf{L}_r - \mathbf{L}_0)\mathbf{A}_r\right]\bar{\boldsymbol{\varepsilon}}. \tag{5.7.6}$$

Comparing (5.7.6) with (5.6.1) leads to

$$\bar{\mathbf{L}} = \mathbf{L}_0 + \sum_{r=1}^{N} c_r(\mathbf{L}_r - \mathbf{L}_0)\mathbf{A}_r. \tag{5.7.7}$$

In other words, the effective stiffness tensor of a composite material can be expressed in terms of the stiffness tensors of the constituent materials, their corresponding volume fraction and the strain concentration tensors. Among these, the strain concentration tensor is the only unknown in typical engineering applications. Much of the effort in micromechanics has been devoted to find ways to compute the concentration tensor. Several of the most commonly used methods to approximate the concentration tensors are discussed in Chapters 7 and 8.

Along a parallel path, it can be shown (Problem 5.2) that under the traction boundary condition $\sigma_{ij}n_j|_S = \overline{\sigma}_{ij}n_j$, the effective compliance tensor of a composite is given by

$$\overline{\mathbf{M}} = \mathbf{M}_0 + \sum_{r=1}^{N} c_r(\mathbf{M}_r - \mathbf{M}_0)\mathbf{B}_r, \tag{5.7.8}$$

where $\mathbf{B}_r$ is the stress concentration tensor for the rth inhomogeneity, that is,

$$\overline{\boldsymbol{\sigma}}_r = \mathbf{B}_r\overline{\boldsymbol{\sigma}}, \tag{5.7.9}$$

where $\overline{\boldsymbol{\sigma}}_r$ is the average stress in the rth inhomogeneity and $\overline{\boldsymbol{\sigma}}$ is the average stress over the entire composite.

The concentration tensors introduced above represent the stress–strain concentration of the rth inhomogeneity with respect to the average strain over the entire sample. It can be regarded as the global concentration tensors. Since the composite has a distinctive matrix phase in which all the other inhomogeneities are embedded, one can also introduce the local concentration tensors through

$$\overline{\boldsymbol{\varepsilon}}_r = \mathbf{G}_r\overline{\boldsymbol{\varepsilon}}_0 \quad \text{and} \quad \overline{\boldsymbol{\sigma}}_r = \mathbf{H}_r\overline{\boldsymbol{\sigma}}_0 \tag{5.7.10}$$

where $\overline{\boldsymbol{\varepsilon}}_0$ and $\overline{\boldsymbol{\sigma}}_0$ are, respectively, the average strain and average stress tensor in the matrix phase. Clearly, $\mathbf{G}_r$ and $\mathbf{H}_r$ are, respectively, the strain and stress concentration tensors of the rth inhomogeneity with respect to the average strain and stress in the matrix immediately surrounding itself.

To find the relationship between the local and global concentration tensors, consider the average strain:

$$\begin{aligned}\overline{\boldsymbol{\varepsilon}} &= \sum_{r=0}^{N} c_r\overline{\boldsymbol{\varepsilon}}_r = c_0\overline{\boldsymbol{\varepsilon}}_0 + \sum_{r=1}^{N} c_r\overline{\boldsymbol{\varepsilon}}_r = c_0\overline{\boldsymbol{\varepsilon}}_0 + \sum_{r=1}^{N} c_r\mathbf{G}_r\overline{\boldsymbol{\varepsilon}}_0 \\ &= \left[c_0\mathbf{I} + \sum_{r=1}^{N} c_r\mathbf{G}_r\right]\overline{\boldsymbol{\varepsilon}}_0. \end{aligned} \tag{5.7.11}$$

This yields

$$\bar{\boldsymbol{\varepsilon}}_0 = \left[c_0\mathbf{I} + \sum_{n=1}^{N} c_n\mathbf{G}_n \right]^{-1} \bar{\boldsymbol{\varepsilon}}. \tag{5.7.12}$$

Substitution of the above equation into (5.7.10) results in

$$\bar{\boldsymbol{\varepsilon}}_r = \mathbf{G}_r \left[c_0\mathbf{I} + \sum_{n=1}^{N} c_n\mathbf{G}_n \right]^{-1} \bar{\boldsymbol{\varepsilon}}. \tag{5.7.13}$$

Comparing this to the definition of global concentration tensor (5.7.3) leads to

$$\mathbf{A}_r = \mathbf{G}_r \left[c_0\mathbf{I} + \sum_{r=1}^{N} c_r\mathbf{G}_r \right]^{-1}. \tag{5.7.14}$$

Similarly, one can show that

$$\mathbf{B}_r = \mathbf{H}_r \left[c_0\mathbf{I} + \sum_{n=1}^{N} c_n\mathbf{H}_n \right]^{-1}. \tag{5.7.15}$$

Because of these relationships, the knowledge of the local concentration tensors is sufficient to determine the effective stiffness (compliant) tensor of the composite.

PROBLEMS

5.1 Prove the equivalency between the two definitions of the effective compliance tensor.

5.2 Assume that a composite is subjected to the traction boundary condition $\sigma_{ij}n_j|_S = \bar{\sigma}_{ij}n_j$. Show Eq. (5.7.8).

5.3 Derive (5.7.15).

SUGGESTED READING

Krajinovic, D. (1996). *Damage Mechanics,* Elsevier, New York, Chapter 1.

Mura, T. (1987). *Micromechanics of Defects in Solids,* Martinus Nijhoff, Boston.

Nemat-Nasser, S. and Hori, M. (1993). *Micromechanics: Overall Properties of Heterogeneous Materials,* North-Holland, New York, Chapter 1.

Torquato, S. (2001). *Random Heterogeneous Materials: Microstructure and Macroscopic Properties,* Springer, New York.

6

BOUNDS FOR EFFECTIVE MODULI

Obtaining the effective modulus tensor of a heterogeneous material is often a very difficult task. In most cases, only approximate solutions can be found. Several of these approximate estimates will be discussed in Chapters 7–9. Although cxact solutions to the effective moduli may not be found easily, for all practical purposes, knowing the bounds for these moduli is enough. In this chapter, some of these bounds are derived based on variation principles.

6.1 CLASSICAL VARIATIONAL THEOREMS IN LINEAR ELASTICITY

To prepare for the study of bounds of the effective properties of composite materials, we state here, without proof, two of the classical variational theorems in linear elasticity.

Minimum Potential Energy Theorem Among all the kinematically admissible displacement fields, the true solution makes the following potential energy minimum:

$$\Pi[u_i] = \frac{1}{2}\int_D L_{ijkl}u_{k,l}u_{i,j}\,dv - \int_{S_\sigma} p_i^0 u_i\,ds, \qquad (6.1.1)$$

where D is the total volume, and S_σ is the portion of the boundary of D where traction p_i^0 is prescribed.

Minimum Complementary Energy Theorem Among all the statically admissible stress fields, the true solution makes the following complementary energy minimum:

$$\Pi_c[\sigma_{ij}] = \frac{1}{2}\int_D M_{ijkl}\sigma_{kl}\sigma_{ij}\, dv - \int_{S_u} u_i^0 \sigma_{ij} n_j\, ds, \tag{6.1.2}$$

where D is the total volume, and S_u is the portion of the boundary of D where displacement u_i^0 is prescribed.

It should be mentioned that the stiffness tensor **L**, and the compliance tensor **M** in the above theorems need not be uniform throughout D, meaning that both theorems are applicable to heterogeneous materials.

In what follows, we will derive, based on the above theorems, some variational principles that are convenient to use for obtaining the bounds of effective properties. To begin, consider the heterogeneous material considered in Section 5.6. Let the displacement be prescribed on the boundary S:

$$u_i|_S = \bar{\varepsilon}_{ij} x_j, \tag{6.1.3}$$

where $\bar{\varepsilon}_{ij}$ is a constant strain tensor, which, according to the average strain theorem (5.4.11), is also the average in the composite under such boundary condition:

$$\bar{\varepsilon}_{ij} = \frac{1}{D}\int_D \varepsilon_{ij}\, dv. \tag{6.1.4}$$

Therefore, the overbar on the strain tensor causes no confusion between the given strain tensor on the boundary and the average strain tensor in D since they are the same. Under the displacement boundary condition (6.1.3), the potential energy of the composite, following (6.1.1), can be written as

$$\Pi[u_i] = \frac{1}{2}\int_D L_{ijkl} u_{k,l} u_{i,j}\, dv = \frac{1}{2}\int_D L_{ijkl}\varepsilon_{kl}\varepsilon_{ij}\, dv, \tag{6.1.5}$$

since no traction is prescribed. To facilitate the following discussions, we will denote the actual displacement field by u_i and any kinematically admissible displacement field by $\hat{u}_i$. Clearly, under the displacement boundary condition, all kinematically admissible displacement (the actual displacement included) fields must satisfy (6.1.3). Let $\overline{U}$ be the average strain energy density in the heterogeneous material when it is subjected to (6.1.3):

$$\overline{U} = \frac{1}{2D}\int_D L_{ijkl}\varepsilon_{kl}\varepsilon_{ij}\,dv = \frac{1}{2}\,\overline{\varepsilon}_{ij}\overline{L}_{ijkl}\overline{\varepsilon}_{kl}, \tag{6.1.6}$$

where $\overline{L}_{ijkl}$ is the effective stiffness tensor of the heterogeneous material and the last equality follows from (5.6.6). Then, according to the minimum potential energy theorem, the following inequality holds for any kinematically admissible displacement field $\hat{u}_i$:

$$\overline{U} \leq \frac{1}{2D}\int_D L_{ijkl}\hat{u}_{k,l}\hat{u}_{i,j}\,dv. \tag{6.1.7}$$

Now, consider the complimentary energy. Again, under the displacement boundary condition (6.1.3), the complimentary energy becomes

$$\Pi_c[\sigma_{ij}] = \frac{1}{2}\int_D M_{ijkl}\sigma_{kl}\sigma_{ij}\,dv - \int_S \overline{\varepsilon}_{ik}x_k\sigma_{ij}n_j\,ds. \tag{6.1.8}$$

Upon the use of the divergence theorem on the second term,

$$\Pi_c[\sigma_{ij}] = \frac{1}{2}\int_D M_{ijkl}\sigma_{kl}\sigma_{ij}\,dv - \int_D (\overline{\varepsilon}_{ik}x_k\sigma_{ij})_{,j}\,ds. \tag{6.1.9}$$

For any statically admissible stress field, that is, $\sigma_{ij,j} = 0$, the above reduces to

$$\Pi_c[\sigma_{ij}] = \frac{1}{2}\int_D M_{ijkl}\sigma_{kl}\sigma_{ij}\,dv - \overline{\varepsilon}_{ij}\int_D \sigma_{ij}\,dv, \tag{6.1.10}$$

where $\overline{\sigma}_{ij}$ is the average stress in D,

$$\overline{\sigma}_{ij} = \frac{1}{D}\int_D \sigma_{ij}\,dv. \tag{6.1.11}$$

Let σ_{ij} be the actual stress field and $\hat{\sigma}_{ij}$ be any kinematically admissible stress field. Then, by making use of Hooke's law, $\varepsilon_{ij} = M_{ijkl}\sigma_{kl}$, we can show, following the steps described in (5.6.6) and (5.6.7), that

$$\Pi_c[\sigma_{ij}] = \frac{1}{2}\int_D M_{ijkl}\sigma_{kl}\sigma_{ij}\,dv - \bar{\varepsilon}_{ij}\int_D \sigma_{ij}\,dv = -D\overline{U}, \qquad (6.1.12)$$

where $\overline{U}$ is the average strain energy density as introduced in (6.1.7). Thus, according to the minimum complimentary energy theorem, the following inequality holds for any statically admissible stress $\hat{\sigma}_{ij}$:

$$\overline{U} \geq \bar{\varepsilon}_{ij}\frac{1}{D}\int_D \hat{\sigma}_{ij}\,dv - \frac{1}{2D}\int_D M_{ijkl}\hat{\sigma}_{kl}\hat{\sigma}_{ij}\,dv. \qquad (6.1.13)$$

Combining (6.1.7) and (6.1.13), we arrive at the following inequalities, in matrix notations:

$$\bar{\boldsymbol{\varepsilon}}\int_D \hat{\boldsymbol{\sigma}}\,dv - \int_D \hat{\boldsymbol{\sigma}}\mathbf{M}\hat{\boldsymbol{\sigma}}\,dv \leq 2D\overline{U} \leq \int_D \hat{\boldsymbol{\varepsilon}}\mathbf{L}\hat{\boldsymbol{\varepsilon}}\,dv, \qquad (6.1.14)$$

where $\hat{\sigma}_{ij}$ can be any stress field that satisfies

$$\hat{\sigma}_{ij,j} = 0 \quad \text{in } D, \qquad (6.1.15)$$

and $\hat{\varepsilon}_{ij} = (\hat{u}_{i,j} + \hat{u}_{j,i})/2$ with $\hat{u}_i$ being any displacement field that satisfies

$$\hat{u}_i = \bar{\varepsilon}_{ij}x_j \quad \text{on } S. \qquad (6.1.16)$$

6.2 VOIGT UPPER BOUND AND REUSS LOWER BOUND

Again, consider the heterogeneous material discussed in Section 5.6. Let the heterogeneous material be subjected to the displacement boundary condition,

$$u_i = \bar{\varepsilon}_{ij}x_j \quad \text{on } S, \qquad (6.2.1)$$

where $\bar{\varepsilon}_{ij}$ is a given constant strain tensor. According to the average strain theorem stated in Section 5.4, the average strain in D is also $\bar{\varepsilon}_{ij}$. Next, we define

$$\hat{u}_i = \bar{\varepsilon}_{ij}x_j \quad \text{everywhere in } D. \qquad (6.2.2)$$

Obviously, the $\hat{u}_i$ so defined is a kinematically admissible displacement field, that is,

$$\hat{u}_i = \bar{\varepsilon}_{ij} x_j \quad \text{on } S. \tag{6.2.3}$$

The kinematical admissible strain field corresponding to $\hat{u}_i$ is simply $\bar{\varepsilon}_{ij}$. Further, for the boundary condition (6.2.1), any constant stress tensor $\hat{\sigma}_{ij}$ would be a statically admissible stress field. Thus, by substituting (6.2.2) and a constant tensor $\hat{\sigma}_{ij}$ into (6.1.14) yields

$$2\bar{\boldsymbol{\varepsilon}}\hat{\boldsymbol{\sigma}} - \hat{\boldsymbol{\sigma}}\overline{\mathbf{M}}^R\hat{\boldsymbol{\sigma}} \leq 2\bar{U} \leq \bar{\boldsymbol{\varepsilon}}\overline{\mathbf{L}}'\bar{\boldsymbol{\varepsilon}}, \tag{6.2.4}$$

where

$$\overline{\mathbf{L}}^V = \frac{1}{D}\int_D \mathbf{L}\, dv, \qquad \overline{\mathbf{M}}^R = \frac{1}{D}\int_D \mathbf{M}\, dv. \tag{6.2.5}$$

Since $\hat{\boldsymbol{\sigma}}$ can be any constant stress tensor, we should try to select one that maximizes the far left-hand side of (6.2.4) to obtain the optimal lower bound. It turns out that

$$\hat{\boldsymbol{\sigma}} = \overline{\mathbf{L}}^R\bar{\boldsymbol{\varepsilon}} = (\overline{\mathbf{M}}^R)^{-1}\bar{\boldsymbol{\varepsilon}} \tag{6.2.6}$$

would be the choice. Substituting (6.2.6) and (6.1.6) into (6.2.4), we obtain the following:

$$\bar{\boldsymbol{\varepsilon}}\overline{\mathbf{L}}^R\bar{\boldsymbol{\varepsilon}} \leq \bar{\boldsymbol{\varepsilon}}\overline{\mathbf{L}}\bar{\boldsymbol{\varepsilon}} \leq \bar{\boldsymbol{\varepsilon}}\overline{\mathbf{L}}^V\bar{\boldsymbol{\varepsilon}}, \tag{6.2.7}$$

where

$$\overline{\mathbf{L}}^R = (\overline{\mathbf{M}}^R)^{-1} = \left[\frac{1}{D}\int_D M\, \mathrm{d}v\right]^{-1}; \tag{6.2.8}$$

These inequalities must hold true for any constant strain tensor $\bar{\boldsymbol{\varepsilon}}$. Therefore, we conclude

$$\overline{\mathbf{L}}^R \leq \overline{\mathbf{L}} \leq \overline{\mathbf{L}}^V, \tag{6.2.9}$$

where $\overline{\mathbf{L}}^V$ defined by the first of (6.2.5) is called the Voigt upper bound, and $\overline{\mathbf{L}}^R$ defined by (6.2.8) is called the Reuss lower bound.

If the stiffness tensor of the rth phase of volume Ω_r is uniform and given by $\mathbf{L}_r$, where $r = 0, 1, \ldots, N$ ($r = 0$ corresponds to the matrix), then the first of (6.2.5) can be simplified to

$$\overline{\mathbf{L}}^V = \sum_{r=0}^{N} c_r \mathbf{L}_r, \tag{6.2.10}$$

where

$$c_r = \frac{\Omega_r}{D}, \qquad r = 1, 2, \ldots, N \tag{6.2.11}$$

are the volume fraction of the inhomogeneities with stiffness tensor $\mathbf{L}_r$. Obviously, we have the relationship

$$\sum_{r=0}^{N} c_r = 1. \tag{6.2.12}$$

Equation (6.2.10) gives the Voigt upper bound for the effective stiffness tensor of a composite with N distinct phases.

Similarly, (6.2.8) can be written as

$$\overline{\mathbf{L}}^R = \left[\sum_{r=0}^{N} c_r \mathbf{M}_r\right]^{-1}, \tag{6.2.13}$$

where $\mathbf{M}_r$ is the compliance tensor for the rth phase.

It is noted that the bounds are derived in terms of the volume faction of each phase only. Therefore, they are independent of the geometry of the phases and their distribution.

Example 6.1 Consider an isotropic matrix of stiffness tensor $\mathbf{L}_0 = (3K_0, 2\mu_0)$ reinforced by particles of stiffness tensor $\mathbf{L}_1 = (3K_1, 2\mu_1)$. We are to find the bounds of the effective elastic constants.

To solve this problem, let us first consider the compliance tensors. Following the rules discussed in Section 1.4:

$$\mathbf{M}_0 = \left(\frac{1}{3K_0}, \frac{1}{2\mu_0}\right), \qquad \mathbf{M}_1 = \left(\frac{1}{3K_1}, \frac{1}{2\mu_1}\right). \tag{6.2.14}$$

Thus,

$$[c_0\mathbf{M}_0 + c_1\mathbf{M}_1]^{-1} = \left(\frac{c_0}{3K_0} + \frac{c_1}{3K_1}, \frac{c_0}{2\mu_0} + \frac{c_1}{2\mu_1}\right)^{-1}$$
$$= \left(\frac{3K_0K_1}{c_0K_1 + c_1K_0}, \frac{2\mu_0\mu_1}{c_0\mu_1 + c_1\mu_0}\right). \quad (6.2.15)$$

According to (6.2.9), the effective stiffness tensor of the composite satisfies the inequality

$$[c_0\mathbf{M}_0 + c_1\mathbf{M}_1]^{-1} \leq \overline{\mathbf{L}} = (3\overline{K}, 2\overline{\mu}) \leq c_0\mathbf{L}_0 + c_1\mathbf{L}_1, \quad (6.2.16)$$

where $\overline{K}$ and $\overline{\mu}$ are the effective bulk and shear moduli of the composite, respectively, and c_1 is the volume fraction of the particles. Making use of (6.2.15) in (6.2.16), we have

$$\frac{K_0K_1}{K_1 - c_1(K_1 - K_0)} \leq \overline{K} \leq K_0 + c_1(K_1 - K_0), \quad (6.2.17)$$

$$\frac{\mu_0\mu_1}{\mu_1 - c_1(\mu_1 - \mu_0)} \leq (\overline{\mu} \leq \mu_0 + c_1(\mu_1 - \mu_0). \quad (6.2.18)$$

These give the upper and lower bounds for the effective bulk and shear moduli of the composite.

6.3 EXTENSIONS OF CLASSICAL VARIATIONAL PRINCIPLES

It is noted that the Voigt upper and the Reuss lower bounds were obtained by specifying a kinematically admissible displacement field in the potential energy and by specifying a statically admissible stress field in the complimentary energy. In other words, the key to find bounds is to find suitable admissible fields. In fact, this approach is rather general. Tighter bounds can be derived by specifying kinematically and statically admissible fields that make some explicit allowance for the microstructure of the composite.

To this end, let us first try to develop an alternative form of the classical variational principles that are better suited for developing bounds of the effective properties of heterogeneous materials. For the heterogeneous material considered in Section 5.6, we assume that the following displacement boundary condition is prescribed:

$$u_i = \bar{\varepsilon}_{ij} x_j \quad \text{on } S. \tag{6.3.1}$$

Our objective is to find a kinematically admissible displacement field (rather than a constant tensor as in the Voigt upper bound) and a statically admissible stress tensor (rather than a constant tensor as in the Reuss lower bound) for this heterogeneous material under the boundary condition (6.3.1).

To this end, let us write the elastic stiffness tensor of the heterogeneous material in the following form:

$$\mathbf{L}(\mathbf{x}) = \mathbf{L}^h + \mathbf{L}^p(\mathbf{x}), \tag{6.3.2}$$

where it is understood that $\mathbf{L}(\mathbf{x})$ is a function of position vector $\mathbf{x}$, $\mathbf{L}^h$ is a constant fourth-order tensor representing a homogeneous material, and $\mathbf{L}^p(\mathbf{x})$, being the difference between $\mathbf{L}(\mathbf{x})$ and $\mathbf{L}^h$, is also a function of position vector $\mathbf{x}$, representing the perturbation from the homogeneous material. The constant stiffness tensor $\mathbf{L}^h$ is introduced to represent a homogeneous "comparison" material. The corresponding actual stress field in the heterogeneous material is then given by

$$\boldsymbol{\sigma}(\mathbf{x}) = \mathbf{L}^h \boldsymbol{\varepsilon}(\mathbf{x}) + \boldsymbol{\tau}(\mathbf{x}), \tag{6.3.3}$$

where

$$\varepsilon_{ij}(\mathbf{x}) = \frac{1}{2}\left(\frac{\partial u_i(\mathbf{x})}{\partial x_j} + \frac{\partial u_j(\mathbf{x})}{\partial x_i}\right) \tag{6.3.4}$$

is the actual strain tensor in D and

$$\boldsymbol{\tau}(\mathbf{x}) = \mathbf{L}^p(\mathbf{x}) \boldsymbol{\varepsilon}(\mathbf{x}) \tag{6.3.5}$$

is called the stress polarization in some of the literature. Clearly, $\boldsymbol{\tau}(\mathbf{x})$ is symmetric, that is, $\tau_{ij}(\mathbf{x}) = \tau_{ji}(\mathbf{x})$, because of the symmetry of $\mathbf{L}^p(\mathbf{x})$.

In light of (6.3.3), the equations of equilibrium in terms of the displacement can be written as

$$L^h_{ijkl} u_{k,lj}(\mathbf{x}) + \tau_{ij,j}(\mathbf{x}) = 0 \quad \text{in } D. \tag{6.3.6}$$

If we treat the second term on the left-hand side as a body force, the solution to the boundary value problem (6.3.1) and (6.3.6) can be attained in terms of Green's function; see (2.6.51):

$$u_i(\mathbf{x}) = \int_S L^h_{pjkl} \frac{\partial G_{ki}(\mathbf{x}, \mathbf{y})}{\partial x_l} \bar{\varepsilon}_{pq} y_q n_j \, dS(\mathbf{y}) + \int_D G_{mi}(\mathbf{x}, \mathbf{y}) \frac{\partial \tau_{mn}(\mathbf{y})}{\partial y_n} \, dV(\mathbf{y}), \tag{6.3.7}$$

where $G_{mi}(\mathbf{x}, \mathbf{y})$ is Green's function in the homogeneous comparison material satisfying

$$L^h_{ijkl} \frac{\partial^2 G_{km,lj}(\mathbf{x}, \mathbf{y})}{\partial x_l \, \partial x_j} + \delta_{im}\delta(\mathbf{x} - \mathbf{y}) = 0 \quad \text{in } V, \tag{6.3.8}$$

$$G_{ij}(\mathbf{x}, \mathbf{y})|_{\mathbf{x} \in S} = 0. \tag{6.3.9}$$

By using the divergence theorem and (6.3.8), the surface integral in (6.3.7) can be simplified to

$$\begin{aligned}
&\int_S L^h_{pjkl} \frac{\partial G_{ki}(\mathbf{x}, \mathbf{y})}{\partial x_i} \bar{\varepsilon}_{pq} y_q n_j \, dS(\mathbf{y}) \\
&= -\int_D L^h_{pjkl} \frac{\partial^2 G_{ki}(\mathbf{x}, \mathbf{y})}{\partial x_l + \partial y_j} \bar{\varepsilon}_{pq} y_q \, dV(\mathbf{y}) + \int_D L^h_{pjkl} \frac{\partial G_{ki}(\mathbf{x}, \mathbf{y})}{\partial x_l} \bar{\varepsilon}_{pj} \, dV(\mathbf{y}) \\
&= \int_D \delta_{pi}\delta(\mathbf{x}, \mathbf{y})\bar{\varepsilon}_{pq} y_q \, dV(\mathbf{y}) + \int_S L^h_{pjkl} G_{ki}(\mathbf{x}, \mathbf{y})\bar{\varepsilon}_{pj} n_j \, dS(\mathbf{y}) \\
&= \bar{\varepsilon}_{iq} x_q.
\end{aligned} \tag{6.3.10}$$

Substituting (6.3.10) into (6.3.7) and applying the divergence theorem to the volume integral in conjunction with (6.3.9), we have

$$\begin{aligned}
u_i(\mathbf{x}) &= \bar{\varepsilon}_{iq} x_q + \int_D G_{mi}(\mathbf{x}, \mathbf{y}) \frac{\partial \tau_{mn}(\mathbf{y})}{\partial y_n} \, dV(\mathbf{y}) \\
&= \bar{\varepsilon}_{iq} x_q - \int_D \frac{\partial G_{mi}(\mathbf{x}, \mathbf{y})}{y_n} \tau_{mn}(\mathbf{y}) \, dV(\mathbf{y}).
\end{aligned} \tag{6.3.11}$$

The displacement gradient is thus given by

$$u_{i,j}(\mathbf{x}) = \bar{\varepsilon}_{ij} - \int_D \frac{\partial^2 G_{mi}(\mathbf{x} - \mathbf{y})}{\partial x_j \, \partial y_n} \tau_{mn}(\mathbf{y}) \, dV(\mathbf{y}). \tag{6.3.12}$$

The strain tensor follows (6.3.12) readily,

$$\varepsilon_{ij}(\mathbf{x}) = \bar{\varepsilon}_{ij} - \frac{1}{2}\int_D \left(\frac{\partial^2 G_{mi}(\mathbf{x}-\mathbf{y})}{\partial x_j\, \partial y_n} + \frac{\partial^2 G_{mj}(\mathbf{x}-\mathbf{y})}{\partial x_i\, \partial y_n} \right) \tau_{mn}(\mathbf{y})\, dV(\mathbf{y}), \tag{6.3.13}$$

Using the symmetry properties of $\tau_{mn}(\mathbf{y})$, we can rewrite the above into

$$\boldsymbol{\varepsilon}(\mathbf{x}) = \bar{\boldsymbol{\varepsilon}} - \int_D \boldsymbol{\Gamma}(\mathbf{x}, \mathbf{y})\boldsymbol{\tau}(\mathbf{y})\, dV(\mathbf{y}), \tag{6.3.14}$$

where

$$\Gamma_{ijkl}(\mathbf{x}, \mathbf{y}) = \frac{1}{4}\left[\frac{\partial^2 G_{ki}(\mathbf{x}-\mathbf{y})}{\partial x_j\, \partial y_l} + \frac{\partial^2 G_{kj}(\mathbf{x}-\mathbf{y})}{\partial x_i\, \partial y_l} + \frac{\partial^2 G_{li}(\mathbf{x}-\mathbf{y})}{\partial x_j\, \partial y_k} + \frac{\partial^2 G_{lj}(\mathbf{x}-\mathbf{y})}{\partial x_i\, \partial y_k} \right]. \tag{6.3.15}$$

Since the integral operator in (6.3.14) will be used many times, we will, for convenience, use the following symbolic notation to indicate the integration operation:

$$\int_D \boldsymbol{\Gamma}(\mathbf{x}, \mathbf{y})\boldsymbol{\tau}(\mathbf{y})\, dV(\mathbf{y}) \equiv \boldsymbol{\Gamma}(\mathbf{x}, \mathbf{y}) * \boldsymbol{\tau}(\mathbf{y}). \tag{6.3.16}$$

Using this notation, (6.3.14) can be awritten as

$$\boldsymbol{\varepsilon}(\mathbf{x}) = \bar{\boldsymbol{\varepsilon}} - \boldsymbol{\Gamma}(\mathbf{x}, \mathbf{y}) * \boldsymbol{\tau}(\mathbf{y}). \tag{6.3.17}$$

Before proceeding forward, we note that the integral operator is self-adjoint, that is,

$$\boldsymbol{\sigma}(\mathbf{x}) * [\boldsymbol{\Gamma}(\mathbf{x}, \mathbf{y}) * \boldsymbol{\tau}(\mathbf{y})] = \boldsymbol{\tau}(\mathbf{x}) * [\boldsymbol{\Gamma}(\mathbf{x}, \mathbf{y}) * \boldsymbol{\sigma}(\mathbf{y})]. \tag{6.3.18}$$

This can be readily proven by noting that

$$\boldsymbol{\Gamma}(\mathbf{x}, \mathbf{y}) = \boldsymbol{\Gamma}(\mathbf{y}, \mathbf{x}). \tag{6.3.19}$$

Further, we have

$$\int_D \mathbf{\Gamma}(\mathbf{x}, \mathbf{y})\, dV(\mathbf{y}) = \mathbf{0}, \qquad \mathbf{\Gamma}(\mathbf{z}, \mathbf{x}) * \mathbf{L}^h \mathbf{\Gamma}(\mathbf{x}, \mathbf{y}) = \mathbf{\Gamma}(\mathbf{z}, \mathbf{y}). \quad (6.3.20)$$

The first of (6.3.20) can be verified by noting, for example,

$$\int_D \frac{\partial^2 G_{ki}(\mathbf{x}, \mathbf{y})}{\partial x_j\, \partial y_l}\, dV(\mathbf{y}) = \frac{\partial}{\partial x_j} \int_D \frac{\partial G_{ki}(\mathbf{x}, \mathbf{y})}{\partial y_l}\, dV(\mathbf{y})$$

$$= \frac{\partial}{\partial x_j} \int_S G_{ki}(\mathbf{x}, \mathbf{y}) n_l\, dS(\mathbf{y}) = 0. \quad (6.3.21)$$

The second of (6.3.20) can be verified by noting, for example,

$$\int_D \frac{\partial^2 G_{ki}(\mathbf{z}, \mathbf{x})}{\partial x_l\, \partial z_j} L^h_{klmn} \frac{\partial^2 G_{pm}(\mathbf{x}, \mathbf{y})}{\partial x_n\, \partial y_q}\, dV(\mathbf{x})$$

$$= \frac{\partial^2}{\partial z_j\, \partial y_q} \int_D \frac{\partial G_{ki}(\mathbf{z}, \mathbf{x})}{\partial x_l} L^h_{klmn} \frac{\partial G_{pm}(\mathbf{x}, \mathbf{y})}{\partial x_n}\, dV(\mathbf{x})$$

$$= \frac{\partial^2}{\partial z_j\, \partial y_q} \int_D \left\{ \frac{\partial}{\partial x_l} \left[G_{ki}(\mathbf{z}, \mathbf{x}) L^h_{klmn} \frac{\partial G_{pm}(\mathbf{x}, \mathbf{y})}{\partial x_n} \right] \right.$$

$$\left. - G_{ki}(\mathbf{z}, \mathbf{x}) L^h_{klmn} \frac{\partial G_{pm}(\mathbf{x}, \mathbf{y})}{\partial x_n\, \partial x_l} \right\} dV(\mathbf{x})$$

$$= -\frac{\partial^2}{\partial z_j\, \partial y_q} \int_D G_{ki}(\mathbf{z}, \mathbf{x}) L^h_{klmn} \frac{\partial G_{pm}(\mathbf{x}, \mathbf{y})}{\partial x_n\, \partial x_l}\, dV(\mathbf{x})$$

$$= \frac{\partial^2}{\partial z_j\, \partial y_q} \int_D G_{ki}(\mathbf{z}, \mathbf{x}) \delta_{kp} \delta(\mathbf{x} - \mathbf{y})\, dV(\mathbf{x})$$

$$= \frac{\partial^2 G_{pi}(\mathbf{z}, \mathbf{y})}{\partial z_j\, \partial y_q}. \quad (6.3.22)$$

Let us now consider (6.3.11). Because of (6.3.9), we have

$$u_i(\mathbf{x})|_{\mathbf{x}\in S} = \bar{\varepsilon}_{iq} x_q|_{\mathbf{x}\in S} + \int_D G_{mi}(\mathbf{x}, \mathbf{y})|_{\mathbf{x}\in S} \frac{\partial \tau_{mn}(\mathbf{y})}{\partial y_n}\, dV(\mathbf{y}) = \bar{\varepsilon}_{iq} x_q|_{\mathbf{x}\in S},$$

regardless of the choice of $\boldsymbol{\tau}(\mathbf{y})$. In other words, for any given $\hat{\boldsymbol{\tau}}(\mathbf{y})$, the displacement derived from (6.3.11), therefore, the corresponding strain derived from (6.3.17),

$$\hat{\boldsymbol{\varepsilon}}(\mathbf{x}) = \overline{\boldsymbol{\varepsilon}} - \boldsymbol{\Gamma}(\mathbf{x}, \mathbf{y}) * \hat{\boldsymbol{\tau}}(\mathbf{y}) \tag{6.3.23}$$

are kinematically admissible displacement and strain fields, respectively. Therefore, the minimum potential energy theorem (6.1.7) leads us to the following inequality:

$$2D\overline{U} \leq \int_D \hat{\boldsymbol{\varepsilon}}\mathbf{L}\hat{\boldsymbol{\varepsilon}}\, dv = \int_D \hat{\boldsymbol{\varepsilon}}\mathbf{L}^h\hat{\boldsymbol{\varepsilon}}\, dv + \int_D \hat{\boldsymbol{\varepsilon}}\mathbf{L}^p\hat{\boldsymbol{\varepsilon}}\, dv. \tag{6.3.24}$$

Substitute (6.3.23) into (6.3.24) and make use of properties of the integral operator $\boldsymbol{\Gamma}(\mathbf{x}, \mathbf{y})$, and (6.3.19) and (6.3.20), we arrive at the following inequality:

$$\begin{aligned}2D(\overline{U} - \overline{U}^h) \leq & \int_D \hat{\boldsymbol{\tau}}\boldsymbol{\Gamma}(\mathbf{x}, \mathbf{y}) * \hat{\boldsymbol{\tau}}(\mathbf{y})\, dv + \int_D \overline{\boldsymbol{\varepsilon}}\mathbf{L}^p\overline{\boldsymbol{\varepsilon}}\, dv \\ & - 2\int_D \overline{\boldsymbol{\varepsilon}}\mathbf{L}^p\boldsymbol{\Gamma}(\mathbf{x}, \mathbf{y}) * \hat{\boldsymbol{\tau}}(\mathbf{y})\, dv \\ & + \int_D \boldsymbol{\Gamma}(\mathbf{x}, \mathbf{y}) * \hat{\boldsymbol{\tau}}(\mathbf{y})\mathbf{L}^p\boldsymbol{\Gamma}(\mathbf{x}, \mathbf{y}) * \hat{\boldsymbol{\tau}}(\mathbf{y})\, dv,\end{aligned} \tag{6.3.25}$$

where

$$\overline{U}^h = \frac{1}{2D}\int_D \overline{\boldsymbol{\varepsilon}}\mathbf{L}^h\overline{\boldsymbol{\varepsilon}}\, dv. \tag{6.3.26}$$

Through a straightforward simplification, one can show that the last term in (6.3.25) can be rewritten as

$$\begin{aligned}&\int_D \boldsymbol{\Gamma}(\mathbf{x}, \mathbf{y}) * \hat{\boldsymbol{\tau}}(\mathbf{y})\mathbf{L}^p\boldsymbol{\Gamma}(\mathbf{x}, \mathbf{y}) * \hat{\boldsymbol{\tau}}(\mathbf{y})\, dv \\ &= \int_D \boldsymbol{\eta}[\hat{\boldsymbol{\tau}}]\mathbf{L}^p\boldsymbol{\eta}[\hat{\boldsymbol{\tau}}]\, dv - \int_D \hat{\boldsymbol{\tau}}(\mathbf{L}^p)^{-1}\hat{\boldsymbol{\tau}}\, dv \\ &\quad - 2\int_D \hat{\boldsymbol{\tau}}\boldsymbol{\Gamma}(\mathbf{x}, \mathbf{y}) * \hat{\boldsymbol{\tau}}(\mathbf{y})\, dv - \int_D \overline{\boldsymbol{\varepsilon}}\mathbf{L}^p\overline{\boldsymbol{\varepsilon}}\, dv \\ &\quad + 2\int_D \overline{\boldsymbol{\varepsilon}}\{\hat{\boldsymbol{\tau}} + \mathbf{L}^p\boldsymbol{\Gamma}(\mathbf{x}, \mathbf{y}) * \hat{\boldsymbol{\tau}}(\mathbf{y})\}\, dv,\end{aligned} \tag{6.3.27}$$

where

$$\boldsymbol{\eta}[\hat{\boldsymbol{\tau}}] = (\mathbf{L}^p)^{-1}\hat{\boldsymbol{\tau}} + \boldsymbol{\Gamma}(\mathbf{x}, \mathbf{y}) * \hat{\boldsymbol{\tau}}(\mathbf{y}) - \overline{\boldsymbol{\varepsilon}}. \tag{6.3.28}$$

Substituting (6.3.27) into (6.3.25), we obtain, for any stress tensor $\hat{\boldsymbol{\tau}}$, the desired inequality,

$$2D(\overline{U} - \overline{U}^h) \leq -H[\hat{\boldsymbol{\tau}}] + \int_D \boldsymbol{\eta}[\hat{\boldsymbol{\tau}}]\mathbf{L}^p\boldsymbol{\eta}[\hat{\boldsymbol{\tau}}]\, dv, \tag{6.3.29}$$

where

$$H[\hat{\boldsymbol{\tau}}] = \int_D \hat{\boldsymbol{\tau}}\boldsymbol{\eta}[\hat{\boldsymbol{\tau}}]\, dv - \int_D \overline{\boldsymbol{\varepsilon}}\hat{\boldsymbol{\tau}}\, dv. \tag{6.3.30}$$

Next, we want to know what happens if the $\hat{\boldsymbol{\tau}}$ in (6.3.29) is indeed the actual stress polarization tensor of the problem, that is, $\hat{\boldsymbol{\tau}} = \boldsymbol{\tau}$. In this case, we have, from (6.3.5) and (6.3.17), that

$$\boldsymbol{\eta}[\boldsymbol{\tau}] = (\mathbf{L}^p)^{-1}\boldsymbol{\tau} + \boldsymbol{\Gamma}(\mathbf{x}, \mathbf{y}) * \boldsymbol{\tau}(\mathbf{y}) - \overline{\boldsymbol{\varepsilon}} = \mathbf{0}, \tag{6.3.31}$$

therefore, the second term on the right-hand side of (6.3.29) would vanish. Further

$$H[\boldsymbol{\tau}] = -\int_D \overline{\boldsymbol{\varepsilon}}\boldsymbol{\tau}\, dv = -\int_D \overline{\boldsymbol{\varepsilon}}(\mathbf{L} - \mathbf{L}^h)\boldsymbol{\varepsilon}\, dv = -2D(\overline{U} - \overline{U}^h). \tag{6.3.32}$$

In other words, the right-hand side of (6.3.29) is minimized if $\hat{\boldsymbol{\tau}} = \boldsymbol{\tau}$ is used, and the minimum value is equal to the left-hand side, $2D(\overline{U} - \overline{U}^h)$.

We can now summarize the above derivations. First, once $\mathbf{L}^h$ is selected, the perturbed part of the stiffness tensor $\mathbf{L}^p$ and Green's function $\boldsymbol{\Gamma}(\mathbf{x}, \mathbf{y})$ can be computed from (6.3.2) and (6.3.8) and (6.3.9). Second, displacement u_i generated from (6.3.11) is a kinematically admissible displacement field, which leads the minimum potential energy theorem to the inequality (6.3.29). Third, the inequality is valid for any stress tensor $\hat{\boldsymbol{\tau}}$. Finally, the equal sign is realized when $\hat{\boldsymbol{\tau}}$ is replaced by the actual solution to the stress polarization tensor of the problem. Therefore, we have just proved the following theorem.

Theorem 6.1 Among all the symmetric second-order tensors, the true solution to the stress polarization tensor $\hat{\boldsymbol{\tau}}$ renders the following functional minimum:

$$I[\hat{\boldsymbol{\tau}}] = \overline{U}^h - \frac{1}{2D} H[\hat{\boldsymbol{\tau}}] + \frac{1}{2D} \int_D \boldsymbol{\eta}[\hat{\boldsymbol{\tau}}]\mathbf{L}^p\boldsymbol{\eta}[\hat{\boldsymbol{\tau}}]\, dv, \qquad (6.3.33)$$

and the minimum value of $I[\hat{\boldsymbol{\tau}}]$ is the strain energy density of the heterogeneous material under the displacement boundary condition (6.3.1), that is,

$$\min\{I[\hat{\boldsymbol{\tau}}]\} = \tfrac{1}{2}\overline{\boldsymbol{\varepsilon}}\,\overline{\mathbf{L}}\,\overline{\boldsymbol{\varepsilon}}. \qquad (6.3.34)$$

The inequality (6.3.29) can be further simplified if one assumes that $\mathbf{L}^p$ is negative semidefinite, that is, for any $\boldsymbol{\eta}$,

$$\int_D \boldsymbol{\eta}[\hat{\boldsymbol{\tau}}]\mathbf{L}^p\boldsymbol{\eta}[\hat{\boldsymbol{\tau}}]\, dv \leq 0. \qquad (6.3.35)$$

In this case, (6.3.29) reduces to

$$2D(\overline{U} - \overline{U}^h) \leq -H[\hat{\boldsymbol{\tau}}]. \qquad (6.3.36)$$

Therefore, we have the following theorem.

Theorem 6.1a (Hashin–Shtrikman Variational Principle) When $\mathbf{L}^p$ is negative semidefinite, among all the symmetric second-order tensors, the true solution to the stress polarization tensor $\hat{\boldsymbol{\tau}}$ renders the following functional minimum:

$$I_h[\hat{\boldsymbol{\tau}}] = \overline{U}^h - \frac{1}{2D} H[\hat{\boldsymbol{\tau}}], \qquad (6.3.37)$$

and the minimum value of $I_h[\hat{\boldsymbol{\tau}}]$ is the strain energy density of the heterogeneous material under the displacement boundary condition (6.3.1), that is,

$$\min\{I_h[\hat{\boldsymbol{\tau}}]\} = \tfrac{1}{2}\overline{\boldsymbol{\varepsilon}}\,\overline{\mathbf{L}}\,\overline{\boldsymbol{\varepsilon}}. \qquad (6.3.38)$$

Let us now turn to the complimentary energy principle. Introducing (6.3.17) to (6.3.3) yields

$$\boldsymbol{\sigma}(\mathbf{x}) = \mathbf{L}^h\bar{\boldsymbol{\varepsilon}} - \mathbf{L}^h \int_D \boldsymbol{\Gamma}(\mathbf{x}, \mathbf{y})\boldsymbol{\tau}(\mathbf{y})\, dV(\mathbf{y}) + \boldsymbol{\tau}(\mathbf{x}). \qquad (6.3.39)$$

Thus, it follows from, for example,

$$\begin{aligned}
&\int_D \left(\frac{\partial}{\partial x_j} L^h_{ijmn} \frac{\partial^2 G_{km}(\mathbf{x}, \mathbf{y})}{\partial x_n\, \partial y_l}\right) \tau_{kl}(\mathbf{y})\, dV(\mathbf{y}) \\
&= \int_D \frac{\partial}{\partial y_l}\left(L^h_{ijmn} \frac{\partial^2 G_{km}(x, \mathbf{y})}{\partial x_n\, \partial x_j}\, \tau_{kl}(\mathbf{y})\right) dV(\mathbf{y}) \\
&\quad - \int_D L^h_{ijmn} \frac{\partial^2 G_{km}(\mathbf{x}, \mathbf{y})}{\partial x_n\, \partial x_j}\, \tau_{kl,l}(\mathbf{y})\, dV(\mathbf{y}) \\
&= \int_S L^h_{ijmn} \frac{\partial^2 G_{km}(\mathbf{x}, \mathbf{y})}{\partial x_n\, \partial x_j}\, n_l\, dS(\mathbf{y}) - \int_D L^h_{ijmn} \frac{\partial^2 G_{km}(\mathbf{x}, \mathbf{y})}{\partial x_n\, \partial x_j}\, \tau_{kl,l}(\mathbf{y})\, dV(\mathbf{y}) \\
&= -\int_S \delta_{ik}\delta(\mathbf{x} - \mathbf{y})n_l\, dS(\mathbf{y}) - \int_D L^h_{ijmn} \frac{\partial^2 G_{km}(\mathbf{x}, \mathbf{y})}{\partial x_n\, \partial x_j}\, \tau_{kl,l}(\mathbf{y})\, dV(\mathbf{y}) \\
&= -\int_D L^h_{ijmn} \frac{\partial^2 G_{km}(\mathbf{x}, \mathbf{y})}{\partial x_n\, \partial x_j}\, \tau_{kl,l}(\mathbf{y})\, dV(\mathbf{y})
\end{aligned}$$

that

$$\operatorname{div}(\boldsymbol{\sigma}(\mathbf{x})) = \mathbf{L}^h \int_D \boldsymbol{\Gamma}(\mathbf{x}, \mathbf{y})\operatorname{div}(\boldsymbol{\tau}(\mathbf{y}))\, dV(\mathbf{y}) + \operatorname{div}(\boldsymbol{\tau}(\mathbf{x})). \qquad (6.3.40)$$

One may then conclude that for any statically admissible stress tensor $\hat{\boldsymbol{\tau}}(\mathbf{x})$, that is, $\operatorname{div}(\boldsymbol{\tau}(\mathbf{x})) = 0$, the stress tensor $\hat{\boldsymbol{\sigma}}(\mathbf{x})$ generated from (6.3.3)

$$\hat{\boldsymbol{\sigma}}(\mathbf{x}) = \mathbf{L}^h\boldsymbol{\varepsilon}(\mathbf{x}) + \hat{\boldsymbol{\tau}}(\mathbf{x}), \qquad (6.3.41)$$

is also a statically admissible stress field, that is,

$$\operatorname{div}(\hat{\boldsymbol{\sigma}}(\mathbf{x})) = \operatorname{div}(\mathbf{L}^h\boldsymbol{\varepsilon}(\mathbf{x}) + \hat{\boldsymbol{\tau}}(\mathbf{x})) = \operatorname{div}(\mathbf{L}^h\boldsymbol{\varepsilon}(\mathbf{x})) = 0. \qquad (6.3.42)$$

Having established that $\hat{\boldsymbol{\sigma}}(\mathbf{x})$ is statically admissible, whenever $\hat{\boldsymbol{\tau}}(\mathbf{x})$ is, we can now consider (6.1.13) for a statically admissible $\hat{\boldsymbol{\sigma}}(\mathbf{x})$,

$$
\begin{aligned}
2D\overline{U} &\geq \overline{\boldsymbol{\varepsilon}} \int_D \hat{\boldsymbol{\sigma}}\, dv - \int_D \hat{\boldsymbol{\sigma}}\mathbf{M}\hat{\boldsymbol{\sigma}}\, dv \\
&= \overline{\boldsymbol{\varepsilon}} \int_D \hat{\boldsymbol{\sigma}}\, dv - \int_D \hat{\boldsymbol{\sigma}}(\mathbf{M}^h + \mathbf{M}^p)\hat{\boldsymbol{\sigma}}\, dv, \qquad (6.3.43)
\end{aligned}
$$

where, analogous to (6.3.2), the compliance tensor is written as

$$
\mathbf{M}(\mathbf{x}) = \mathbf{M}^h + \mathbf{M}^p(\mathbf{x}), \qquad (6.3.44)
$$

and the superscripts on the compliance tensors carry the same meaning as those for the stiffness tensors. Substituting (6.3.41) into (6.3.43), following similar steps in deriving (6.3.29) and making use of the identity,

$$
\mathbf{L}^h\mathbf{M}^p + \mathbf{L}^p(\mathbf{M}^h + \mathbf{M}^p) = \mathbf{0}, \qquad (6.3.45)
$$

we conclude that

$$
2D(\overline{U} - \overline{U}^h) \geq -H[\hat{\boldsymbol{\tau}}] - \int_D L^h\boldsymbol{\eta}[\hat{\boldsymbol{\tau}}]\mathbf{M}^p\mathbf{L}^h\boldsymbol{\eta}[\hat{\boldsymbol{\tau}}]\, dv, \qquad (6.3.46)
$$

for any statically admissible stress tensor $\hat{\boldsymbol{\tau}}(\mathbf{x})$. As shown earlier, if $\hat{\boldsymbol{\tau}}(\mathbf{x})$ happens to be the true stress polarization to the deformation problem considered here, the equal sign in (6.3.46) holds. We have just proved the following variational theorem.

Theorem 6.2 Among all the statically admissible stress fields, the true solution renders the following functional maximum,

$$
I^c[\hat{\boldsymbol{\tau}}] = \overline{U}^h - \frac{1}{2D} H[\hat{\boldsymbol{\tau}}] - \frac{1}{2D} \int_D \mathbf{L}^h\boldsymbol{\eta}[\hat{\boldsymbol{\tau}}]\mathbf{M}^p\mathbf{L}^h\boldsymbol{\eta}[\hat{\boldsymbol{\tau}}]\, dv, \qquad (6.3.47)
$$

and the maximum value of $I^c[\hat{\boldsymbol{\tau}}]$ is the strain energy density of the heterogeneous material under the displacement boundary condition (6.3.1), that is,

$$
\min\{I^c[\hat{\boldsymbol{\tau}}]\} = \tfrac{1}{2}\overline{\boldsymbol{\varepsilon}}\overline{\mathbf{L}}\overline{\boldsymbol{\varepsilon}}. \qquad (6.3.48)
$$

If $\mathbf{M}^p$ is negative semidefinite (or, equivalently, $\mathbf{L}^p$ is positive semidefinite), that is,

$$\int_D \mathbf{L}^h\boldsymbol{\eta}[\hat{\boldsymbol{\tau}}]\mathbf{M}^p\mathbf{L}^h\boldsymbol{\eta}[\hat{\boldsymbol{\tau}}]\, dv \le 0,$$

then (6.3.46) implies

$$2D(\overline{U} - \overline{U}^h) \ge -H[\hat{\boldsymbol{\tau}}]. \tag{6.3.49}$$

This leads to another form of the Hashin–Shtrikman variational principle.

Theorem 6.2a (Hashin–Shtrikman Variational Principle) Among all the statically admissible stress fields, the true solution renders the following functional maximum:

$$I_h^c[\hat{\boldsymbol{\tau}}] = 2D\overline{U}^h - H[\hat{\boldsymbol{\tau}}], \tag{6.3.50}$$

and the maximum value of $I_h^c[\hat{\boldsymbol{\tau}}]$ is the strain energy density of the heterogeneous material under the displacement boundary condition (6.3.1), that is,

$$\min\{I_h^c[\hat{\boldsymbol{\tau}}]\} = \tfrac{1}{2}\overline{\boldsymbol{\varepsilon}}\,\overline{\mathbf{L}}\,\overline{\boldsymbol{\varepsilon}}. \tag{6.3.51}$$

6.4 HASHIN–SHTRIKMAN BOUNDS

As can be seen from (6.3.36) and (6.3.49), the key to developing the bounds is to find an appropriate statically admissible stress tensor $\hat{\boldsymbol{\tau}}$. This is what we will do next. Remembering that the Reuss and Voigt bounds were obtained by assuming uniform admissible fields throughout the composite, let us try a piecewise uniform field for $\hat{\boldsymbol{\tau}}$,

$$\hat{\boldsymbol{\tau}}(\mathbf{x})\big|_{\mathbf{x}\in\Omega_r} = \hat{\boldsymbol{\tau}}_r, \qquad r = 0, 1, 2, \ldots, N. \tag{6.4.1}$$

In other words, we assume that the stress polarization tensor is a constant tensor on each inhomogeneity including the matrix. Introducing (6.4.1) into (6.3.28) yields

$$\boldsymbol{\eta}[\hat{\boldsymbol{\tau}}] = (\mathbf{L}^p)^{-1}\hat{\boldsymbol{\tau}} + D\sum_{s=0}^{N}\frac{c_r}{\Omega_s}\int_{\Omega_s}\boldsymbol{\Gamma}(\mathbf{x},\mathbf{y})\,dV(\mathbf{y})\boldsymbol{\tau}_s - \bar{\boldsymbol{\varepsilon}}. \quad (6.4.2)$$

It thus follows from (6.3.30) that

$$H[\hat{\boldsymbol{\tau}}] = \sum_{r=0}^{N}\int_{\Omega_r}\boldsymbol{\tau}_r\left[(\mathbf{L}^p)^{-1}\boldsymbol{\tau} + D\sum_{s=0}^{N}\frac{c_r}{\Omega_s}\int_{\Omega_s}\boldsymbol{\Gamma}(\mathbf{x},\mathbf{y})\,dV(\mathbf{y})\boldsymbol{\tau}_s - \bar{\boldsymbol{\varepsilon}}\right]dv$$

$$- \sum_{r=0}^{N}\int_{\Omega_r}\bar{\boldsymbol{\varepsilon}}\boldsymbol{\tau}_r\,dv$$

$$= D\sum_{r=0}^{N}c_r\boldsymbol{\tau}_r(\bar{\mathbf{L}}_r^p)^{-1}\boldsymbol{\tau}_r + D^2\sum_{r=0}^{N}\sum_{s=0}^{N}c_rc_s\boldsymbol{\tau}_r\mathbf{P}_{rs}\boldsymbol{\tau}_s - 2D\bar{\boldsymbol{\varepsilon}}\sum_{r=0}^{N}c_r\boldsymbol{\tau}_r, \quad (6.4.3)$$

where c_r is the volume fraction of the rth inhomogeneity and

$$(\bar{\mathbf{L}}_r^p)^{-1} = \frac{1}{\Omega_r}\int_{\Omega_r}(\mathbf{L}^p)^{-1}\,dv, \quad (6.4.4)$$

$$\mathbf{P}_{rs} = \frac{1}{\Omega_r\Omega_s}\int_{\Omega_r}\int_{\Omega_s}\boldsymbol{\Gamma}(\mathbf{x},\mathbf{y})\,dV(\mathbf{y})\,dV(\mathbf{x}). \quad (6.4.5)$$

The validity of (6.4.4) is based on the fact that the stiffness tensor is uniform on each inhomogeneity.

Clearly, $H[\hat{\boldsymbol{\tau}}]$ is now a quadratic function of the constant tensor $\boldsymbol{\tau}_r$. The specific $\boldsymbol{\tau}_r$ that gives the extreme value of $H[\hat{\boldsymbol{\tau}}]$ can be determined from $\partial H[\hat{\boldsymbol{\tau}}]/\partial\boldsymbol{\tau}_r = 0$,

$$(\bar{\mathbf{L}}_r^p)^{-1}\boldsymbol{\tau}_r + D\sum_{s=0}^{N}c_s\mathbf{P}_{rs}\boldsymbol{\tau}_s = \bar{\boldsymbol{\varepsilon}}, \qquad r = 0, 1, \ldots, N. \quad (6.4.6)$$

This is a system of linear algebraic equations for $\boldsymbol{\tau}_r$. The solution can be written formally as

$$\boldsymbol{\tau}_r = \mathbf{R}_r\bar{\boldsymbol{\varepsilon}}. \quad (6.4.7)$$

This leads to the following extreme value of $H[\hat{\boldsymbol{\tau}}]$:

$$H[\hat{\boldsymbol{\tau}}]|_{\boldsymbol{\tau}_r=\mathbf{R}_r\bar{\boldsymbol{\varepsilon}}} = -D\bar{\boldsymbol{\varepsilon}} \sum_{r=0}^{N} c_r \mathbf{R}_r \bar{\boldsymbol{\varepsilon}} = D\bar{\boldsymbol{\varepsilon}}(\mathbf{L}^h - \tilde{\mathbf{L}})\bar{\boldsymbol{\varepsilon}}, \tag{6.4.8}$$

where

$$\tilde{\mathbf{L}} = \mathbf{L}^h + \sum_{r=0}^{N} c_r \mathbf{R}_r. \tag{6.4.9}$$

For $\mathbf{L}^p$ being negative semidefinite, substituting (6.4.8) into (6.3.36) yields

$$2\overline{U} = \bar{\boldsymbol{\varepsilon}}\overline{\mathbf{L}}\bar{\boldsymbol{\varepsilon}} \le \bar{\boldsymbol{\varepsilon}}\tilde{\mathbf{L}}\bar{\boldsymbol{\varepsilon}}, \tag{6.4.10}$$

or

$$\overline{\mathbf{L}} \le \tilde{\mathbf{L}} = \mathbf{L}^h + \sum_{r=0}^{N} c_r \mathbf{R}_r, \tag{6.4.11}$$

with $\tilde{\mathbf{L}}$ being given by (6.4.9). Dually, for $\mathbf{L}^p$ being positive semidefinite, substituting (6.4.8) into (6.3.49) yields

$$\overline{\mathbf{L}} \ge \tilde{\mathbf{L}} = \mathbf{L}^h + \sum_{r=0}^{N} c_r \mathbf{R}_r. \tag{6.4.12}$$

The inequalities (6.4.11) and (6.4.12) are the Hashin–Shtrikman bounds for the effective stiffness tensor of a heterogeneous material containing N-phase inhomogeneities. In general, the Hashin–Shtrikman bounds are more restrictive than Voigt and Reuss bounds.

It is noted that, in order to evaluate the Hashin–Shtrikman bounds, the double integral in (6.4.5) needs to be evaluated. Although, theoretically speaking, the integral can be carried out for a given heterogeneous material, the large number of inhomogeneities and the statistical nature of their distributions in a heterogeneous material render the integration impractical. We need to find a practical way to evaluate it, albeit, approximately.

Recall from (6.3.15) that the integrand consists of the second derivatives of Green's function $\mathbf{G}_{ij}(\mathbf{x}, \mathbf{y})$ in the domain D, which satisfies zero displacement boundary condition; see (6.3.8) and (6.3.9). One of the proposals of simplifying (6.4.5) (see Willis, 1982) is to replace

Green's function $G_{ij}(\mathbf{x}, \mathbf{y})$ in the finite domain D by its counterpart in the infinite domain,

$$\mathbf{P}_{rs} \approx \frac{1}{\Omega_r \Omega_s} \int_{\Omega_r} \int_{\Omega_s} \mathbf{\Gamma}^{\infty}(\mathbf{x}, \mathbf{y})\, dV(\mathbf{y})\, dV(\mathbf{x}), \tag{6.4.13}$$

where $\mathbf{\Gamma}^{\infty}(\mathbf{x}, \mathbf{y})$ is given by (3.3.18), which is related to Green's function in an unbounded domain, $G_{ij}^{\infty}(\mathbf{x}, \mathbf{y})$ derived in Section 2.6:

$$G_{ij}^{\infty}(\mathbf{x}, \mathbf{y}) = \frac{1}{(2\pi)^3} \int_{-\infty}^{\infty} N_{ij}(\boldsymbol{\xi}) D^{-1}(\boldsymbol{\xi}) e^{i\boldsymbol{\xi}\cdot(\mathbf{x}-\mathbf{y})}\, d\boldsymbol{\xi}. \tag{6.4.14}$$

For statistically homogeneous and isotropic distribution of inhomogeneities, the integral in (6.4.13) can be evaluated; see Appendix 6.A:

$$\mathbf{P}_{rs} \approx \frac{1}{\Omega_r \Omega_s} \int_{\Omega_r} \int_{\Omega_s} \mathbf{\Gamma}^{\infty}(\mathbf{x}, \mathbf{y})\, dV(\mathbf{y})\, dV(\mathbf{x}) = \frac{1}{\Omega_s} (\delta_{rs} - c_s)\mathbf{P}, \tag{6.4.15}$$

where

$$\mathbf{P}(\mathbf{x}) = \int_{\hat{S}^2} \Gamma^{\infty}(\mathbf{x}, \mathbf{y})\, dV(\mathbf{y}) \quad \text{for} \quad \mathbf{x} \in \hat{S}^2, \tag{6.4.16}$$

and the integral is over a unit sphere $\hat{S}^2$. It follows from (4.3.19) that the tensor $\mathbf{P}$ is related to the Eshelby tensor $\mathbf{S}$ through

$$\mathbf{P} = \mathbf{S}(\mathbf{L}^h)^{-1}. \tag{6.4.17}$$

Substituting (6.4.15) into (6.4.6) yields

$$(\overline{\mathbf{L}}_r^p)^{-1}\boldsymbol{\tau}_r + \mathbf{P}\boldsymbol{\tau}_r - \mathbf{P} \sum_{s=0}^{N} c_s \boldsymbol{\tau}_s = \bar{\boldsymbol{\varepsilon}}, \qquad r = 0, 1, \ldots, N, \tag{6.4.18}$$

which can be rewritten as

$$\boldsymbol{\tau}_r = [(\overline{\mathbf{L}}_r^p)^{-1} + \mathbf{P}]^{-1} \left(\bar{\boldsymbol{\varepsilon}} + \mathbf{P} \sum_{s=0}^{N} c_s \boldsymbol{\tau}_s \right), \qquad r = 0, 1, \ldots, N. \tag{6.4.19}$$

Further,

$$\sum_{r=0}^{N} c_r \boldsymbol{\tau}_r = \sum_{r=0}^{N} c_r\,[(\overline{\mathbf{L}}_r^p)^{-1} + \mathbf{P}]^{-1}\left(\bar{\boldsymbol{\varepsilon}} + \mathbf{P}\sum_{s=0}^{N} c_s \boldsymbol{\tau}_s\right)$$

$$= \sum_{r=0}^{N} c_r[(\overline{\mathbf{L}}_r^p)^{-1} + \mathbf{P}]^{-1}\bar{\boldsymbol{\varepsilon}} + \sum_{r=0}^{N} [(\overline{\mathbf{L}}_r^p)^{-1} + \mathbf{P}]^{-1}\,\mathbf{P}\sum_{s=0}^{N} c_s \boldsymbol{\tau}_s. \tag{6.4.20}$$

Finally, making use of (6.4.7), we have

$$\sum_{s=0}^{N} c_s \mathbf{R}_s = \left\{\mathbf{I} - \sum_{r=0}^{N} c_r[(\overline{\mathbf{L}}_r^p)^{-1} + \mathbf{P}]^{-1}\mathbf{P}\right\}^{-1} \sum_{r=0}^{N} c_r[(\overline{\mathbf{L}}_r^p)^{-1} + \mathbf{P}]^{-1}$$

$$= \left[\sum_{r=0}^{N} c_r(\mathbf{I} + \overline{\mathbf{L}}_r^p\mathbf{P})^{-1}\right]^{-1} \sum_{r=0}^{N} c_r(\mathbf{I} + \overline{\mathbf{L}}_r^p\mathbf{P})^{-1}\mathbf{L}_r^p. \tag{6.4.21}$$

Thus, it follows from (6.4.9) that

$$\tilde{\mathbf{L}} = \mathbf{L}^h + \left[\sum_{r=0}^{N} c_r(\mathbf{I} + \overline{\mathbf{L}}_r^p\mathbf{P})^{-1}\right]^{-1} \sum_{r=0}^{N} c_r(\mathbf{I} + \overline{\mathbf{L}}_r^p\mathbf{P})^{-1}\overline{\mathbf{L}}_r^p. \tag{6.4.22}$$

Example 6.2 Let a composite material be made of a matrix with elastic stiffness tensor $\mathbf{L}_0$ containing uniformly distributed identical spherical inhomogeneities with elastic stiffness tensor $\mathbf{L}_1$. Assume that both $\mathbf{L}_0$ and $\mathbf{L}_1$ are isotropic and given by

$$\mathbf{L}_0 = (3K_0, 2\mu_0), \qquad \mathbf{L}_1 = (3K_1, 2\mu_1), \tag{6.4.23}$$

where $K_1 > K_0$, $\mu_1 > \mu_0$. If the volume fraction of the inhomogeneities is c_1, find the bounds of its effective stiffness tensor.

To find the bounds for the effective stiffness tensor of the composites, we need to know the tensor $\tilde{\mathbf{L}}$ as given by (6.4.22). To this end, let us compute the relevant tensors one at a time. First, we choose (it will be seen later that the choice is arbitrary)

$$\mathbf{L}^h = \mathbf{L}_0. \tag{6.4.24}$$

Consequently,

$$\mathbf{L}^p(\mathbf{x}) = \begin{cases} \mathbf{L}_1 - \mathbf{L}_0 & \mathbf{x} \in \Omega_1 \\ \mathbf{0} & \mathbf{x} \in \Omega_0 \end{cases}. \tag{6.4.25}$$

So,

$$\overline{\mathbf{L}}_0^p = \frac{1}{\Omega_0}\int_{\Omega_0} \mathbf{L}^p \, dv = \mathbf{0}, \qquad \overline{\mathbf{L}}_1^p = \frac{1}{\Omega_1}\int_{\Omega_1} \mathbf{L}^p \, dv = \mathbf{L}_1 - \mathbf{L}_0. \tag{6.4.26}$$

Introducing the above to (6.4.22) leads to

$$\begin{aligned} \tilde{\mathbf{L}} &= \mathbf{L}_0 + \left[\sum_{r=0}^{1} c_r(\mathbf{I} + \overline{\mathbf{L}}_r^p\mathbf{P})^{-1}\right]^{-1} \sum_{r=0}^{1} c_r(\mathbf{I} + \overline{\mathbf{L}}_r^p\mathbf{P})^{-1}(\mathbf{L}_r - \mathbf{L}_0) \\ &= \left[\sum_{r=0}^{1} c_r(\mathbf{I} + \overline{\mathbf{L}}_r^p\mathbf{P})^{-1}\right]^{-1} \sum_{r=0}^{1} c_r\,(\mathbf{I} + \overline{\mathbf{L}}_r^p\mathbf{P})^{-1}\mathbf{L}_r \\ &= \{c_0\mathbf{I} + c_1[\mathbf{I} + (\mathbf{L}_1 - \mathbf{L}_0)\mathbf{P}]^{-1}\}^{-1} \\ &\quad \times \{c_0\mathbf{L}_0 + c_1[\mathbf{I} + (\mathbf{L}_1 - \mathbf{L}_0)\mathbf{P}]^{-1}\mathbf{L}_1\}. \end{aligned} \tag{6.4.27}$$

For spherical inhomogeneities, the Eshelby tensor is given by (see Appendix 4.B)

$$\mathbf{S} = (3\gamma_0, 2\delta_0), \tag{6.4.28}$$

where

$$\gamma_0 = \frac{K_0}{3K_0 + 4\mu_0} = \frac{1 + \nu_0}{9(1 - \nu_0)}, \qquad \delta_0 = \frac{3(K_0 + 2\mu_0)}{5(3K_0 + 4\mu_0)} = \frac{4 - 5\nu_0}{15(1 - \nu_0)}. \tag{6.4.29}$$

Therefore, making use of (6.4.17), we have

$$\begin{aligned} &\mathbf{I} + (\mathbf{L}_1 - \mathbf{L}_0)\mathbf{P} \\ &\quad = (1,1) + (3K_1 - 3K_0, 2\mu_1 - 2\mu_0)(3\gamma_0, 2\delta_0)\left(\frac{1}{3K_0}, \frac{1}{2\mu_0}\right) \\ &\quad = \left(\frac{K_0 + 3\gamma_0(K_1 - K_0)}{K_0}, \frac{\mu_0 + 2\gamma_0(\mu_1 - \mu_0)}{\mu_0}\right). \end{aligned} \tag{6.4.30}$$

Thus,

$$[\mathbf{I} + (\mathbf{L}_1 - \mathbf{L}_0)\mathbf{P}]^{-1} = \left(\frac{K_0}{K_0 + 3\gamma_0(K_1 - K_0)}, \frac{\mu_0}{\mu_0 + 2\gamma_0(\mu_1 - \mu_0)} \right). \tag{6.4.31}$$

This leads to

$$\tilde{\mathbf{L}} = \{c_0\mathbf{I} + c_1[\mathbf{I} + (\mathbf{L}_1 - \mathbf{L}_0)\mathbf{P}]^{-1}\}^{-1}\{c_0\mathbf{L}_0 + c_1[\mathbf{I} + (\mathbf{L}_1 - \mathbf{L}_0)\mathbf{P}]^{-1}\mathbf{L}_1\}$$
$$= (3\breve{K}, 2\breve{\mu}),$$

where

$$\breve{K} = \frac{K_0\{(1 - c_1)[K_0 + 3\gamma_0(K_1 - K_0)] + c_1K_1\}}{K_0 + 3\gamma_0(1 - c_1)(K_1 - K_0)},$$
$$\breve{\mu} = \frac{\mu_0\{(1 - c_1)[\mu_0 + 2\delta_0(\mu_1 - \mu_0)] + c_1\mu_1\}}{\mu_0 + 2\delta_0(1 - c_1)(\mu_1 - \mu_0)}. \tag{6.4.32}$$

Introducing (6.4.29) into the above expressions,

$$\breve{K} = \frac{K_0(3K_1 + 4\mu_0) + 4c_1\mu_0(K_1 - K_0)}{3K_0 + 4\mu_0 + 3(1 - c_1)(K_1 - K_0)},$$
$$\breve{\mu} = \frac{6\mu_0\mu_1(K_0 + 2\mu_0) + \mu_0(9K_0 + 8\mu_0)[c_1\mu_1 + (1 - c_1)\mu_0]}{\mu_0(9K_0 + 8\mu_0) + 6(K_0 + 2\mu_0)[c_1\mu_0 + (1 - c_1)\mu_1]}. \tag{6.4.33}$$

Since $K_1 > K_0$, $\mu_1 > \mu_0$, one can see that the tensor $\mathbf{L}^p(\mathbf{x})$ defined by (6.4.25) is positive semidefinite. It thus follows from (6.4.12) that

$$\overline{\mathbf{L}} = (3\overline{K}, 2\overline{\mu}) \geq \tilde{\mathbf{L}} = (3\breve{K}, 2\breve{\mu}),$$

or equivalently,

$$\overline{K} \geq \frac{K_0(3K_1 + 4\mu_0) + 4c_1\mu_0(K_1 - K_0)}{3K_0 + 4\mu_0 + 3(1 - c_1)(K_1 - K_0)}, \tag{6.4.34}$$

$$\overline{\mu} \geq \frac{6\mu_0\mu_1(K_0 + 2\mu_0) + \mu_0(9K_0 + 8\mu_0)[c_1\mu_1 + (1 - c_1)\mu_0]}{\mu_0(9K_0 + 8\mu_0) + 6(K_0 + 2\mu_0)[c_1\mu_0 + (1 - c_1)\mu_1]}. \tag{6.4.35}$$

These are the Hashin–Shtrikman lower bounds.

To find the Hashin–Shtrikman upper bounds, we choose

$$\mathbf{L}^h = \mathbf{L}_1. \tag{6.4.36}$$

Consequently,

$$\mathbf{L}^p(\mathbf{x}) = \begin{cases} \mathbf{L}_0 - \mathbf{L}_1 & \mathbf{x} \in \Omega_0 \\ \mathbf{0} & \mathbf{x} \in \Omega_1 \end{cases}. \tag{6.4.37}$$

So,

$$\overline{\mathbf{L}}_0^p = \frac{1}{\Omega_0}\int_{\Omega_0} \mathbf{L}^p\, d\nu = \mathbf{L}_0 - \mathbf{L}_1, \qquad \overline{\mathbf{L}}_1^p = \frac{1}{\Omega_1}\int_{\Omega_1} \mathbf{L}^p\, d\nu = \mathbf{0}. \tag{6.4.38}$$

Thus, we have

$$\begin{aligned}\tilde{\mathbf{L}} &= \mathbf{L}_1 + \left[\sum_{r=0}^{1} c_r(\mathbf{I} + \overline{\mathbf{L}}_r^p\mathbf{P})^{-1}\right]^{-1} \sum_{r=0}^{1} c_r(\mathbf{I} + \overline{\mathbf{L}}_r^p\mathbf{P})^{-1}(\mathbf{L}_r - \mathbf{L}_1) \\ &= \left[\sum_{r=0}^{1} c_r(\mathbf{I} + \overline{\mathbf{L}}_r^p\mathbf{P})^{-1}\right]^{-1} \sum_{r=0}^{1} c_r\,(\mathbf{I} + \overline{\mathbf{L}}_r^p\mathbf{P})^{-1}\mathbf{L}_r \\ &= \{c_0[\mathbf{I} + (\mathbf{L}_0 - \mathbf{L}_1)\mathbf{P}]^{-1} + c_1\mathbf{I}\}^{-1} \\ &\quad \times \{c_0[\mathbf{I} + (\mathbf{L}_0 - \mathbf{L}_1)\mathbf{P}]^{-1}\mathbf{L}_0 + c_1\mathbf{L}_1\}. \end{aligned} \tag{6.4.39}$$

Since

$$\begin{aligned}&\mathbf{I} + (\mathbf{L}_0 - \mathbf{L}_1)\mathbf{P} \\ &\quad = (1,1) + (3K_0 - 3K_1, 2\mu_0 - 2\mu_1)(3\gamma_0, 2\delta_0)\left(\frac{1}{3K_0}, \frac{1}{2\mu_0}\right) \\ &\quad = \left(\frac{K_0 + 3\gamma_0(K_0 - K_1)}{K_0}, \frac{\mu_0 + 2\gamma_0(\mu_0 - \mu_1)}{\mu_0}\right). \end{aligned} \tag{6.4.40}$$

Thus,

$$[\mathbf{I} + (\mathbf{L}_0 - \mathbf{L}_1)\mathbf{P}]^{-1} = \left(\frac{K_0}{K_0 + 3\gamma_0(K_0 - K_1)}, \frac{\mu_0}{\mu_0 + 2\gamma_0(\mu_0 - \mu_1)}\right). \tag{6.4.41}$$

This leads to

$$\tilde{\mathbf{L}} = \{c_0[\mathbf{I} + (\mathbf{L}_0 - \mathbf{L}_1)\mathbf{P}]^{-1} + c_1\mathbf{I}\}^{-1}$$
$$\times \{c_0[\mathbf{I} + (\mathbf{L}_0 - \mathbf{L}_1)\mathbf{P}]^{-1}\mathbf{L}_0 + c_1\mathbf{L}_1\} = (3\widehat{K}, 2\widehat{\mu}), \quad (6.4.42)$$

where

$$\widehat{K} = \frac{K_0^2 - c_1(K_1 - K_0)(3\gamma_0 K_1 - K_0)}{K_0 - 3\gamma_0 c_1(K_1 - K_0)},$$
$$\widehat{\mu} = \frac{\mu_0^2 - c_1(\mu_1 - \mu_0)(2\delta_0\mu_1 - \mu_0)}{\mu_0 - 2\delta_0 c_1(\mu_1 - \mu_0)}. \quad (6.4.43)$$

Introducing (6.4.29) into the above expressions,

$$\widehat{K} = \frac{K_0(3K_0 + 4\mu_0) + c_1(K_1 - K_0)[4\mu_0 - 3(K_1 - K_0)]}{3K_0 + 4\mu_0 - 3c_1(K_1 - K_0)},$$
$$\widehat{\mu} = \frac{5\mu_0^2(3K_0 + 4\mu_0) + c_1(\mu_1 - \mu_0)[5\mu_0(3K_0 + 4\mu_0) - 6\mu_1(K_0 + 2\mu_0)]}{5\mu_0(3K_0 + 4\mu_0) - 6c_1(K_0 + 2\mu_0)(\mu_1 - \mu_0)}.$$
$$(6.4.44)$$

Since $K_1 > K_0$, $\mu_1 > \mu_0$, one can see that the tensor $\mathbf{L}^p(\mathbf{x})$ defined by (6.4.37) is negative semidefinite. It thus follows from (6.4.11) that

$$\overline{\mathbf{L}} = (3\overline{K}, 2\overline{\mu}) \leq \tilde{\mathbf{L}} = (3\widehat{K}, 2\widehat{\mu}),$$

or equivalently,

$$\overline{K} \leq \frac{K_0(3K_0 + 4\mu_0) + c_1(K_1 - K_0)[4\mu_0 - 3(K_1 - K_0)]}{3K_0 + 4\mu_0 - 3c_1(K_1 - K_0)}, \quad (6.4.45)$$

$$\overline{\mu} \leq \frac{5\mu_0^2(3K_0 + 4\mu_0) + c_1(\mu_1 - \mu_0)[5\mu_0(3K_0 + 4\mu_0) - 6\mu_1(K_0 + 2\mu_0)]}{5\mu_0(3K_0 + 4\mu_0) - 6c_1(K_0 + 2\mu_0)(\mu_1 - \mu_0)}. \quad (6.4.46)$$

These are the Hashin–Shtrikman upper bounds.

PROBLEMS

6.1 Show that

$$2\bar{\boldsymbol{\varepsilon}}\hat{\boldsymbol{\sigma}} - \hat{\boldsymbol{\sigma}}\overline{\mathbf{M}}^R\hat{\boldsymbol{\sigma}} \le \bar{\boldsymbol{\varepsilon}}(\overline{\mathbf{M}}^R)^{-1}\bar{\boldsymbol{\varepsilon}}$$

for any constant tensor $\hat{\sigma}_{ij}$.

6.2 Show that the integral operator

$$\int_D \boldsymbol{\Gamma}(\mathbf{x}, \mathbf{y})\boldsymbol{\tau}(\mathbf{y})\, dV(\mathbf{y}) \equiv \boldsymbol{\Gamma}(\mathbf{x}, \mathbf{y}) * \boldsymbol{\tau}(\mathbf{y})$$

is self-adjoint, that is,

$$\boldsymbol{\sigma}(\mathbf{x}) * (\boldsymbol{\Gamma}(\mathbf{x}, \mathbf{y}) * \boldsymbol{\tau}(\mathbf{y})) = \boldsymbol{\tau}(\mathbf{x}) * (\boldsymbol{\Gamma}(\mathbf{x}, \mathbf{y}) * \boldsymbol{\sigma}(\mathbf{y})).$$

6.3 Show the identity $\mathbf{L}^h\mathbf{M}^p + \mathbf{L}^p(\mathbf{M}^h + \mathbf{M}^p) = \mathbf{0}$.

6.4 Show that, for a composite comprising isotropic matrix and N-phase isotropic, spherical inhomogeneities, the Hashin–Shtrikman bounds are given by

$$\left(\sum_{r=0}^{N} \frac{c_r}{K_r + K_S^*}\right)^{-1} - K_S^* \le \overline{K} \le \left(\sum_{r=0}^{N} \frac{c_r}{K_r + K_L^*}\right)^{-1} - K_L^*,$$

$$\left(\sum_{r=0}^{N} \frac{c_r}{\mu_r + \mu_S^*}\right)^{-1} - \mu_S^* \le \overline{\mu} \le \left(\sum_{r=0}^{N} \frac{c_r}{\mu_r + \mu_L^*}\right)^{-1} - \mu_L^*,$$

where

$$K_L^* = \frac{4\mu_L}{3}, \qquad \mu_L^* = \frac{3}{2[1/\mu_L + 10/(9K_L + 8\mu_L)]},$$

$$K_S^* = \frac{4\mu_S}{3}, \qquad \mu_S^* = \frac{3}{2[1/\mu_S + 10/(9K_S + 8\mu_S)]},$$

$$\mu_L = \max\{\mu_0, \mu_1, \ldots, \mu_N\}, \qquad \mu_S = \min\{\mu_0, \mu_1, \ldots, \mu_N\},$$

$$K_L = \max\{K_0, K_1, \ldots, K_N\}, \qquad K_S = \min\{K_0, K_1, \ldots, K_N\}.$$

6.5 Assume that the traction boundary condition $\boldsymbol{\sigma}\mathbf{n}|_{\mathbf{x}\in S} = \overline{\boldsymbol{\sigma}}\mathbf{n}$ is prescribed on the entire boundary, where $\overline{\boldsymbol{\sigma}}$ is a constant stress tensor (the average stress). Use the minimum potential energy theorem to prove that for any continuously differentiable displacement field $\hat{u}_i$,

$$2DU \geq -\int_D \hat{\boldsymbol{\varepsilon}}\mathbf{L}\hat{\boldsymbol{\varepsilon}}\, dv + 2\overline{\boldsymbol{\sigma}} \int_D \hat{\boldsymbol{\varepsilon}}\, dv,$$

where

$$U = \frac{1}{2D}\int_D \boldsymbol{\varepsilon}\mathbf{L}\boldsymbol{\varepsilon}\, dv = \frac{1}{2}\,\overline{\boldsymbol{\varepsilon}}\mathbf{L}\overline{\boldsymbol{\varepsilon}},$$

$$\hat{\varepsilon}_{ij} = \tfrac{1}{2}(\hat{u}_{i,j} + \hat{u}_{j,i}).$$

(Hint: Since there is no displacement boundary, any continuously differentiable displacement field would be kinematically admissible.)

6.6 Show that under the traction boundary condition $\boldsymbol{\sigma}\mathbf{n}|_{\mathbf{x}\in S} = \overline{\boldsymbol{\sigma}}\mathbf{n}$,

$$-\int_D L_{ijkl}\hat{u}_{i,j}\hat{u}_{k,l}\, dv + 2\overline{\sigma}_{ij}\int_D \hat{u}_{i,j}\, dv \leq 2DU \leq \int_D \hat{\sigma}_{ij}M_{ijkl}\hat{\sigma}_{kl}\, dv,$$

where $\hat{u}_i$ is any continuously differentiable displacement field and $\hat{\sigma}_{ij}$ is any statically admissible stress fields, that is, $\hat{\sigma}_{ij,i} = 0$ in D and $\sigma_{ij}n_j|_{\mathbf{x}\in S} = \overline{\sigma}_{ij}n_j$ on S.

6.7 Following the notations introduced in Appendix 6.A, please show that the following inequality holds for any eigenstrain $\hat{\boldsymbol{\varepsilon}}$:

$$2D(\overline{U} - \overline{U}_c^h) \geq -\tilde{H}(\hat{\boldsymbol{\varepsilon}}) - \int_D \boldsymbol{\gamma}(\mathbf{x})\mathbf{L}^p(\mathbf{x})\boldsymbol{\gamma}(\mathbf{x})\, dV(\mathbf{x}).$$

6.8 Prove Theorems 6.4 and 6.4a in Appendix 6A.

6.9 Prove $L_{pqij}\int_S \dfrac{\partial \tilde{G}_{mi}(\mathbf{x}, \mathbf{y})}{\partial x_j}\,\overline{\sigma}_{ms}n_s\, dS(\mathbf{y}) = \overline{\sigma}_{pq}$.

APPENDIX 6.A

In Sections 6.3 and 6.4, the variational theorems have been established for heterogeneous material under the displacement boundary conditions. Let us now consider the traction boundary condition,

$$\boldsymbol{\sigma}(\mathbf{x})\mathbf{n}\big|_{\mathbf{x}\in S} = \bar{\boldsymbol{\sigma}}\mathbf{n}, \tag{6.A.1}$$

where $\bar{\boldsymbol{\sigma}}$ is a constant stress tensor. Analogous to (6.3.5), we introduce a strain polarization $\boldsymbol{\varepsilon}^*$ (or eigenstrain) through

$$\boldsymbol{\varepsilon} = \mathbf{M}^h\boldsymbol{\sigma} + \boldsymbol{\varepsilon}^* \quad \text{or} \quad \boldsymbol{\sigma} = (\mathbf{M}^h)^{-1}(\boldsymbol{\varepsilon} - \boldsymbol{\varepsilon}^*), \tag{6.A.2}$$

where $\mathbf{M}^h$ is defined in (6.3.44), and $\boldsymbol{\varepsilon}$ is the actual total strain in the heterogeneous material D, which is related to the displacements through (6.3.4). Using Hooke's law $\boldsymbol{\varepsilon} = \mathbf{M}\boldsymbol{\sigma}$ in the first of (6.A.2) gives

$$\boldsymbol{\varepsilon}^* = \mathbf{M}^p\boldsymbol{\sigma}. \tag{6.A.3}$$

Substituting the second of (6.A.2) into the equations of equilibrium,

$$L^h_{ijkl}u_{k,lj}(\mathbf{x}) - L^h_{ijkl}\varepsilon^*_{kl,j}(\mathbf{x}) = 0 \quad \text{in } D, \tag{6.A.4}$$

where L^h_{ijkl} is to denote the inverse of M^h_{ijkl}. To solve (6.A.4) under the traction boundary condition (6.A.1), let us consider the following Green's function:

$$L^h_{ijkl}\frac{\partial^2\tilde{G}_{km}(\mathbf{x}, \mathbf{y})}{\partial x_l\,\partial x_j} + \delta_{im}\delta(\mathbf{x} - \mathbf{y}) = 0 \quad \text{in } V, \tag{6.A.5}$$

$$L^h_{ijkl}\frac{\partial\tilde{G}_{km}(\mathbf{x}, \mathbf{y})}{\partial x_l}\,n_j\big|_{\mathbf{x}\in S} = 0. \tag{6.A.6}$$

It then follows from (2.6.51) that

$$u_i(\mathbf{x}) = \int_S \tilde{G}_{mi}(\mathbf{x}, \mathbf{y})\bar{\sigma}_{ms}n_s\,dS(\mathbf{y}) - \int_V \tilde{G}_{mi}(\mathbf{x}, \mathbf{y})\frac{\partial L^h_{mnkl}\varepsilon^*_{kl}(\mathbf{y})}{\partial y_n}\,dV(\mathbf{y}). \tag{6.A.7}$$

The corresponding displacement gradient tensor follows immediately:

$$u_{i,j}(\mathbf{x}) = \int_S \frac{\partial \tilde{G}_{mi}(\mathbf{x}, \mathbf{y})}{\partial x_j} \bar{\sigma}_{ms} n_s \, dS(\mathbf{y}) - \int_V \frac{\partial \tilde{G}_{mi}(\mathbf{x}, \mathbf{y})}{x_j} \frac{\partial L^h_{mnkl} \varepsilon^*_{kl}(\mathbf{y})}{\partial y_n} \, dV(\mathbf{y}). \tag{6.A.8}$$

The use of Hooke's law leads us to the corresponding stress tensor:

$$\sigma_{pq}(\mathbf{x}) = L_{pqij} \int_S \frac{\partial \tilde{G}_{mi}(\mathbf{x}, \mathbf{y})}{\partial x_j} \bar{\sigma}_{ms} n_s \, dS(\mathbf{y}) - L_{pqij} \int_D \frac{\partial \tilde{G}_{mi}(\mathbf{x}, \mathbf{y})}{\partial x_j} \frac{\partial L^h_{mnkl} \varepsilon^*_{kl}(\mathbf{y})}{\partial y_n} \, dV(\mathbf{y}). \tag{6.A.9}$$

It is easy to show that

$$L_{pqij} \int_S \frac{\partial \tilde{G}_{mi}(\mathbf{x}, \mathbf{y})}{\partial x_j} \bar{\sigma}_{ms} n_s \, dS(\mathbf{y}) = \bar{\sigma}_{pq}. \tag{6.A.10}$$

Thus, the stress tensor is simplified to

$$\sigma_{pq}(\mathbf{x}) = \bar{\sigma}_{pq} - L_{pqij} \int_D \frac{\partial \tilde{G}_{mi}(\mathbf{x}, \mathbf{y})}{\partial x_j} \frac{\partial L^h_{mnkl} \varepsilon^*_{kl}(\mathbf{y})}{\partial y_n} \, dV(\mathbf{y}). \tag{6.A.11}$$

Note that $\bar{\sigma}_{pq}$ is a constant stress tensor (the average stress in the composite). Therefore, by making use of the properties of Green's function, one can show that

$$\begin{aligned}
\sigma_{pq,q}(\mathbf{x}) &= -L_{pqij} \int_D \frac{\partial \tilde{G}_{mi}(\mathbf{x}, \mathbf{y})}{\partial x_j \, \partial x_q} \frac{\partial L^h_{mnkl} \varepsilon^*_{kl}(\mathbf{y})}{\partial y_n} \, dV(\mathbf{y}) \\
&= \int_D \delta_{mp} \delta(\mathbf{x} - \mathbf{y}) \frac{\partial L^h_{mnkl} \varepsilon^*_{kl}(\mathbf{y})}{\partial y_n} \, dV(\mathbf{y}) \\
&= \frac{\partial L^h_{pnkl} \varepsilon^*_{kl}(\mathbf{x})}{\partial x_n} = L^h_{pnkl} \varepsilon^*_{kl,n}(\mathbf{x})
\end{aligned} \tag{6.A.12}$$

and

$$\sigma_{pq}(\mathbf{x})n_q\big|_{\mathbf{x}\in S} = -L_{pqij}\int_D \frac{\partial \tilde{G}_{mi}(\mathbf{x}, \mathbf{y})}{\partial x_j} n_q \frac{\partial L^h_{mnkl}\varepsilon^*_{kl}(\mathbf{y})}{\partial y_n} dV(\mathbf{y}) + \overline{\sigma}_{pq}n_q$$

$$= -L_{pqij}\int_D \frac{\partial \tilde{G}_{mi}(\mathbf{x}, \mathbf{y})}{\partial x_j} n_q \frac{\partial L^h_{mnkl}\varepsilon^*_{kl}(\mathbf{y})}{\partial y_n} dV(\mathbf{y}) + \overline{\sigma}_{pq}n_q$$

$$= \overline{\sigma}_{pq}n_q. \tag{6.A.13}$$

In other words, regardless of the choice of $\varepsilon^*_{kl}(\mathbf{y})$, the stress tensor generated from (6.A.11) is always a statically admissible stress field, that is, it satisfies (6.A.4) and (6.A.1).

For convenience, we define the following operator to represent the volume integral in (6.A.11):

$$L_{pqij}\int_D \frac{\partial \tilde{G}_{mi}(\mathbf{x}, y)}{\partial x_j} \frac{\partial L^h_{mjkl}\varepsilon^*_{kl}(\mathbf{y})}{\partial y_j} dV(\mathbf{y}) \equiv \int_D \tilde{\Gamma}_{pqkl}(\mathbf{x}, \mathbf{y})\varepsilon^*_{kl}(\mathbf{y})\, dV(\mathbf{y}). \tag{6.A.14}$$

Analogues to (6.3.16), the operator can also be written symbolically,

$$\int_D \tilde{\mathbf{\Gamma}}(\mathbf{x}, \mathbf{y})\boldsymbol{\varepsilon}^*(\mathbf{y})\, dV(\mathbf{y}) \equiv \tilde{\mathbf{\Gamma}}(\mathbf{x}, \mathbf{y}) * \boldsymbol{\varepsilon}^*(\mathbf{y}). \tag{6.A.15}$$

Thus, the stress tensor can be written formally as

$$\boldsymbol{\sigma}(\mathbf{x}) = \overline{\boldsymbol{\sigma}} - \tilde{\mathbf{\Gamma}}(\mathbf{x}, \mathbf{y}) * \boldsymbol{\varepsilon}^*(\mathbf{y}). \tag{6.A.16}$$

It then follows from (6.A.3) that

$$(\mathbf{M}^p)^{-1}\boldsymbol{\varepsilon}^*(\mathbf{x}) + \tilde{\mathbf{\Gamma}}(\mathbf{x}, \mathbf{y}) * \boldsymbol{\varepsilon}^* = \overline{\boldsymbol{\sigma}}. \tag{6.A.17}$$

Let $\hat{\boldsymbol{\varepsilon}}(\mathbf{y})$ be any strain polarization, and

$$\hat{\boldsymbol{\sigma}}(\mathbf{x}) = \overline{\boldsymbol{\sigma}} - \tilde{\mathbf{\Gamma}}(\mathbf{x}, \mathbf{y}) * \boldsymbol{\varepsilon}^*(\mathbf{y}) \tag{6.A.18}$$

the corresponding statically admissible stress field. The minimum complementary energy theorem takes the following form:

$$2D\overline{U}_c \leq \int_D \hat{\boldsymbol{\sigma}}(\mathbf{x})\mathbf{M}(\mathbf{x})\hat{\boldsymbol{\sigma}}(\mathbf{x})\, dV(\mathbf{x}), \tag{6.A.19}$$

where

$$\overline{U}_c = \frac{1}{2D}\int_D M_{ijkl}\sigma_{kl}\sigma_{ij}\, dv = \frac{1}{2}\,\overline{\boldsymbol{\sigma}}\overline{\mathbf{M}}\overline{\boldsymbol{\sigma}}. \tag{6.A.20}$$

Substituting (6.A.18) into (6.A.19), after some simplifications, yields

$$2D(\overline{U} - \overline{U}_c^h) \leq -\tilde{H}(\hat{\boldsymbol{\varepsilon}}) + \int_D \boldsymbol{\gamma}(\mathbf{x})\mathbf{M}^p(\mathbf{x})\boldsymbol{\gamma}(\mathbf{x})\, dV(\mathbf{x}). \tag{6.A.21}$$

where the compliance tensor has been decomposed as in (6.3.44) and

$$\overline{U}^h = \tfrac{1}{2}\overline{\boldsymbol{\sigma}}\overline{\mathbf{M}}^h\overline{\boldsymbol{\sigma}}, \tag{6.A.22}$$

$$\mathbf{L}^h\boldsymbol{\gamma}(\mathbf{x}) - \tilde{\boldsymbol{\Gamma}}(\mathbf{x}, \mathbf{y}) * \hat{\boldsymbol{\varepsilon}}(\mathbf{y}) + (\overline{\mathbf{M}}^h)^{-1}\hat{\boldsymbol{\varepsilon}}(\mathbf{x}) - \overline{\boldsymbol{\sigma}},$$

$$\tilde{H}(\hat{\boldsymbol{\varepsilon}}) = -\int_D \hat{\boldsymbol{\varepsilon}}(\mathbf{x})\mathbf{L}^h\boldsymbol{\gamma}(\mathbf{x})\, dV(\mathbf{x}) + \int_D \hat{\boldsymbol{\varepsilon}}(\mathbf{x})\overline{\boldsymbol{\sigma}}\, dv. \tag{6.A.23}$$

If $\hat{\boldsymbol{\varepsilon}}$ happens to be the actual strain polarization $\boldsymbol{\varepsilon}^*$ for the problem at hand, then the use of (6.A.17) leads to

$$\boldsymbol{\gamma}(\mathbf{x}) = \mathbf{0},$$

$$\tilde{H}(\boldsymbol{\varepsilon}^*) = \int_D \boldsymbol{\varepsilon}^*(\mathbf{x})\overline{\boldsymbol{\sigma}}\, dv = \int_D \overline{\boldsymbol{\sigma}}(\mathbf{M}(\mathbf{x}) - \mathbf{M}^h)\overline{\boldsymbol{\sigma}}\, dv = 2D(\overline{U}_c - \overline{U}_c^h).$$

In other words, the equal sign in (6.A.21) is attained when the actual strain polarization $\boldsymbol{\varepsilon}^*$ for the problem at hand is used. Therefore, we can state the following variational principle.

Theorem 6.3 Among all the symmetric second-order tensors, the true solution to the strain polarization tensor $\boldsymbol{\varepsilon}^*$ renders the following functional minimum:

$$J^c[\hat{\boldsymbol{\varepsilon}}] = \frac{1}{2}\,\overline{\boldsymbol{\sigma}}\,\overline{\mathbf{M}}^h\overline{\boldsymbol{\sigma}} - \frac{1}{2D}\,\tilde{H}(\hat{\boldsymbol{\varepsilon}}) + \frac{1}{2D}\int_D \boldsymbol{\gamma}(\mathbf{x})\mathbf{M}^p(\mathbf{x})\boldsymbol{\gamma}(\mathbf{x})\,dV(\mathbf{x}), \tag{6.A.24}$$

and the minimum value of $J^c[\hat{\boldsymbol{\varepsilon}}]$ is the strain energy of the heterogeneous material under the displacement boundary condition (6.3.1), that is,

$$\min\{J^c[\hat{\boldsymbol{\varepsilon}}]\} = \tfrac{1}{2}\overline{\boldsymbol{\sigma}}\,\overline{\mathbf{M}}\overline{\boldsymbol{\sigma}}. \tag{6.A.25}$$

The inequality (6.A.21) can be further simplified if one assumes that $\mathbf{M}^p(\mathbf{x})$ is negative semidefinite, that is, for any $\hat{\boldsymbol{\varepsilon}}$,

$$\int_D \boldsymbol{\gamma}(\mathbf{x})\mathbf{M}^p(\mathbf{x})\boldsymbol{\gamma}(\mathbf{x})\,dv \leq 0. \tag{6.A.26}$$

In this case, (6.A.21) reduces to

$$2D(\overline{U} - \overline{U}^h) \leq -H[\hat{\boldsymbol{\tau}}]. \tag{6.A.27}$$

Therefore, we have the following theorem.

Theorem 6.3a When $\mathbf{M}^p(\mathbf{x})$ is negative semidefinite, among all the symmetric second-order tensors, the true solution to the strain polarization tensor $\boldsymbol{\varepsilon}^*$ renders the following functional minimum:

$$J^c_h[\hat{\boldsymbol{\varepsilon}}] = \frac{1}{2}\,\overline{\boldsymbol{\sigma}}\,\overline{\mathbf{M}}^h\overline{\boldsymbol{\sigma}} - \frac{1}{2D}\,\tilde{H}(\hat{\boldsymbol{\varepsilon}}), \tag{6.A.28}$$

and the minimum value of $J^c_h[\hat{\boldsymbol{\varepsilon}}]$ is the strain energy of the composite under the displacement boundary condition (6.3.1), that is,

$$\min\{J^c_h[\hat{\boldsymbol{\varepsilon}}]\} = \frac{1}{2D}\,\overline{\boldsymbol{\sigma}}\,\overline{\mathbf{M}}\overline{\boldsymbol{\sigma}}. \tag{6.A.29}$$

This is another form of the Hashin–Shtrikman variational principle.

The duals of the above theorems can be easily proven by using the minimum potential energy theorem; see Problem 6.7.

Theorem 6.4 Among all the symmetric second-order tensors, the true solution to the strain polarization tensor $\boldsymbol{\varepsilon}^*$ renders the following functional maximum:

$$J[\hat{\boldsymbol{\varepsilon}}] = \frac{1}{2}\,\overline{\boldsymbol{\sigma}}\,\overline{\mathbf{M}^h}\,\overline{\boldsymbol{\sigma}} - \frac{1}{2D}\,\tilde{H}(\hat{\boldsymbol{\varepsilon}}) - \frac{1}{2D}\int_D \boldsymbol{\gamma}(\mathbf{x})\mathbf{L}^p(\mathbf{x})\boldsymbol{\gamma}(\mathbf{x})\,dV(\mathbf{x}), \quad (6.A.30)$$

and the maximum value of $J[\hat{\boldsymbol{\varepsilon}}]$ is the strain energy of the composite under the displacement boundary condition (6.3.1), that is,

$$\min\{J[\hat{\boldsymbol{\varepsilon}}]\} = \tfrac{1}{2}\overline{\boldsymbol{\sigma}}\,\overline{\mathbf{M}}\,\overline{\boldsymbol{\sigma}}. \quad (6.A.31)$$

The inequality (6.A.21) can be further simplified if one assumes that $\mathbf{M}^p(\mathbf{x})$ is negative semidefinite, that is, for any $\hat{\boldsymbol{\varepsilon}}$,

$$\int_D \boldsymbol{\gamma}(\mathbf{x})\mathbf{L}^p(\mathbf{x})\boldsymbol{\gamma}(\mathbf{x})\,dv \leq 0. \quad (6.A.32)$$

In this case, we have the following theorem.

Theorem 6.4a When $\mathbf{M}^p(\mathbf{x})$ is negative semidefinite, among all the symmetric second-order tensors, the true solution to the strain polarization tensor $\boldsymbol{\varepsilon}^*$ renders the following functional maximum:

$$J_h[\hat{\boldsymbol{\varepsilon}}] = \frac{1}{2}\,\overline{\boldsymbol{\sigma}}\,\overline{\mathbf{M}^h}\,\overline{\boldsymbol{\sigma}} - \frac{1}{2D}\,\tilde{H}(\hat{\boldsymbol{\varepsilon}}), \quad (6.A.33)$$

and the maximum value of $J_h[\hat{\boldsymbol{\varepsilon}}]$ is the strain energy of the composite under the displacement boundary condition (6.3.1), that is,

$$\min\{J_h[\hat{\boldsymbol{\varepsilon}}]\} = \tfrac{1}{2}\overline{\boldsymbol{\sigma}}\,\overline{\mathbf{M}}\,\overline{\boldsymbol{\sigma}}. \quad (6.A.34)$$

This is another form of the Hashin–Shtrikman variational principle.

REFERENCE

Willis, J. R. (1982). Elasticity Theory of Composites, in *Mechanics of Solids—The Rodney Hill 60th Anniversary Volume,* H. G. Hopkins and M. J. Sewell, eds., Pergamon Press, Oxford. pp. 635–686.

SUGGESTED READING

Hashin, Z. and S. Shtrikman. (1961). Note on a Variational Approach to the Theory of Composite Elastic Materials, *J. Franklin Inst.*, Vol. 271, pp. 336–341.

Hashin, Z. and S. Shtrikman. (1963). A Variational Approach to the Theory of the Elastic Behavior of Multiphase Materials, *J. Mech. Phys. Solids,* Vol. 11, pp. 127–140.

Mura, T. and T. Koya. (1992). *Variational Methods in Mechanics,* Oxford University Press, New York, Chapter 21.

Nemat-Nasser, S. and M. Hori. (1993). *Micromechanics: Overall Properties of Heterogeneous Materials,* North-Holland, New York, Chapter 3, Section 9.

Willis, J. R. (1977). Bounds and Self-Consistent Estimates for the Overall Properties of Anisotropic Composites, *J. Mech. Phys. Solids*, Vol. 25, pp. 185–202.

Willis, J. R. (1981). Variational and Related Methods for the Overall Properties of Composites, *Adv. Appl. Mech.*, Vol. 21, pp. 1–78.

7

DETERMINATION OF EFFECTIVE MODULI

In this chapter, we develop several methods to evaluate approximately the effective properties of heterogeneous materials. All the approaches presented are based on the Eshelby single-inclusion solution. In the next chapter, we will present other methods of evaluating the effective properties that are based on multiple-inclusion approaches.

7.1 BASIC IDEAS OF MICROMECHANICS FOR EFFECTIVE PROPERTIES

Consider a heterogeneous material D consisting of a matrix phase Ω_0 and ellipsoidal inhomogeneities, $\Omega_1, \Omega_2, \ldots, \Omega_N$, as shown in Figure 7.1. The volume fraction of Ω_r is $c_r = \Omega_r/\Omega$. Clearly, one has

$$\sum_{r=0}^{N} c_r = 1.$$

Let the stiffness tenor of Ω_r be $\mathbf{L}_r$ ($r = 0, 1, \ldots, N$). Our objective is to determine the effective elastic modulus of the composite materials in terms of the stiffness tensors of the matrix and the inhomogeneities and their respective volume fractions. We learned from Section 5.7 that the effective elastic stiffness tensor can be easily computed once the local or the global concentration tensors are known. However, the exact expressions of the concentration tensors are rather difficult to obtain. So, the various methods discussed in this chapter are aimed at obtaining approximate solutions to these concentration tensors.

The multiphase composite shown in Figure 7.1 is a very general case. Most engineering composites are two-phase materials, namely,

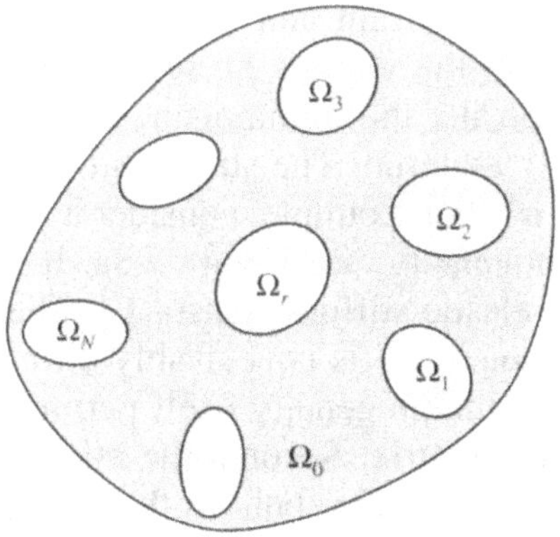

Figure 7.1 Multiphase composite.

matrix phase and the reinforcement phase. The most common types of reinforcement phases are either particles (particulate-reinforced composites) or fibers (fiber-reinforced composites). The results derived below for the multiphase composite can be easily specified to the case of two-phase composite by assuming that all the inhomogeneities are the same.

The basic idea of micromechanics is to develop solutions of either the global or local concentration tensors. Once such concentration tensors are known, the effective properties can be easily obtained per the discussions in Section 5.7. One common approach of developing the concentration tensors is to use the Eshelby solution in conjunction with the equivalent inclusion method.

To this end, we assume that the composite is subjected to either the displacement boundary condition

$$\mathbf{u}|_S = \boldsymbol{\varepsilon}^0\mathbf{x} \tag{7.1.1}$$

or the traction boundary condition

$$\boldsymbol{\sigma} \bullet \mathbf{n}|_S = \boldsymbol{\sigma}^0\mathbf{n}, \tag{7.1.2}$$

where $\boldsymbol{\varepsilon}^0$ and $\boldsymbol{\sigma}^0$ are constant second-order tensors and are related by

$$\boldsymbol{\sigma}^0 = L_0\boldsymbol{\varepsilon}^0.$$

If all the inhomogeneities are absent and the entire volume D is filled with the matrix material $\mathbf{L}_0$, or equivalently, $\mathbf{L}_0 = \mathbf{L}_1 = \mathbf{L}_2 = \cdots =$

$\mathbf{L}_N$, it is obvious that the strain and stress field would be $\boldsymbol{\varepsilon}^0$ and $\boldsymbol{\sigma}^0$, respectively, throughout the volume D; see Figure 7.2.

In the actual composite, the inhomogeneities have distinctive properties from the matrix material. The strain and stress in the matrix will no longer be $\boldsymbol{\varepsilon}^0$ and $\boldsymbol{\sigma}^0$. For example, consider a typical inhomogeneity called the rth inhomogeneity; see Figure 7.3a. It is surrounded by the matrix material with elastic stiffness tensor $\mathbf{L}_0$. The strain in the matrix around the rth inhomogeneity is conceivably different from $\boldsymbol{\varepsilon}^0$ for two reasons. First, the rth inhomogeneity itself perturbs the uniform distribution of strains in the matrix. Second, the other inhomogeneities, especially those near by, also contribute to the strain perturbation. Thus, as far as the rth inhomogeneity is concerned, it might be viewed as an isolated inhomogeneity placed in a uniform matrix that possessed certain strain before the inhomogeneity is embedded. For convenience, we will denote this strain by $\hat{\boldsymbol{\varepsilon}}^0$. Furthermore, it is also plausible that the influence of other inhomogeneities can be accounted for by assuming that the rth inhomogeneity is embedded in a matrix material that is somewhat different from the actual matrix. We will use $\hat{\mathbf{L}}_0$ to denote the stiffness tensor of this fictitious matrix material; see Figure 7.3b.

Now the companion problem can be restated as follows. An ellipsoidal inhomogeneity Ω_r with stiffness tensor $\mathbf{L}_r$ is placed within a uniform matrix of stiffness tensor $\hat{\mathbf{L}}_0$, which had been subjected to the uniform strain $\hat{\boldsymbol{\varepsilon}}^0$ before the inhomogeneity was embedded; see Figure 7.3b. Our objective is to find proper values for $\hat{\boldsymbol{\varepsilon}}^0$ and $\hat{\mathbf{L}}_0$ so that the stress (and strain) on Ω_r is the same as that occurring on the rth inhomogeneity the actual composite.

To find such $\hat{\boldsymbol{\varepsilon}}^0$ and $\hat{\mathbf{L}}_0$, let us pretend at the moment that they are known. We will first establish how the stresses and strains of the companion problem are related to $\hat{\boldsymbol{\varepsilon}}^0$ and $\hat{\mathbf{L}}_0$. This can be done by solving the Eshelby inclusion problem. The stress and strain on the inhomogeneity can be obtained by the equivalent inclusion method discussed in Section 4.4. The equivalent inclusion equation in this case is

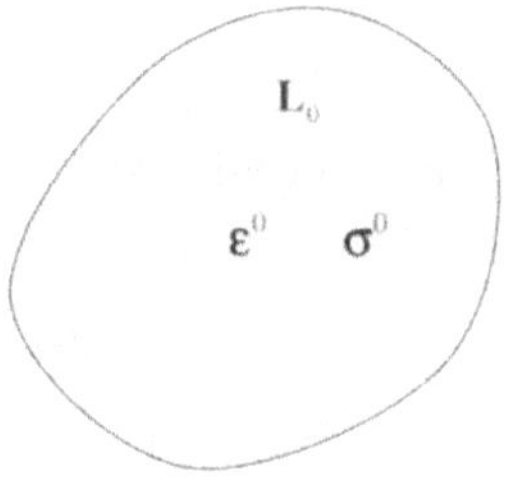

Figure 7.2 Homogeneous material.

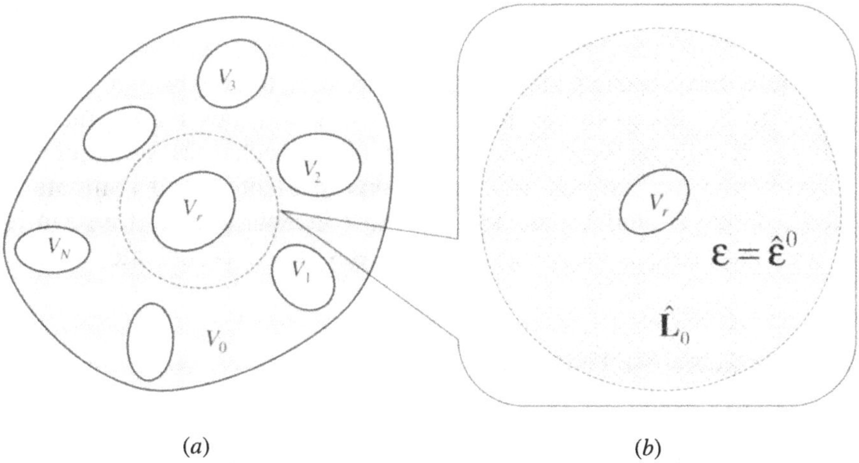

Figure 7.3 (*a*) The *r*th inhomogeneity in the composite; (*b*) The *r*th inhomogeneity in a uniform matrix of stiffness $\hat{\mathbf{L}}_0$, which had been subjected to the uniform strain $\hat{\boldsymbol{\varepsilon}}^0$ before the inhomogeneity was embedded.

$$\mathbf{L}_r\,(\hat{\boldsymbol{\varepsilon}}^0 + \boldsymbol{\varepsilon}_r^{pt}) = \hat{\mathbf{L}}_0\,(\hat{\boldsymbol{\varepsilon}}^0 + \boldsymbol{\varepsilon}_r^{pt} - \boldsymbol{\varepsilon}_r^*), \tag{7.1.3}$$

where, according to the Eshelby solution given in Section 4.3,

$$\boldsymbol{\varepsilon}_r^{pt} = \hat{\mathbf{S}}_r \boldsymbol{\varepsilon}_r^*, \tag{7.1.4}$$

where $\hat{\mathbf{S}}_r$ is the Eshelby tensor computed using the elastic constants of $\hat{\mathbf{L}}_0$ and the geometry of the rth inhomogeneity Ω_r. Substitution of the Eshelby solution into the equivalent inclusion equation yields the eigenstrain

$$\boldsymbol{\varepsilon}_r^* = [(\mathbf{L}_r - \hat{\mathbf{L}}_0)\hat{\mathbf{S}}_r + \hat{\mathbf{L}}_0]^{-1}(\mathbf{L}_r - \mathbf{L}_0)\hat{\boldsymbol{\varepsilon}}^0. \tag{7.1.5}$$

The total strain in the rth inhomogeneity can then be written as

$$\boldsymbol{\varepsilon}_r = \hat{\boldsymbol{\varepsilon}}^0 + \boldsymbol{\varepsilon}_r^{pt} = \hat{\boldsymbol{\varepsilon}}^0 + \hat{\mathbf{S}}_r \boldsymbol{\varepsilon}^* = \hat{\mathbf{T}}_r \hat{\boldsymbol{\varepsilon}}^0, \tag{7.1.6}$$

where

$$\hat{\mathbf{T}}_r = [\mathbf{I} + \hat{\mathbf{S}}_r \hat{\mathbf{L}}_0^{-1}(\mathbf{L}_r - \hat{\mathbf{L}}_0)]^{-1}. \tag{7.1.7}$$

The derivation of the last equality in (7.1.6) is similar to that of (4.4.15). The corresponding stress on the rth inhomogeneity is

$$\boldsymbol{\sigma}_r = \mathbf{L}_r \boldsymbol{\varepsilon}_r = \mathbf{L}_r \hat{\mathbf{T}}_r \hat{\boldsymbol{\varepsilon}}^0. \tag{7.1.8}$$

We now have related the strain and stress fields in the rth inhomogeneity to an unknown strain field $\hat{\boldsymbol{\varepsilon}}^0$ and a fictitious matrix material with $\hat{\mathbf{L}}_0$. It will be seen later that the micromechanics schemes discussed below use various approximations to identify and compute $\hat{\boldsymbol{\varepsilon}}^0$ and $\hat{\mathbf{L}}_0$. Once $\hat{\boldsymbol{\varepsilon}}^0$ and $\hat{\mathbf{L}}_0$ are known, the concentration tensors can be obtained and so can the effective properties of the composite.

7.2 ESHELBY METHOD

If the inhomogeneities in the composite shown in Figure 7.1 are far apart from each other, their interactions may be neglected. In other words, each inhomogeneity can be treated as if it exists in a homogeneous matrix without the interference by other inhomogeneities. Therefore, a typical inhomogeneity, for example, the rth inhomogeneity, can be treated as an ellipsoidal inhomogeneity in an otherwise uniform matrix of stiffness $\mathbf{L}_0$, which was subjected to a uniform strain $\boldsymbol{\varepsilon}^0$ before the inhomogeneity was imbedded. Therefore, we choose in (7.1.3),

$$\hat{\mathbf{L}}_0 = \mathbf{L}_0 \quad \text{and} \quad \hat{\boldsymbol{\varepsilon}}^0 = \boldsymbol{\varepsilon}^0 \tag{7.2.1}$$

and consequently,

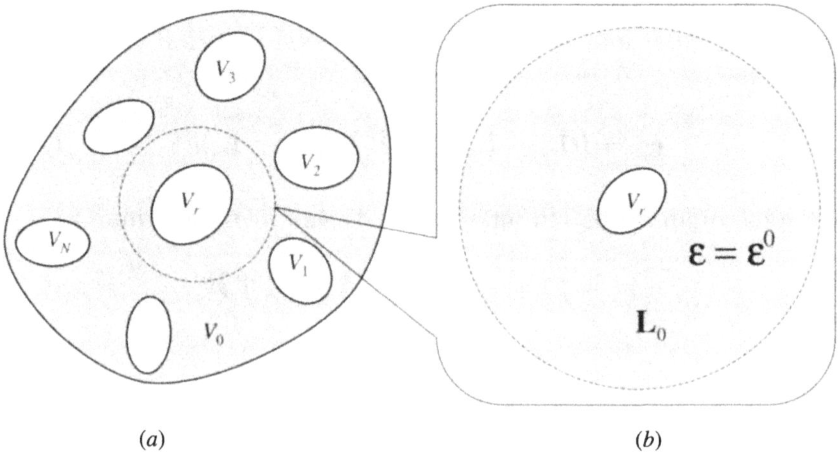

Figure 7.4 (*a*) The rth inhomogeneity in the composite. (*b*) The rth inhomogeneity in the original matrix of stiffness $\mathbf{L}_0$, which had been subjected to the uniform strain $\boldsymbol{\varepsilon}^0$ before the inhomogeneity was embedded.

$$\hat{\mathbf{S}}_r = \mathbf{S}_r, \tag{7.2.2}$$

where $\mathbf{S}_r$ is the Eshelby tensor computed using the matrix material property $\mathbf{L}_0$ and the geometry of the rth inhomogeneity Ω_r. These equations simply imply that other inhomogeneities have no influence on the rth inhomogeneity at all. So, the interactions between inhomogeneities are completely ignored. In this case, the strain in the rth inhomogeneity follows directly from (7.1.6):

$$\boldsymbol{\varepsilon}_r = \boldsymbol{\varepsilon}^0 + \boldsymbol{\varepsilon}_r^{pt} = \boldsymbol{\varepsilon}^0 + \mathbf{S}_r\boldsymbol{\varepsilon}^* = \mathbf{T}_r\boldsymbol{\varepsilon}^0, \tag{7.2.3}$$

where

$$\mathbf{T}_r = [\mathbf{I} + \mathbf{S}_r\mathbf{L}_0^{-1}(\mathbf{L}_r - \mathbf{L}_0)]^{-1}. \tag{7.2.4}$$

For future reference, we want to mention that although $\mathbf{T}_r$ is defined for $r > 0$, it is nevertheless true that $\mathbf{T}_0 = \mathbf{I}$ as seen from (7.2.4).

We now assume that the composite is subjected to the displacement boundary condition

$$\mathbf{u}|_S = \boldsymbol{\varepsilon}^0 \cdot \mathbf{x}. \tag{7.2.5}$$

It follows from the average strain theorem discussed in Section 5.4 that the average strain tensor of the composite is equal to $\boldsymbol{\varepsilon}^0$, that is,

$$\bar{\boldsymbol{\varepsilon}} = \frac{1}{V}\int_V \boldsymbol{\varepsilon}\, dV = \boldsymbol{\varepsilon}^0. \tag{7.2.6}$$

Therefore, Eq. (7.2.3) can be rewritten as

$$\boldsymbol{\varepsilon}_r = \mathbf{T}_r\bar{\boldsymbol{\varepsilon}}. \tag{7.2.7}$$

By comparing (5.7.3) and (7.2.7), it is clear that the global strain concentration tensor for the rth inhomogeneity is

$$\mathbf{A}_r = \mathbf{T}_r. \tag{7.2.8}$$

Thus, it follows from (5.7.7) that the effective stiffness tensor of the composite is given by

$$\bar{\mathbf{L}} = \mathbf{L}_0 + \sum_{r=1}^{N} c_r (\mathbf{L}_r - \mathbf{L}_0)\mathbf{T}_r$$

$$= \mathbf{L}_0 + \sum_{r=1}^{N} c_r(\mathbf{L}_r - \mathbf{L}_0)[\mathbf{I} + \mathbf{S}_r(\mathbf{M}_0\mathbf{L}_r - \mathbf{I})]^{-1}. \qquad (7.2.9)$$

This is the Eshelby estimate of the effective stiffness tensor. Recall that this effective stiffness tensor was derived based on the assumption made earlier that the inhomogeneities in the composite are so far apart that they do not interfere with each other. Thus, the Eshelby estimate is valid only for very low volume fraction of inhomogeneities, or the dilute case.

Instead of the displacement boundary conditions, a traction boundary condition

$$\boldsymbol{\sigma} \cdot \mathbf{n}|_S = \boldsymbol{\sigma}^0\mathbf{n} \qquad (7.2.10)$$

can be prescribed. In this case, the average strain of the composite is no longer equal to $\boldsymbol{\varepsilon}^0$, that is,

$$\bar{\boldsymbol{\varepsilon}} = \frac{1}{V}\int_V \boldsymbol{\varepsilon}\, dV \neq \boldsymbol{\varepsilon}^0, \qquad (7.2.11)$$

where

$$\boldsymbol{\varepsilon}^0 = \mathbf{M}_0\boldsymbol{\sigma}^0. \qquad (7.2.12)$$

Instead, we have, according to the average stress theorem discussed in Section 5.4,

$$\bar{\sigma} = \frac{1}{V}\int_V \sigma\, dV = \boldsymbol{\sigma}^0. \qquad (7.2.13)$$

Substituting (7.2.12) and (7.2.13) into (7.2.3) and making use of $\boldsymbol{\sigma}_r = \mathbf{L}_r\boldsymbol{\varepsilon}_r$ yield

$$\boldsymbol{\sigma}_r = \mathbf{L}_r\mathbf{T}_r\mathbf{M}_0\bar{\boldsymbol{\sigma}}. \qquad (7.2.14)$$

The global stress concentration tensor is thus given by

$$\mathbf{B}_r = \mathbf{L}_r\mathbf{T}_r\mathbf{M}_0 = [\mathbf{M}_r - \mathbf{S}_r (\mathbf{M}_r - \mathbf{M}_0)]^{-1} \mathbf{M}_0. \qquad (7.2.15)$$

It then follows from (5.7.8) that the effective compliant tensor of the composite is given by

$$\overline{\mathbf{M}} = \mathbf{M}_0 + \sum_{r=1}^{N} c_r(\mathbf{M}_r - \mathbf{M}_0)\mathbf{B}_r = \mathbf{M}_0 + \sum_{r=1}^{N} c_r(\mathbf{M}_r - \mathbf{M}_0)\mathbf{L}_r\mathbf{T}_r\mathbf{M}_0. \tag{7.2.16}$$

It can be further simplified to

$$\overline{\mathbf{M}} = \mathbf{M}_0 - \sum_{r=1}^{N} c_r[(\mathbf{L}_r - \mathbf{L}_0)\mathbf{S}_r + \mathbf{L}_0]^{-1}(\mathbf{L}_r - \mathbf{L}_0)\mathbf{M}_0. \tag{7.2.17}$$

This is the Eshelby estimate of the effective compliance of the composite in consideration.

Finally, let us check the consistency of the Eshelby estimates by considering

$$\begin{aligned} \overline{\mathbf{M}}\,\overline{\mathbf{L}} &= \left[\mathbf{M}_0 + \sum_{r=1}^{N} c_r\,(\mathbf{M}_r - \mathbf{M}_0)\mathbf{L}_r\mathbf{T}_r\mathbf{M}_0\right]\left[\mathbf{L}_0 + \sum_{r=1}^{N} c_r\,(\mathbf{L}_r - \mathbf{L}_0)\mathbf{T}_r\right] \\ &= \mathbf{I} - \left[\sum_{r=1}^{N} c_r\,(\mathbf{I} - \mathbf{M}_0\mathbf{L}_r)\mathbf{T}_r\right]\left[\sum_{r=1}^{N} c_r(\mathbf{I} - \mathbf{M}_0\mathbf{L}_r)\mathbf{T}_r\right] \\ &= \mathbf{I} + O(c_r^2). \end{aligned} \tag{7.2.18}$$

It is seen that $\overline{\mathbf{M}}\,\overline{\mathbf{L}} = \mathbf{I}$ only when the terms with order of c_r^2 are neglected. In other words, the Eshelby estimates are consistent only up to the first order of the volume fraction of the inhomogeneities.

To close this section, we note that both the volume fraction and the geometry of the inhomogeneities are taken into consideration in the Eshelby estimate. However, the distribution of the inhomogeneities is neglected. More importantly, the interaction between the inhomogeneities is not taken into consideration because the eigenstrain in each inclusion is calculated by assuming that the other inclusions are not present. Therefore, this method is limited to low (dilute) concentration of inhomogeneities. The method is also called by some authors the dilute concentration method.

Example 7.1 Consider a particulate-reinforced composite material. Assume that all particles are spherical and randomly dispersed throughout the matrix. Further, assume that both the particle and the matrix are made of isotropic elastic materials. Their elastic stiffness tensors are

$$\mathbf{L}_0 = (3K_0, 2\mu_0) \quad \text{for the matrix,} \tag{7.2.19}$$

and

$$\mathbf{L}_1 = (3K_1, 2\mu_1) \quad \text{for the particles,} \tag{7.2.20}$$

where the matrix symbolic notation for the fourth-order tensors introduced in Section 1.4 has been used for the stiffness tensors. The isotropic elastic constants K_r and μ_r are the bulk and shear moduli, respectively, for the matrix ($r = 0$) and the particles ($r = 1$). Finally, we assume that the volume fraction of the particles in the composite is c_1.

To find the effective stiffness tensor of the composites, we can use (7.2.9) directly by specifying $N = 1$. To this end, let us compute the relevant tensors one at a time. First, the Eshelby tensor for an isotropic material with $\mathbf{L}_0 = (3K_0, 2\mu_0)$ can be written as (see Appendix 4.B)

$$\mathbf{S}_1 = (3\gamma_0, 2\delta_0), \tag{7.2.21}$$

where

$$\gamma_0 = \frac{K_0}{3K_0 + 4\mu_0} = \frac{1 + v_0}{9(1 - v_0)}, \qquad \delta_0 = \frac{3(K_0 + 2\mu_0)}{5(3K_0 + 4\mu_0)} = \frac{4 - 5v_0}{15(1 - v_0)}. \tag{7.2.22}$$

Next, let us compute the fourth-order tensor $\mathbf{T}_1$:

$$\begin{aligned}
\mathbf{T}_1 &= [\mathbf{I} + \mathbf{S}_1\mathbf{L}_0^{-1}(\mathbf{L}_1 - \mathbf{L}_0)]^{-1} \\
&= \{(1,1) + (3\gamma_0, 2\delta_0)[(3K_0, 2\mu_0)^{-1}(3K_1, 2\mu_1) - (1,1)]\}^{-1} \\
&= \left[(1, 1) + (3\gamma_0, 2\delta_0)\left(\frac{K_1 - K_0}{K_0}, \frac{\mu_1 - \mu_0}{\mu_0}\right)\right]^{-1} \\
&= \left[\frac{K_0 + 3\gamma_0(K_1 - K_0)}{K_0}, \frac{\mu_0 + 2\delta_0(\mu_1 - \mu_0)}{\mu_0}\right]^{-1} \\
&= \left[\frac{K_0}{K_0 + 3\gamma_0(K_1 - K_0)}, \frac{\mu_0}{\mu_0 + 2\delta_0(\mu_1 - \mu_0)}\right].
\end{aligned} \tag{7.2.23}$$

Finally,

$$\begin{aligned}\overline{\mathbf{L}} &= \mathbf{L}_0 + c_1\,(\mathbf{L}_1 - \mathbf{L}_0)\mathbf{T}_1 \\ &= (3K_0,\, 2\mu_0) + \left[\frac{3c_1K_0\,(K_1 - K_0)}{K_0 + 3\gamma_0\,(K_1 - K_0)},\ \frac{2c_1\mu_0(\mu_1 - \mu_0)}{\mu_0 + 2\delta_0(\mu_1 - \mu_0)}\right].\end{aligned} \tag{7.2.24}$$

Thus, we have

$$\overline{K} = K_0 + \frac{c_1K_0\,(K_1 - K_0)}{K_0 + 3\gamma_0\,(K_1 - K_0)},$$

$$\overline{\mu} = \mu_0 + \frac{c_1\mu_0\,(\mu_1 - \mu_0)}{\mu_0 + 2\delta_0(\mu_1 - \mu_0)}. \tag{7.2.25}$$

Making use of (7.2.22) in (7.2.25) yields alternative expressions

$$\overline{K} = K_0 + \frac{c_1(K_1 - K_0)(3K_0 + 4\mu_0)}{3K_1 + 4\mu_0}, \tag{7.2.26}$$

$$\overline{\mu} = \mu_0 + \frac{5c_1\mu_0(\mu_1 - \mu_0)(3K_0 + 4\mu_0)}{3K_0(3\mu_0 + 2\mu_1) + 4\mu_0\,(2\mu_0 + 3\mu_1)}. \tag{7.2.27}$$

This completes the solution. Note that in the above derivations, we have used the symbolic notations of fourth-order isotropic tensors and followed the rules of operation on using such symbolic notations, as introduced in Section 1.4. The advantages of such operation are clearly seen.

7.3 MORI–TANAKA METHOD

Before presenting the Mori–Tanaka method, let us recall the average strain and average stress of a composite as defined in Section 5.4:

$$\overline{\boldsymbol{\varepsilon}} = \frac{1}{D}\int_D \boldsymbol{\varepsilon}\, dV = \sum_{r=0}^{N} \overline{\boldsymbol{\varepsilon}}_r, \qquad \overline{\boldsymbol{\sigma}} = \frac{1}{D}\int_D \boldsymbol{\sigma}\, dV = \sum_{r=0}^{N} \overline{\boldsymbol{\sigma}}_r, \tag{7.3.1}$$

where

$$\bar{\boldsymbol{\varepsilon}}_r = \frac{1}{\Omega_r}\int_{\Omega_r} \boldsymbol{\varepsilon}\, dV \quad \text{and} \quad \bar{\boldsymbol{\sigma}}_r = \frac{1}{\Omega_r}\int_{\Omega_r} \boldsymbol{\sigma}\, dV \tag{7.3.2}$$

are the strain and stress tensors, respectively, averaged over the rth inhomogeneity. For example, $\bar{\boldsymbol{\varepsilon}}_1$ is the average strain in the inhomogeneity with stiffness tensor $\mathbf{L}_1$, and $\bar{\boldsymbol{\varepsilon}}_0$ is the average strain in the matrix phase, see Fig. 7.5. Furthermore,

$$\bar{\boldsymbol{\sigma}}_r = \mathbf{L}_r \bar{\boldsymbol{\sigma}}_r.$$

Now consider the composite material shown in Figure 7.1. For a typical inhomogeneity $\mathbf{L}_r$ $(r > 0)$ in the composite, the effects (or the existence) of other inhomogeneities are communicated to it through the strain and stress fields in its surrounding matrix material. Although the strain and stress fields are different from one location to another in the matrix, the averages ($\bar{\boldsymbol{\varepsilon}}_0$ and $\bar{\boldsymbol{\sigma}}_0$) represent good approximations of the actual fields in the matrix surrounding each inhomogeneity, when a large number of inhomogeneities exit and are randomly distributed in the matrix (which is the case for most engineering composites). Also, it would be reasonable to assume that taking only one inhomogeneity out will not affect the overall elastic behavior of the composite. In other words, when the rth inhomogeneity is removed and replaced by the matrix material (or equivalently, let $\mathbf{L}_r = \mathbf{L}_0$), the averages ($\bar{\boldsymbol{\varepsilon}}_0$ and $\bar{\boldsymbol{\sigma}}_0$) will remain the same. Therefore, as far as the rth inhomogeneity is concerned, it can be viewed as an ellipsoidal inhomogeneity with stiffness tensor $\mathbf{L}_r$ placed within a uniform matrix of stiffness tensor

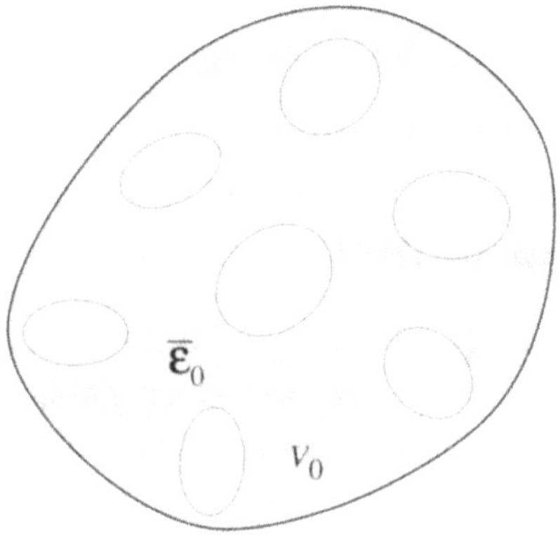

Figure 7.5 Average strain in matrix.

$\mathbf{L}_0$, which had been subjected to the uniform strain $\overline{\boldsymbol{\varepsilon}}_0$ before the inhomogeneity was embedded, see Fig. 7.6. Thus, we choose in (7.1.3),

$$\hat{\mathbf{L}}_0 = \mathbf{L}_0 \quad \text{and} \quad \hat{\boldsymbol{\varepsilon}}^0 = \overline{\boldsymbol{\varepsilon}}_0 \tag{7.3.3}$$

and consequently

$$\hat{\mathbf{S}}_r = \mathbf{S}_r. \tag{7.3.4}$$

Comparing these with (7.2.1) and (7.2.2), it is seen that the only difference between the Eshelby and the Mori–Tanaka methods is the use of $\hat{\boldsymbol{\varepsilon}}^0$. Note that $\hat{\boldsymbol{\varepsilon}}^0$ represents how the effects of other inhomogeneities are accounted for. Because the Eshelby method assumes no interaction between the inhomogeneities, it takes $\hat{\boldsymbol{\varepsilon}}^0 = \boldsymbol{\varepsilon}^0$, which is the strain in the matrix when none of the inhomogeneities is present. The Mori–Tanaka method also assumes the absence of all inhomogeneities, but it includes certain effects of the inhomogeneity by taking $\hat{\boldsymbol{\varepsilon}}^0 = \overline{\boldsymbol{\varepsilon}}_0$, because $\overline{\boldsymbol{\varepsilon}}_0$ is the average strain in the matrix phase when all inhomogeneities are present. For this reason, the Mori–Tanaka method provides better estimate of the effective modulus than the Eshelby method does for composite with nondilute reinforcement phases.

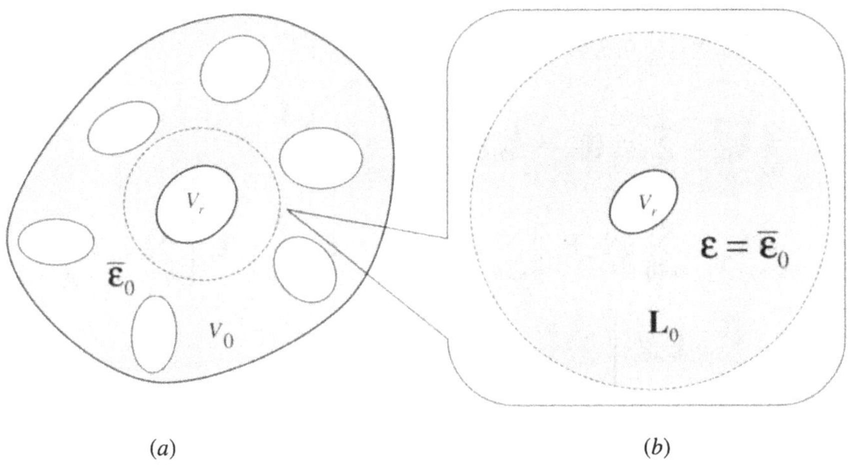

Figure 7.6 (*a*) The *r*th inhomogeneity in the composite. (*b*) The *r*th inhomogeneity in the original matrix of stiffness $\mathbf{L}_0$, which had been subjected to the uniform strain $\overline{\boldsymbol{\varepsilon}}_0$ before the inhomogeneity was embedded.

The use of (7.3.3) and (7.3.4) in (7.1.6) yields the strain in the rth inhomogeneity

$$\boldsymbol{\varepsilon}_r = \bar{\boldsymbol{\varepsilon}}_0 + \boldsymbol{\varepsilon}_r^{pt} = \bar{\boldsymbol{\varepsilon}}_0 + \mathbf{S}_r\boldsymbol{\varepsilon}_r^* = \mathbf{T}_r\bar{\boldsymbol{\varepsilon}}_0, \tag{7.3.5}$$

where $\mathbf{T}_r$ is the same as that of (7.2.4). Although (7.2.7) and (7.3.5) have similar appearance, they are very different equations in that (7.2.7) relates the total strain on the inhomogeneity to the average strain over the entire composite, while (7.3.5) relates the total strain on the inhomogeneity to the average strain on the matrix surrounding the inhomogeneity. Therefore, the tensor $\mathbf{T}_r$ in (7.2.7) represents the global strain concentration, while the tensor $\mathbf{T}_r$ in (7.3.5) represents local strain concentration tensor introduced in (5.7.9), that is,

$$\mathbf{G}_r = \mathbf{T}_r. \tag{7.3.6}$$

Thus, it follows from (5.7.14) that the global strain concentration tensor is

$$\mathbf{A}_r = \mathbf{G}_r\left[c_0\mathbf{I} + \sum_{n=1}^{N} c_n\mathbf{G}_n\right]^{-1} = \mathbf{T}_r\left[\sum_{n=0}^{N} c_n\mathbf{T}_n\right]^{-1}, \tag{7.3.7}$$

where the fact that $\mathbf{T}_0 = \mathbf{I}$ has been used. Consequently, the effective stiffness tensor of the composite follows directly from (5.7.7):

$$\begin{aligned}
\bar{\mathbf{L}} &= \mathbf{L}_0 + \sum_{r=1}^{N} c_r\,(\mathbf{L}_r - \mathbf{L}_0)\mathbf{A}_r \\
&= \mathbf{L}_0 + \sum_{r=0}^{N} c_r\,(\mathbf{L}_r - \mathbf{L}_0)\mathbf{T}_r\left[\sum_{n=0}^{N} c_n\mathbf{T}_n\right]^{-1} \\
&= \mathbf{L}_0 + \sum_{r=0}^{N} c_r\mathbf{L}_r\mathbf{T}_r\left[\sum_{n=0}^{N} c_n\mathbf{T}_n\right]^{-1} - \mathbf{L}_0\sum_{r=0}^{N} c_r\mathbf{T}_r\left[\sum_{n=0}^{N} c_n\mathbf{T}_n\right]^{-1} \\
&= \sum_{r=0}^{N} c_r\mathbf{L}_r\mathbf{T}_r\left[\sum_{n=0}^{N} c_n\mathbf{T}_n\right]^{-1}
\end{aligned} \tag{7.3.8}$$

or

$$\bar{\mathbf{L}} = \sum_{r=0}^{N} c_r\mathbf{L}_r\mathbf{T}_r\left[\sum_{n=0}^{N} c_n\mathbf{T}_n\right]^{-1}. \tag{7.3.9}$$

This is the Mori–Tanaka estimate of the effective stiffness tensor.

To obtain the Mori–Tanaka estimate of the effective complaint tensor, we substitute

$$\bar{\boldsymbol{\varepsilon}}_0 = \mathbf{M}_0 \bar{\boldsymbol{\sigma}}_0, \qquad \boldsymbol{\sigma}_r = \mathbf{L}_r \boldsymbol{\varepsilon}_r \tag{7.3.10}$$

into (7.3.5), which yields

$$\boldsymbol{\sigma}_r = \mathbf{L}_r \mathbf{T}_r \mathbf{M}_0 \bar{\boldsymbol{\sigma}}_0. \tag{7.3.11}$$

The local stress concentration tensor is thus given by

$$\mathbf{H}_r = \mathbf{L}_r \mathbf{T}_r \mathbf{M}_0. \tag{7.3.12}$$

It then follows from (5.7.15) that the global stress concentration tensor is given by

$$\begin{aligned}
\mathbf{B}_r = \mathbf{H}_r \left[c_0 \mathbf{I} + \sum_{n=1}^{N} c_n \mathbf{H}_n \right]^{-1} \\
= \mathbf{L}_r \mathbf{T}_r \mathbf{M}_0 \left[\sum_{n=0}^{N} c_n \mathbf{L}_n \mathbf{T}_n \mathbf{M}_0 \right]^{-1},
\end{aligned} \tag{7.3.13}$$

where the fact that $\mathbf{H}_0 = \mathbf{L}_0 \mathbf{T}_0 \mathbf{M}_0 = \mathbf{I}$ has been used. The Mori–Tanaka effective compliant tensor thus follows from (5.7.8):

$$\begin{aligned}
\overline{\mathbf{M}} &= \mathbf{M}_0 + \sum_{r=1}^{N} c_r (\mathbf{M}_r - \mathbf{M}_0) \mathbf{B}_r \\
&= \mathbf{M}_0 + \sum_{r=0}^{N} c_r (\mathbf{M}_r - \mathbf{M}_0) \mathbf{L}_r \mathbf{T}_r \mathbf{M}_0 \left[\sum_{n=0}^{N} c_n \mathbf{L}_n \mathbf{T}_n \mathbf{M}_0 \right]^{-1} \\
&= \mathbf{M}_0 + \sum_{r=0}^{N} c_r \mathbf{T}_r \mathbf{M}_0 \left[\sum_{n=0}^{N} c_n \mathbf{L}_n \mathbf{T}_n \mathbf{M}_0 \right]^{-1} \\
&\quad - \mathbf{M}_0 \sum_{r=0}^{N} c_r \mathbf{L}_r \mathbf{T}_r \mathbf{M}_0 \left[\sum_{n=0}^{N} c_n \mathbf{L}_n \mathbf{T}_n \mathbf{M}_0 \right]^{-1} \\
&= \sum_{r=0}^{N} c_r \mathbf{T}_r \mathbf{M}_0 \left[\sum_{n=0}^{N} c_n \mathbf{L}_n \mathbf{T}_n \mathbf{M}_0 \right]^{-1} \\
&= \sum_{r=0}^{N} c_r \mathbf{T}_r \left[\sum_{n=0}^{N} c_n \mathbf{L}_n \mathbf{T}_n \right]^{-1}
\end{aligned} \tag{7.3.14}$$

or

$$\overline{\mathbf{M}} = \sum_{r=0}^{N} c_r \mathbf{T}_r \left[\sum_{n=0}^{N} c_n \mathbf{L}_n \mathbf{T}_n \right]^{-1} \tag{7.3.15}$$

This is the Mori–Tanaka estimate of the effective compliance tensor of the composite in consideration.

It is easy to see from (7.3.9) and (7.3.15) that, unlike in the Eshelby method, the Mori–Tanaka method yields $\overline{\mathbf{M}}\overline{\mathbf{L}} = \overline{\mathbf{L}}\overline{\mathbf{M}} = \mathbf{I}$ for any c_r.

Example 7.2 Let us use the Mori–Tanaka method to estimate the effective elastic tensor and the effective compliant tensor for the composite considered in Example 7.1. First, consider (7.3.9) for $r = 1$,

$$\begin{aligned} \overline{\mathbf{L}} &= (c_0 \mathbf{L}_0 \mathbf{T}_0 + c_1 \mathbf{L}_1 \mathbf{T}_1)(c_0 \mathbf{T}_0 + c_1 \mathbf{T}_1)^{-1} \\ &= (c_0 \mathbf{L}_0 + c_1 \mathbf{L}_1 \mathbf{T}_1)(c_0 \mathbf{I} + c_1 \mathbf{T}_1)^{-1}, \end{aligned} \tag{7.3.16}$$

where

$$\mathbf{T}_1 = [\mathbf{I} + \mathbf{S}_1 (\mathbf{M}_0 \mathbf{L}_1 - \mathbf{I})]^{-1}. \tag{7.3.17}$$

By substituting (7.3.17) into (7.3.16) and making use of (7.2.19)–(7.2.22), we obtain using the symbolic notation,

$$\overline{\mathbf{L}} = (3\overline{K}, 2\overline{\mu}), \tag{7.3.18}$$

where

$$\overline{K} = K_0 + \frac{c_1 K_0 (K_1 - K_0)}{K_0 + 3\gamma_0 (1 - c_1)(K_1 - K_0)}, \tag{7.3.19}$$

$$\overline{\mu} = \mu_0 + \frac{c_1 \mu_0 (\mu_1 - \mu_0)}{\mu_0 + 2\delta_0 (1 - c_1)(\mu_1 - \mu_0)}. \tag{7.3.20}$$

Eliminating γ_0 and δ_0 by using (7.2.22), we can rewrite (7.3.20) as

$$\overline{K} = K_0 + \frac{c_1 (K_1 - K_0)(3K_0 + 4\mu_0)}{3K_0 + 4\mu_0 + 3(1 - c_1)(K_1 - K_0)}, \tag{7.3.21}$$

$$\overline{\mu} = \mu_0 + \frac{5 c_1 \mu_0 (\mu_1 - \mu_0)(3K_0 + 4\mu_0)}{5\mu_0 (3K_0 + 4\mu_0) + 6(1 - c_1)(\mu_1 - \mu_0)(K_0 + 2\mu_0)}. \tag{7.3.22}$$

These are the Mori–Tanaka estimates of the effective elastic constants for a composite with spherical particles.

7.4 SELF-CONSISTENT METHODS FOR COMPOSITE MATERIALS

Again, consider the composite material stated in Section 7.1. Let us pretend at this point that we already knew the effective stiffness tensor $\overline{\mathbf{L}}$ and its inverse $\overline{\mathbf{M}}$. Since there are numerous inhomogeneities in the composite, we know that the effective properties ($\overline{\mathbf{L}}$ and $\overline{\mathbf{M}}$) will not be affected if one inhomogeneity is removed from the composite. Now, focus our attention to a typical inhomogeneity, for example, the rth inhomogeneity. When the composite is subjected to either the displacement boundary condition or the traction boundary condition, one may envision that the effects of the applied loads (through the boundary condition) and the interaction with other inhomogeneity can be accounted for by assuming that the rth inhomogeneity is placed within a homogeneous matrix of stiffness $\overline{\mathbf{L}}$ that had been subjected to the strain tensor $\overline{\boldsymbol{\varepsilon}}$ before the inhomogeneity is embedded; see Figure 7.7. With this understanding, we can choose in (7.1.3),

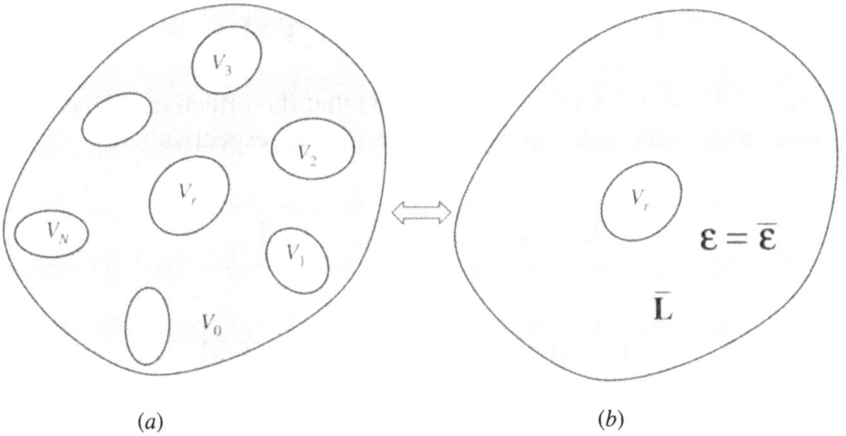

Figure 7.7 (*a*) The rth inhomogeneity in the actual composite. (*b*) The rth inhomogeneity in a homogeneous matrix with (as-yet-unknown) effective stiffness $\overline{\mathbf{L}}$ that has been subjected to the uniform strain $\overline{\boldsymbol{\varepsilon}}$ before the inhomogeneity was embedded.

$$\hat{\mathbf{L}}_0 = \overline{\mathbf{L}} \quad \text{and} \quad \hat{\boldsymbol{\varepsilon}}^0 = \overline{\boldsymbol{\varepsilon}}, \tag{7.4.1}$$

and consequently,

$$\hat{\mathbf{S}}_r = \overline{\mathbf{S}}_r, \tag{7.4.2}$$

where the overbar on the Eshelby tensor indicates that its elements should be computed based on the effective elastic constants of the composite.

Substituting (7.4.1) and (7.4.2) into (7.1.6) yields the total strain in the rth inhomogeneity:

$$\boldsymbol{\varepsilon}_r = \overline{\boldsymbol{\varepsilon}} + \boldsymbol{\varepsilon}_r^{pt} = \overline{\boldsymbol{\varepsilon}} + \overline{\mathbf{S}}\boldsymbol{\varepsilon}_r^* = \overline{\mathbf{T}}_r\overline{\boldsymbol{\varepsilon}}, \tag{7.4.3}$$

where

$$\overline{\mathbf{T}}_r = [\mathbf{I} = \overline{\mathbf{S}}_r\overline{\mathbf{L}}^{-1}(\mathbf{L}_r - \overline{\mathbf{L}})]^{-1}. \tag{7.4.4}$$

The corresponding stress on the rth inhomogeneity is

$$\boldsymbol{\sigma}_r = \mathbf{L}_r\boldsymbol{\varepsilon}_r = \mathbf{L}_r\overline{\mathbf{T}}_r\overline{\boldsymbol{\varepsilon}} = \mathbf{L}_r\overline{\mathbf{T}}_r\overline{\mathbf{M}}\overline{\boldsymbol{\sigma}}. \tag{7.4.5}$$

It is seen that the global strain and stress concentration tensors for the rth inhomogeneity are, respectively,

$$\mathbf{A}_r = \overline{\mathbf{T}}_r \quad \text{and} \quad \mathbf{B}_r = \mathbf{L}_r\overline{\mathbf{T}}_r\overline{\mathbf{M}}. \tag{7.4.6}$$

Thus, it follows from (5.7.7) and (5.7.8) that the effective stiffness and compliant tensors of the composite are given, respectively, by

$$\overline{\mathbf{L}} = \mathbf{L}_0 + \sum_{r=1}^{N} c_r\,(\mathbf{L}_r - \mathbf{L}_0)\overline{\mathbf{T}}_r, \tag{7.4.7}$$

$$\overline{\mathbf{M}} = \mathbf{M}_0 + \sum_{r=1}^{N} c_r\,(\mathbf{M}_r - \mathbf{M}_0)\mathbf{L}_r\overline{\mathbf{T}}_r\overline{\mathbf{M}}, \tag{7.4.8}$$

where $\overline{\mathbf{T}}_r$ is given in (7.4.4). These are the self-consistent estimates of the effective stiffness and compliant tensors of the composite. Unlike the Eshelby and Mori–Tanaka methods, the self-consistent method yields implicit equations for the effective properties. Solutions to these implicit equations typically require numerical iterations using computers.

To prove the consistency of the self-consistent estimates, let us consider (7.4.7). Premultiplying by $\mathbf{M}_0$ and postmultiplying by $\overline{\mathbf{L}}^{-1}$, we arrive at

$$\begin{aligned} \mathbf{M}_0 &= \overline{\mathbf{L}}^{-1} + \sum_{r=1}^{N} c_r \mathbf{M}_0 (\mathbf{L}_r - \mathbf{L}_0)\overline{\mathbf{T}}_r\overline{\mathbf{L}}^{-1} \\ &= \overline{\mathbf{L}}^{-1} + \sum_{r=1}^{N} c_r (\mathbf{M}_0 - \mathbf{M}_r)\mathbf{L}_r\overline{\mathbf{T}}_r\overline{\mathbf{L}}^{-1}. \end{aligned} \tag{7.4.9}$$

By rearranging the terms, we obtain

$$\overline{\mathbf{L}}^{-1} = \mathbf{M}_0 + \sum_{r=1}^{N} c_r (\mathbf{M}_r - \mathbf{M}_0)\mathbf{L}_r\overline{\mathbf{T}}_r\overline{\mathbf{L}}^{-1}. \tag{7.4.10}$$

Comparing (7.4.10) with (7.4.8), we observe that

$$\overline{\mathbf{M}} = \overline{\mathbf{L}}^{-1}. \tag{7.4.11}$$

Similarly, one can show that

$$\overline{\mathbf{L}} = \overline{\mathbf{M}}^{-1}. \tag{7.4.12}$$

Thus, we have $\overline{\mathbf{M}}\overline{\mathbf{L}} = \overline{\mathbf{L}}\overline{\mathbf{M}} = \mathbf{I}$, that is, the self-consistent estimates are indeed consistent.

Example 7.3 Consider again the composite described in Example 7.1. To find the effective stiffness tensor of the composites using the self-consistent method, we can use (7.4.7) directly by specifying $N = 1$. To this end, let us compute the relevant tensors one at a time. First, the Eshelby tensor for an isotropic material with $\mathbf{L} = (3\overline{K}, 2\overline{\mu})$ can be written as (see Appendix 4.B)

$$\overline{\mathbf{S}} = (3\overline{\gamma}, 2\overline{\delta}), \tag{7.4.13}$$

where

$$\bar{\gamma} = \frac{\bar{K}}{3\bar{K} + 4\bar{\mu}} = \frac{1 + \bar{\nu}}{9(1 - \bar{\nu})}, \quad \bar{\delta} = \frac{3\bar{K} + 6\bar{\mu}}{15\bar{K} + 20\bar{\mu}} = \frac{4 - 5\bar{\nu}}{15(1 - \bar{\nu})}. \tag{7.4.14}$$

Next, let us compute the fourth-order tensor $\bar{\mathbf{T}}_1$,

$$\begin{aligned}
\bar{\mathbf{T}}_1 &= [\mathbf{I} + \bar{\mathbf{S}}(\bar{\mathbf{L}}^{-1}\mathbf{L}_1 - \mathbf{I})]^{-1} \\
&= \{(1,1) + (3\bar{\gamma}, 2\bar{\delta})[(3\bar{K}, 2\bar{\mu})^{-1}(3K_1, 2\mu_1) - (1,1)]\}^{-1} \\
&= \left[(1,1) + (3\bar{\gamma}, 2\bar{\delta})\left(\frac{K_1 - \bar{K}}{\bar{K}}, \frac{\mu_1 - \bar{\mu}}{\bar{\mu}}\right)\right]^{-1} \\
&= \left[\frac{\bar{K} + 3\bar{\gamma}(K_1 - \bar{K})}{\bar{K}}, \frac{\bar{\mu} + 2\bar{\delta}(\mu_1 - \bar{\mu})}{\bar{\mu}}\right]^{-1} \\
&= \left[\frac{\bar{K}}{\bar{K} + 3\bar{\gamma}(K_1 - \bar{K})}, \frac{\bar{\mu}}{\bar{\mu} + 2\bar{\delta}(\mu_1 - \bar{\mu})}\right].
\end{aligned} \tag{7.4.15}$$

Finally

$$\begin{aligned}
\bar{\mathbf{L}} &= \mathbf{L}_0 + c_1 (\mathbf{L}_1 - \mathbf{L}_0)\bar{\mathbf{T}}_1 \\
&= (3K_0, 2\mu_0) + c_1 \left[\frac{3\bar{K}(K_1 - K_0)}{\bar{K} + 3\bar{\gamma}(K_1 - \bar{K})}, \frac{2\bar{\mu}(\mu_1 - \mu_0)}{\bar{\mu} + 2\bar{\delta}(\mu_1 - \bar{\mu})}\right] \\
&= \left[3K_0 + \frac{3c_1\bar{K}(K_1 - K_0)}{\bar{K} + 3\bar{\gamma}(K_1 - \bar{K})}, 2\mu_0 + \frac{2c_1\bar{\mu}(\mu_1 - \mu_0)}{\bar{\mu} + 2\bar{\delta}(\mu_1 - \bar{\mu})}\right].
\end{aligned} \tag{7.4.16}$$

Thus, we have

$$\bar{K} = K_0 + \frac{c_1\bar{K}(K_1 - K_0)}{\bar{K} + 3\bar{\gamma}(K_1 - \bar{K})}, \quad \bar{\mu} = \mu_0 + \frac{c_1\bar{\mu}(\mu_1 - \mu_0)}{\bar{\mu} + 2\bar{\delta}(\mu_1 - \bar{\mu})}. \tag{7.4.17}$$

This is a pair of nonlinear algebraic equations. The roots of these equations are the effective elastic constants of the composite. Because of the uniqueness of the problem, selections must be made judiciously based on physical ground if there are multiple roots. A common method to find the roots is iteration.

Making use of (7.4.14) in (7.4.17) yields a pair of alternative expressions:

$$\overline{K} = K_0 + \frac{c_1(K_1 - K_0)(3\overline{K} + 4\overline{\mu})}{3K_1 + 4\overline{\mu}}, \tag{7.4.18}$$

$$\overline{\mu} = \mu_0 + \frac{5c_1\overline{\mu}(\mu_1 - \mu_0)(3\overline{K} + 4\overline{\mu})}{3\overline{K}(3\overline{\mu} + 2\mu_1) + 4\overline{\mu}(2\overline{\mu} + 3\mu_1)}. \tag{7.4.19}$$

7.5 SELF-CONSISTENT METHODS FOR POLYCRYSTALLINE MATERIALS

The derivations used to develop the methods of determining the effective properties described in previous sections including the self-consistent method rely on the fact that the composite has a distinct matrix in which other inhomogeneities are embedded. For certain heterogeneous materials, such as polycrystalline materials, there is no distinct matrix phase. In a polycrystal, each grain can be viewed as an inhomogeneity embedded in the remaining grains, and, hence, all grains have the same significance. Clearly, there is a complete symmetry in treating each grain as an inhomogeneity. The concept of a matrix with embedded inhomogeneities is thus no longer meaningful. In this section, we will describe other self-consistent methods that treat all inhomogeneities (one may consider the matrix as one of the inhomogeneities) on an equal footing.

To begin, let us consider a polycrystalline material comprising N randomly distributed grains. The different grains can be either different materials (in the case of alloys) or grains with different crystallographic orientations (in the case of single-element polycrystals). We assume that the rth grain has volume fraction of c_r, $r = 1, 2, 3, \ldots, N$. It then follows from (5.4.1) that the average strain and average stress in the polycrystal is given by

$$\overline{\boldsymbol{\varepsilon}} = \sum_{r=1}^{N} c_r \overline{\boldsymbol{\varepsilon}}_r, \qquad \overline{\boldsymbol{\sigma}} = \sum_{r=1}^{N} c_r \overline{\boldsymbol{\sigma}}_r, \tag{7.5.1}$$

where $\overline{\boldsymbol{\varepsilon}}_r$ and $\overline{\boldsymbol{\sigma}}_r$ are the average strain and stress tensors, respectively, on the rth grain. Note that $\overline{\varepsilon}$ is the average strain of the entire polycrystal and $\overline{\varepsilon}_r$ is the average strain of the rth grain. The global strain concentration tensor $\mathbf{A}_r$ is defined for the rth grain through (5.7.3):

$$\bar{\boldsymbol{\varepsilon}}_r = \mathbf{A}_r \bar{\boldsymbol{\varepsilon}}. \tag{7.5.2}$$

Similarly, the global stress concentration tensor is defined for the rth grain through (5.7.9):

$$\bar{\boldsymbol{\sigma}}_r = \mathbf{B}_r \bar{\boldsymbol{\sigma}}. \tag{7.5.3}$$

It then follows from (7.5.2) and (7.5.3) and the use of Hooke's law that

$$\bar{\boldsymbol{\sigma}} = \sum_{r=1}^{N} c_r \bar{\boldsymbol{\sigma}}_r = \sum_{r=1}^{N} c_r \mathbf{L}_r \bar{\boldsymbol{\varepsilon}}_r = \sum_{r=1}^{N} c_r \mathbf{L}_r \mathbf{A}_r \bar{\boldsymbol{\varepsilon}}, \tag{7.5.4}$$

$$\bar{\boldsymbol{\varepsilon}} = \sum_{r=1}^{N} c_r \bar{\boldsymbol{\varepsilon}}_r = \sum_{r=1}^{N} c_r \mathbf{M}_r \bar{\boldsymbol{\sigma}}_r = \sum_{r=1}^{N} c_r \mathbf{M}_r \mathbf{B}_r \bar{\boldsymbol{\sigma}}. \tag{7.5.5}$$

These lead to

$$\overline{\mathbf{L}} = \sum_{r=1}^{N} c_r \mathbf{L}_r \mathbf{A}_r, \qquad \overline{\mathbf{M}} = \sum_{r=1}^{N} c_r \mathbf{M}_r \mathbf{B}_r. \tag{7.5.6}$$

Clearly, once the strain concentration tensor $\mathbf{A}_r$ is known as a function of the effective stiffness tensor $\overline{\mathbf{L}}$ for the polycrystal, the first of (7.5.6) can be used to obtain the effective stiffness tensor $\overline{\mathbf{L}}$. Similarly, if the stress concentration tensor $\mathbf{B}_r$ is known as a function of the effective compliance tensor $\overline{\mathbf{M}}$, the second of (7.5.6) can be used to obtain the effective compliance tensor $\overline{\mathbf{M}}$. Generally speaking, the effective compliance and stiffness tensor obtained from these two equations may not be consistent, unless the exact expressions of $\mathbf{A}_r$ and $\mathbf{B}_r$ are used.

One way to obtain an approximate solution to $\mathbf{A}_r$ is to view each grain as an ellipsoidal inhomogeneity embedded in a homogeneous matrix (consisting of all the other grains) of elastic stiffness tensor $\overline{\mathbf{L}}$. Under this assumption, when the polycrystal is subjected to displacement boundary condition $\mathbf{u}|_{\mathbf{x}\in S} = \bar{\boldsymbol{\varepsilon}}\mathbf{x}$, the strain concentration tensor $\mathbf{A}_r$ is given by (7.2.4) and (7.2.7) with $\mathbf{M}_0$ and $\mathbf{L}_0$ replaced by $\overline{\mathbf{M}}$ and $\overline{\mathbf{L}}$, respectively, because in this case the rth inhomogeneity is embedded in a homogeneous matrix with $\overline{\mathbf{L}}$, instead of $\mathbf{L}_0$,

$$\mathbf{A}_r = [\mathbf{I} + \overline{\mathbf{S}}_r \overline{\mathbf{M}}(\mathbf{L}_r - \overline{\mathbf{L}}]^{-1}, \tag{7.5.7}$$

where $\bar{\mathbf{S}}_r$ is the Eshelby tensor computed based on $\bar{\mathbf{L}}$. Substituting (7.5.7) into the first of (7.5.6) results in a self-consistent scheme for $\bar{\mathbf{L}}$:

$$\bar{\mathbf{L}} = \sum_{r=1}^{N} c_r \mathbf{L}_r [\mathbf{I} + \bar{\mathbf{S}}_r \bar{\mathbf{L}}^{-1}(\mathbf{L}_r - \bar{\mathbf{L}})]^{-1}. \tag{7.5.8}$$

To obtain an approximate solution to $\mathbf{B}_r$, let us consider the traction boundary condition (7.2.10). In this case, it follows from (7.2.15) that

$$\mathbf{B}_r = [\mathbf{M}_r - \bar{\mathbf{S}}_r(\mathbf{M}_r - \bar{\mathbf{M}})]^{-1}\, \bar{\mathbf{M}}. \tag{7.5.9}$$

Substituting (7.5.9) into the second of (7.5.6) yields a self-consistent scheme for $\bar{\mathbf{M}}$,

$$\mathbf{I} = \sum_{r=1}^{N} c_r \mathbf{M}_r\, [\mathbf{M}_r - \bar{\mathbf{S}}_r(\mathbf{M}_r - \bar{\mathbf{M}})]^{-1}. \tag{7.5.10}$$

It can be shown (see Problem 7.8) that the concentration tensors can be written in alternative forms:

$$\mathbf{A}_r = (\mathbf{L}_r + \bar{\mathbf{H}}_r)^{-1}(\bar{\mathbf{L}} + \bar{\mathbf{H}}_r), \qquad \mathbf{B}_r = (\mathbf{M}_r + \bar{H}_r^{-1})^{-1}(\bar{\mathbf{M}} + \bar{\mathbf{H}}_r^{-1}), \tag{7.5.11}$$

where

$$\bar{\mathbf{H}}_r = \bar{\mathbf{L}}(\bar{\mathbf{S}}_r^{-1} - \mathbf{I}) \tag{7.5.12}$$

is the Hill's constraint tensor (4.5.13). Thus, the self-consistent schemes can be recast into more symmetric forms:

$$\bar{\mathbf{L}} = \sum_{r=1}^{N} c_r \mathbf{L}_r(\mathbf{L}_r + \bar{\mathbf{H}}_r)^{-1}(\bar{\mathbf{L}} + \bar{\mathbf{H}}_r), \tag{7.5.13}$$

$$\bar{\mathbf{M}} = \sum_{r=1}^{N} c_r \mathbf{M}_r(\mathbf{M}_r + \bar{\mathbf{H}}_r^{-1})^{-1}(\bar{\mathbf{M}} + \bar{\mathbf{H}}_r^{-1}). \tag{7.5.14}$$

These are implicit algebraic equations for the effective stiffness tensor $\overline{\mathbf{L}}$. Solutions to these algebraic equations are the effective properties of the polycrystal. Making use of the relationships (see Problem 7.9)

$$\mathbf{L}_r\mathbf{A}_r = \mathbf{B}_r\overline{\mathbf{L}}, \qquad \mathbf{M}_r\mathbf{B}_r = \mathbf{A}_r\overline{\mathbf{M}},$$

one can easily show that the stiffness and compliance tensors predicted from (7.5.13) and (7.5.14), respectively, are indeed consistent, that is, $\overline{\mathbf{L}}\overline{\mathbf{M}} = \overline{\mathbf{M}}\overline{\mathbf{L}} = \mathbf{I}$.

Let us now consider a single-phase polycrystalline material. Single-phase polycrystal is an assembly of grains, all of which are comprised of the same material. Although all the grains are of the same material, their physical orientations are different. By physical orientation we mean rotations of the grain with respect to a fixed reference frame, or equivalently, the rotation of the stiffness tensor with respect to a fixed coordinate system. For example, let $\mathbf{L}$ be the elastic stiffness tensor of a single crystal in a fixed coordinate system (x_1, x_2, x_3). Then, the stiffness tensor of the same crystal in a different orientation can be represented by a rotation of $\mathbf{L}$, that is,

$$L_{ijkl}^{(r)} = \alpha_{ip}\alpha_{jq}\alpha_{kr}\alpha_{ls}L_{pqrs}, \tag{7.5.15}$$

where the rotation matrix can be expressed, for example, in terms of the Euler angles φ, θ, and ψ:

$$\begin{aligned}
\alpha_{11} &= \cos\theta\cos\psi\cos\varphi - \sin\psi\sin\varphi, \\
\alpha_{12} &= -\cos\theta\cos\psi\sin\varphi - \sin\psi\cos\varphi, \\
\alpha_{13} &= \sin\theta\cos\psi, \\
\alpha_{21} &= \cos\theta\sin\psi\cos\varphi + \cos\psi\sin\varphi, \\
\alpha_{22} &= -\cos\theta\sin\psi\sin\varphi + \cos\psi\cos\varphi, \\
\alpha_{13} &= \sin\theta\sin\psi, \\
\alpha_{31} &= -\sin\theta\cos\varphi, \\
\alpha_{32} &= \sin\theta\sin\varphi, \\
\alpha_{32} &= \cos\theta,
\end{aligned} \tag{7.5.16}$$

where

$$0 \leq \varphi \leq 2\pi, \qquad 0 \leq \theta \leq \pi, \qquad 0 \leq \psi \leq 2\pi. \tag{7.5.17}$$

Since all the grains are comprised of the same material, only with different orientations, we can write, symbolically, that,

$$\mathbf{L}_r = \tilde{\mathbf{L}}(\mathbf{L}, \varphi, \theta, \psi), \qquad \mathbf{M}_r = \tilde{\mathbf{M}}(\mathbf{M}, \varphi, \theta, \psi). \tag{7.5.18}$$

To characterize the physical orientation of the grains in a given single-phase polycrystal, we use a distribution function $f(\varphi, \theta, \psi)$, that is,

$$\frac{1}{8\pi^2}\int_0^{2\pi} d\psi \int_0^{\pi} \sin\theta \, d\theta \int_0^{2\pi} f(\varphi, \theta, \psi)\, d\varphi = 1. \tag{7.5.19}$$

The value $f(\varphi_0, \theta_0, \psi_0)$ represents the percent of grains that are all oriented in the direction $(\varphi_0, \theta_0, \psi_0)$. Clearly, $f(\varphi, \theta, \psi) = \delta(\varphi - \varphi_0)\delta(\theta - \theta_0)\delta(\psi - \psi_0)$ would mean all the grains are orientated in the $(\varphi_0, \theta_0, \psi_0)$ direction.

In addition to physical orientation, each grain (ellipsoidal inhomogeneity) in the polycrystal has its own size, shape, and orientation, all of which may not be the same for all grains. Since these geometric attributes come into the self-consistent equations through only the Eshelby tensor, which is independent of the size of the inhomogeneity, one may conclude that the effective stiffness tensor predicted by the self-consistent methods (7.5.13) and (7.5.14) is invariant to the grain size.

Because of the large number of grains involved, distribution functions are typically used to characterize the geometric shapes and orientations of grains. For example, we can use the Euler angles to indicate the geometric orientation of a grain, so that

$$\overline{S}^{(r)}_{ijkl} = \alpha_{ip}\alpha_{jq}\alpha_{kr}\alpha_{ls}\overline{S}_{pqrs}, \tag{7.5.20}$$

where $\overline{S}_{pqrs}$ is the Eshelby tensor for an ellipsoidal inclusion written in the same fixed coordinate system (x_1, x_2, x_3). For convenience, one may choose the ellipsoidal inclusion in such a way that its major axes coincide with the axes of the rectangular coordinate system. Therefore, if all grains have similar shape, the Eshelby tensor for each grain can be written as

$$\mathbf{S}_r = \tilde{\tilde{\mathbf{S}}}(\overline{\mathbf{L}}, \varphi, \theta, \psi), \tag{7.5.21}$$

where (φ, θ, ψ) follows a distribution function $g(\varphi, \theta, \psi)$.

Replacing the summation by integration and the volume fraction by the distribution function, we can change the self-consistent equations (7.5.13) and (7.5.14) to their integral counterparts that are more suited for polycrystalline materials:

$$\langle g \langle f \tilde{\mathbf{L}}[\mathbf{I} + \tilde{\tilde{\mathbf{S}}}\overline{\mathbf{L}}^{-1}(\tilde{\mathbf{L}} - \overline{\mathbf{L}})]^{-1} \rangle\rangle = \overline{\mathbf{L}}, \tag{7.5.22}$$

$$\langle g \langle f \tilde{\mathbf{M}}[\tilde{\mathbf{M}} - \tilde{\tilde{\mathbf{S}}}(\tilde{\mathbf{M}} - \overline{\mathbf{M}})]^{-1} \rangle\rangle = \mathbf{I}, \tag{7.5.23}$$

where the symbol $\langle\ \rangle =$ is defined as

$$\langle\ \rangle \equiv \int_0^{2\pi} d\varphi \int_0^{\pi} \sin\theta\, d\theta \int_0^{2\pi} d\psi. \tag{7.5.24}$$

Let us, for brevity, use only (7.5.22) to consider some special cases:

1. Both geometric and physical orientations arc random. In this case, $f(\varphi, \theta, \psi) = g(\varphi, \theta, \psi) = 1$. Thus, (7.5.22) reduces to

$$\langle\langle \tilde{\mathbf{L}}\, [\mathbf{I} + \tilde{\tilde{\mathbf{S}}}\overline{\mathbf{L}}^{-1}(\tilde{\mathbf{L}} - \overline{\mathbf{L}})]^{-1} \rangle\rangle = \overline{\mathbf{L}}. \tag{7.5.25}$$

2. Both geometric and physical orientations are aligned. In this case, $f(\varphi, \theta, \psi) = g(\varphi, \theta, \psi) = \delta(\varphi)\delta(\theta)\delta(\psi)$. Thus, (7.5.22) reduces to

$$\tilde{\mathbf{L}}(\mathbf{L}, 0, 0, 0)[\mathbf{I} + \tilde{\tilde{\mathbf{S}}}(\overline{\mathbf{L}}, 0, 0, 0)\overline{\mathbf{M}}(\tilde{\mathbf{L}}(\mathbf{L}, 0, 0, 0) - \overline{\mathbf{L}})]^{-1} = \overline{\mathbf{L}}. \tag{7.5.26}$$

It can be easily seen from (7.5.20) and (7.5.15) that

$$\tilde{\tilde{\mathbf{S}}}(\overline{\mathbf{L}}, 0, 0, 0) = \overline{\mathbf{S}}, \qquad \tilde{\mathbf{L}}(\mathbf{L}, 0, 0, 0) = \mathbf{L}. \tag{7.5.27}$$

Thus, we have

$$\mathbf{L}[\mathbf{I} + \overline{\mathbf{S}}\overline{\mathbf{M}}(\mathbf{L} - \overline{\mathbf{L}})]^{-1} = \overline{\mathbf{L}}. \tag{7.5.28}$$

Equivalently, this leads to

$$\overline{\mathbf{L}} = \mathbf{L}, \tag{7.5.29}$$

that is, the effective stiffness tensor is the stiffness tensor of the single crystal. This is certainly expected since all the grains are aligned both geometrically and physically. In other words, the polycrystal in this case is actually a single crystal!

7.6 DIFFERENTIAL SCHEMES

The differential scheme in determining the effective modulus of a composite takes a somewhat different approach from the methods introduced in previous sections. The differential approach was motivated by the fact that the effective properties of a dilute composite can be calculated with relatively high accuracy. Therefore, one may imagine that a composite with a finite concentration of inhomogeneities can be constructed through the following process. First, start with the homogeneous matrix material with elastic stiffness tensor $\mathbf{L}_0$. Then, add a small amount (volume fraction Δc) inhomogeneities with elastic stiffness tensor $\mathbf{L}_1$. We now have a composite with dilute concentration. The effective elastic stiffness tensor $\overline{\mathbf{L}}^{(1)}(\Delta c, \mathbf{L}_0)$ can be obtained accurately, where the fact that the effective stiffness tensor depends on the volume fraction of the inhomogeneities and the matrix material is explicitly indicated. In the next step, we will start with a homogenous matrix of elastic stiffness tensor $\overline{\mathbf{L}}^{(1)}(\Delta c, \mathbf{L}_0)$. Then add a small amount (volume fraction Δc) of inhomogeneities of stiffness tensor $\mathbf{L}_1$. Again, we have created a new composite. As far as this composite is concerned, it still has a dilute concentration of $\mathbf{L}_1$. Therefore, the effective stiffness tensor of this new composite can be obtained accurately using the same method used for the previous step. The only difference here is that the matrix material in the current step is $\overline{\mathbf{L}}^{(1)}(\Delta c, \mathbf{L}_0)$. Therefore, one can symbolically write the effective stiffness tensor at the current step as $\overline{\mathbf{L}}^{(2)}(\Delta c, \overline{\mathbf{L}}^{(1)})$. This process can continue. Each time, the newly obtained composite will be viewed as a homogeneous matrix, and a small amount of inhomogeneities of $\mathbf{L}_1$ is added to obtain the next composite, until the desired volume fraction is achieved. Such a repetitive process leads to differential equations for the effective properties as functions of the concentrations (or volume fractions of the inhomogeneities). This is the basic idea of all the differential schemes.

To illustrate how such thought process can be translated into differential equations, let us consider a two-phase composite ($N = 1$). According to (5.7.7), the effective stiffness tensor of a two-phase composite can be written as

$$\bar{\mathbf{L}}(c_1) = \mathbf{L}_0 + c_1 (\mathbf{L}_1 - \mathbf{L}_0):\mathbf{A}_1(\mathbf{L}_0), \tag{7.6.1}$$

where the dependence of $\bar{\mathbf{L}}$ on the volume fraction c_1 and $\mathbf{A}_1$ on $\mathbf{L}_0$ is explicitly indicated, and the strain concentration tensor $\mathbf{A}_1$ is defined by

$$\bar{\boldsymbol{\varepsilon}}_1 = \mathbf{A}_1 (\mathbf{L}_0):\bar{\boldsymbol{\varepsilon}}. \tag{7.6.2}$$

In the above, the symbol : has been used to indicate the tensor product. This is to make it clear that $\mathbf{A}_1 (\mathbf{L}_0)$ means $\mathbf{A}_1$ is a function of $\mathbf{L}_0$.

Let us assume that the effective stiffness tensor $\bar{\mathbf{L}}(c_1)$ is known at a volume fraction

$$c_1 = \frac{\Omega_1}{\Omega_0 + \Omega_1}, \tag{7.6.3}$$

where Ω_0 and Ω_1 are, respectively, the volumes of the matrix and inhomogeneities. If the composite with stiffness tensor $\bar{\mathbf{L}}(c_1)$ is viewed as a homogeneous "matrix," a new composite can be made by adding $\Delta\Omega_1$ amount of new inhomogeneities to the "matrix." The volume fraction of the newly added inhomogeneities is

$$\frac{\Delta\Omega_1}{\Omega_0 + \Omega_1 + \Delta\Omega_1}. \tag{7.6.4}$$

For such small-volume fractions, the effective elastic stiffness tensor can be obtained from (7.6.1) by replacing $\mathbf{L}_0$ by the new matrix property $\bar{\mathbf{L}}(c_1)$, that is,

$$\bar{\mathbf{L}}(c_1 + \Delta c_1) = \bar{\mathbf{L}}(c_1) + \frac{\Delta\Omega_1}{\Omega_0 + \Omega_1 + \Delta\Omega_1} (\mathbf{L}_1 - \bar{\mathbf{L}}(c_1)):\mathbf{A}_1(\bar{\mathbf{L}}(c_1)), \tag{7.6.5}$$

where Δc is the increment of the volume fraction of $\mathbf{L}_1$ in the composite due to the addition of $\Delta\Omega_1$ amount of $\mathbf{L}_1$. Clearly, the total volume fraction of $\mathbf{L}_1$ in this newly constructed composite is

$$c_1 + \Delta c_1 = \frac{\Omega_1 + \Delta\Omega_1}{\Omega_0 + \Omega_1 + \Delta\Omega_1}. \tag{7.6.6}$$

Subtracting (7.6.3) from (7.6.6) yields

$$\begin{aligned}\Delta c_1 &= \frac{\Omega_0 \Delta\Omega_1}{(\Omega_0 + \Omega_1)(\Omega_0 + \Omega_1 + \Delta\Omega_1)} \\ &= \frac{(\Omega_0 + \Omega_1 - \Omega_1)\Delta\Omega_1}{(\Omega_0 + \Omega_1)(\Omega_0 + \Omega_1 + \Delta\Omega_1)} \\ &= \frac{(1 - c_1)\Delta\Omega_1}{\Omega_0 + \Omega_1 + \Delta\Omega_1}\end{aligned} \tag{7.6.7}$$

or

$$\frac{\Delta\Omega_1}{\Omega_0 + \Omega_1 + \Delta\Omega_1} = \frac{\Delta c_1}{(1 - c_1)}. \tag{7.6.8}$$

Substitution of (7.6.8) into (7.6.5) yield

$$\frac{\overline{\mathbf{L}}(c_1 + \Delta c_1) - \overline{\mathbf{L}}(c_1)}{\Delta c_1} = \frac{1}{1 - c_1}(\mathbf{L}_1 - \overline{\mathbf{L}}(c_1)):\mathbf{A}_1(\overline{\mathbf{L}}(c_1)). \tag{7.6.9}$$

In the limit of $\Delta c_1 \rightarrow 0$, a first-order differential equation is derived:

$$\frac{d\overline{\mathbf{L}}(c_1)}{dc_1} = \frac{1}{(1 - c_1)}(\mathbf{L}_1 - \overline{\mathbf{L}}(c_1)):\mathbf{A}_1(\overline{\mathbf{L}}(c_1)). \tag{7.6.10}$$

The initial condition to compliment the differential equation is obviously

$$\overline{\mathbf{L}}(c_1)|_{c1=0} = \mathbf{L}_0. \tag{7.6.11}$$

It is straightforward to derive the dual of (7.6.10) and (7.6.11),

$$\frac{d\overline{\mathbf{M}}(c_1)}{dc_1} = \frac{1}{(1 - c_1)}(\mathbf{M}_1 - \overline{\mathbf{M}}(c_1)):\mathbf{B}_1(\overline{\mathbf{M}}(c_1)), \tag{7.6.12}$$

$$\overline{\mathbf{M}}(c_1)|_{c1=0} = \mathbf{M}_0, \tag{7.6.13}$$

where $\mathbf{B}_1$ is the stress concentration tensor defined by (5.7.8). These initial value problems are the starting point for the various differential schemes, depending on how the strain concentration tensors $\mathbf{A}_1$ and $\mathbf{B}_1$ are computed.

Let us consider first the Eshelby approach discussed in Section 7.2. It follows from (7.2.8) and (7.2.4) that

$$\mathbf{A}_1(\overline{\mathbf{L}}(c_1)) = \mathbf{T}_1|_{\mathbf{L}_0=\overline{\mathbf{L}}(c_1)} = [\mathbf{I} + \overline{\mathbf{S}}_1(\overline{\mathbf{L}}^{-1}\mathbf{L}_1 - \mathbf{I})]^{-1}, \qquad (7.6.14)$$

where

$$\overline{\mathbf{L}}^{-1} = \left(\frac{1}{3\overline{K}}, \frac{1}{2\overline{\mu}}\right), \qquad (7.6.15)$$

$$\overline{\mathbf{S}}_1 = (3\overline{\gamma}, 2\overline{\delta}) = \left(\frac{\overline{K}}{3\overline{K} + 4\overline{\mu}}, \frac{3(\overline{K} + 2\overline{\mu})}{5(3\overline{K} + 4\overline{\mu})}\right). \qquad (7.6.16)$$

Note that it goes without saying that all the quantities with an overbar are functions of the volume fraction c.

Substituting the above into (7.6.10) and (7.6.11) yields

$$\frac{d\overline{\mathbf{L}}}{dc_1} = \frac{1}{1 - c_1}(\mathbf{L}_1 - \overline{\mathbf{L}})[\mathbf{I} + \mathbf{S}_1(\overline{\mathbf{L}}^{-1}\mathbf{L}_1 - \mathbf{I})]^{-1}, \qquad (7.6.17)$$

$$\overline{\mathbf{L}}(0) = \mathbf{L}_0. \qquad (7.6.18)$$

For the composite described in Example 7.2, we have

$$\begin{aligned}\mathbf{T}_1|_{\mathbf{L}_0=\overline{\mathbf{L}}(c)} &= [\mathbf{I} + \mathbf{S}_1(\overline{\mathbf{L}}^{-1}\mathbf{L}_1 - \mathbf{I})]^{-1} \\ &= \left(\frac{\overline{K}}{\overline{K} + 3\overline{\gamma}(K_1 - \overline{K})}, \frac{\overline{\mu}}{\overline{\mu} + 2\overline{\delta}(\mu_1 - \overline{\mu})}\right). \end{aligned} \qquad (7.6.19)$$

Making use of the above in (7.6.17), we have

$$\frac{d\overline{K}}{dc} + \frac{(\overline{K} - K_1)(3\overline{K} + 4\overline{\mu})}{(1 - c)(3K_1 + 4\overline{\mu})} = 0, \qquad (7.6.20)$$

$$\frac{d\overline{\mu}}{dc} + \frac{5\overline{\mu}(\overline{\mu} - \mu_1)(3\overline{K} + 4\overline{\mu})}{(1 - c)[3\overline{K}(3\overline{\mu} + 2\mu_1) + 4\overline{\mu}(2\overline{\mu} + 3\mu_1)]} = 0. \qquad (7.6.21)$$

The solution must satisfy the initial conditions

$$\overline{K}(0) = K_0 \quad \text{and} \quad \overline{\mu}(0) = \mu_0. \qquad (7.6.22)$$

Clearly, this is a set of highly nonlinear ordinary differential equations. Numerical procedures are typically required to obtain the solution.

Another version of the differential scheme is attained by following the Mori–Tanaka approach in obtaining the strain concentration tensor. It follows from (7.3.7) that

$$\mathbf{A}_1(\overline{\mathbf{L}}(c_1)) = \mathbf{T}_1[(1 - c_1)\mathbf{I} + c_1\mathbf{T}_1]^{-1}|_{\mathbf{L}_0=\overline{\mathbf{L}}(c_1)}.$$

The differential equation now becomes

$$\frac{d\overline{\mathbf{L}}}{dc_1} = \frac{1}{1 - c_1}(\mathbf{L}_1 - \overline{\mathbf{L}})\mathbf{T}_1[(1 - c_1)\mathbf{I} + c_1\mathbf{T}_1]^{-1}|_{\mathbf{L}_0=\overline{\mathbf{L}}(c_1)}. \qquad (7.6.23)$$

7.7 COMPARISON OF DIFFERENT METHODS

In the previous sections, we have introduced several methods of estimating the effective stiffness (compliance) tensors for a given composite, namely, the Eshelby method (or the dilute concentration method), the Mori–Tanaka method, the self-consistent method, and the differential method. In this section, we will use these methods to obtain the effective moduli of several idealized composite materials where explicit analytical solutions can be obtained. We will compare the results to assess the validity and accuracy of these methods.

We first state that, based on the assumptions made in deriving them, all these methods ignore the spatial distribution of the inhomogeneities, that is, they all assume uniform distribution. However, the shapes and orientations of the ellipsoidal inhomogeneities are taken into account through the Eshelby tensor $\mathbf{S}_r$. Note that the Eshelby tensor is shape dependent but not size dependent. Thus, the effective modulus tensor predicted by these methods will not depend on the size of the inhomogeneities. Interactions among the inhomogeneities are taken into consideration differently by different methods. In general, the Eshelby method works only for very dilute concentration, and the other methods are applicable to somewhat higher concentration. We will see some numerical examples later in this section.

Now consider the composite material described in Example 7.1. For convenience, we list below the equations for the effective moduli of this composite obtained by all these methods and consider a few special cases.

The Eshelby method:

$$\overline{K} = K_0 + \frac{c_1(K_1 - K_0)(3K_0 + 4\mu_0)}{3K_1 + 4\mu_0}, \tag{7.7.1}$$

$$\overline{\mu} = \mu_0 + \frac{5c_1\mu_0(\mu_1 - \mu_0)(3K_0 + 4\mu_0)}{3K_0(3\mu_0 + 2\mu_1) + 4\,\mu_0(2\mu_0 + 3\mu_1)}. \tag{7.7.2}$$

The Mori–Tanaka method:

$$\overline{K} = K_0 + \frac{c_1(K_1 - K_0)(3K_0 + 4\mu_0)}{3K_0 + 4\mu_0 + 3(1 - c_1)(K_1 - K_0)}, \tag{7.7.3}$$

$$\overline{\mu} = \mu_0 + \frac{5c_1\mu_0(\mu_1 - \mu_0)(3K_0 + 4\mu_0)}{5\mu_0(3K_0 + 4\mu_0) + 6(1 - c_1)(\mu_1 - \mu_0)(K_0 + 2\mu_0)}. \tag{7.7.4}$$

The self-consistent method:

$$\overline{K} = K_0 + \frac{c_1(K_1 - K_0)(3\overline{K} + 4\overline{\mu})}{3K_1 + 4\overline{\mu}}, \tag{7.7.5}$$

$$\overline{\mu} = \mu_0 + \frac{5c_1\overline{\mu}(\mu_1 - \mu_0)(3\overline{K} + 4\overline{\mu})}{3\overline{K}(3\overline{\mu} + 2\mu_1) + 4\overline{\mu}(2\overline{\mu} + 3\mu_1)}. \tag{7.7.6}$$

The differential method:

$$\frac{d\overline{K}}{dc_1} + \frac{(\overline{K} - K_1)(3\overline{K} + 4\overline{\mu})}{(1 - c_1)(3K_1 + 4\overline{\mu})} = 0, \tag{7.7.7}$$

$$\frac{d\overline{\mu}}{dc_1} + \frac{5\overline{\mu}(\overline{\mu} - \mu_1)(3\overline{K} + 4\overline{\mu})}{(1 - c_1)[3\overline{K}(3\overline{\mu} + 2\mu_1) + 4\overline{\mu}(2\overline{\mu} + 3\mu_1)]} = 0, \tag{7.7.8}$$

$$\overline{K}|_{c_1=0} = K_0 \quad \text{and} \quad \overline{\mu}|_{c_1=0} = \mu_0. \tag{7.7.9}$$

Dilute Concentration

For dilute concentration of inhomogeneities, we can assume $c_1 << 1$. Therefore, by expanding (7.7.3) and (7.7.4) into power series of c_1 and neglecting high-order terms, we have

$$\overline{K} = K_0 + \frac{c_1(K_1 - K_0)(3K_0 + 4\mu_0)}{3K_1 + 4\mu_0}, \tag{7.7.10}$$

$$\overline{\mu} = \mu_0 + \frac{5c_1\mu_0(\mu_1 - \mu_0)(3K_0 + 4\mu_0)}{3K_0(3\mu_0 + 2\mu_1) + 4\mu_0(2\mu_0 + 3\mu_1)}. \tag{7.7.11}$$

Comparing the above with the Eshelby estimate, it is seen that at the dilute concentration, the Mori–Tanaka estimate reduces to the Eshelby results.

As for the self-consistent estimate, we can first expand $\overline{K}$ and $\overline{\mu}$ into power series of c_1, that is,

$$\overline{K} = K_0 + c_1K_0^{(1)} + c_1^2K_0^{(1)} \ldots, \tag{7.7.12}$$

$$\overline{\mu} = \mu_0 + c_1\mu_0^{(1)} + c_1^2\mu_0^{(1)} \ldots. \tag{7.7.13}$$

Substituting these expansions into (7.7.5) and (7.7.6) yields immediately that the self-consistent estimate, in the dilute case, also reduces to the Eshelby estimate.

Following similar steps, one can show that the differential method predicts the same results as well (as expected, since the differential method was based on the solution to the dilute case). In other words, all four methods yield the same results when the volume fraction of the inhomogeneity is very low. In the limit of $c_1 \to 0$, the effective modulus of the composite as predicted by all these methods reduce to the modulus of the matrix, as it should.

High-Concentration Asymptotes

For this case, we are interested in the limit $c_1 \to 1$, meaning the entire "composite" consists of particles only. Certainly, this would require multiple size particles; see Problem 7.6. Since the composite contains particles only, it is expected that the effective modulus of the composite should be the modulus of the particles. Taking the limit $c_1 \to 1$ in (7.7.1)–(7.7.6), we find that the Eshelby method predicts

$$\bar{K} = \frac{K_0(2K_1 - K_0) + 4\mu_0 K_1}{3K_1 + 4\mu_0}, \tag{7.7.14}$$

$$\bar{\mu} = \frac{\mu_0[3K_0(7\mu_1 - 2\mu_0) + 4\mu_0(8\mu_1 - 3\mu_0)]}{3K_0(3\mu_0 + 2\mu_1) + 4\mu_0(2\mu_0 + 3\mu_1)}. \tag{7.7.15}$$

It is not surprising that the Eshelby method does not predict the correct values because the Eshelby method is valid only for dilute concentrations.

It is very easy to see by setting $c_1 = 1$ in (7.7.3) and (7.7.4) that the Mori–Tanaka method does predict the proper limiting values for the effective moduli:

$$\bar{K} = K_1, \qquad \bar{\mu} = \mu_1. \tag{7.7.16}$$

It can also be easily verified that (7.7.16) represent a set of solutions to the self-consistent equations (7.7.5) and (7.7.6). In other words, the self-consistent method also predicts the property high-concentration limit.

Rigid Particles

To mimic rigid particles, one may consider the limiting case of

$$\frac{\mu_0}{\mu_1} \to 0, \qquad \frac{K_0}{K_1} \to 0. \tag{7.7.17}$$

Also, on physical grounds, we assume that

$$\frac{\bar{\mu}}{\mu_1} \to 0, \qquad \frac{\bar{K}}{K_1} \to 0 \tag{7.7.18}$$

for any $0 \le c_1 < 1$.

Making use of the above in (7.7.1)–(7.7.4), we obtain

$$\bar{K} = K_0 + \frac{c_1(3K_0 + 4\mu_0)}{3}, \qquad \bar{\mu} = \mu_0 + \frac{5c_1\mu_0(3K_0 + 4\mu_0)}{6(K_0 + 2\mu_0)}, \tag{7.7.19}$$

from the Eshelby method, and

$$\bar{K} = K_0 + \frac{c_1(3K_0 + 4\mu_0)}{3(1 - c_1)}, \qquad \bar{\mu} = \mu_0 + \frac{5c_1\mu_0(3K_0 + 4\mu_0)}{6(K_0 + 2\mu_0)(1 - c_1)} \tag{7.7.20}$$

from the Mori–Tanaka method. Again, it is seen that both methods predict correctly that $\bar{K} = K_0$ and $\bar{\mu} = \mu_0$ at the low-concentration limit $c_1 \to 0$. At the high-concentration limit, $c_1 \to 1$, the Mori–Tanaka method predicts that the effective modulus increases without bound (as it should because the composite is full of rigid particles), while the Eshelby method yields a finite value. This is expected, for we know that the Eshelby method is applicable only for the dilute case.

Making use of (7.7.17) and (7.7.18) in the self-consistent estimate, we obtain

$$\bar{K} = K_0 + \frac{c_1(3\bar{K} + \bar{\mu})}{3}, \qquad \bar{\mu} = \mu_0 + \frac{5c_1\bar{\mu}(3\bar{K} + 4\bar{\mu})}{6(\bar{K} + 2\bar{\mu})}. \tag{7.7.21}$$

The effective bulk modulus $\bar{K}$ can be solved from (7.7.21) to yield

$$\bar{K} = \frac{3K_0 + 4c_1\bar{\mu}}{3(1 - c_1)}. \tag{7.7.22}$$

Substituting (7.7.22) into the second of (7.7.21) yields

$$12(1 - 2c_1)\left(\frac{\bar{\mu}}{\mu_0}\right)^2 + \left[3(2 - 5c_1)\frac{K_0}{\mu_0} - 4(3 - c_1)\right]\frac{\bar{\mu}}{\mu_0} - \frac{6K_0}{\mu_0} = 0. \tag{7.7.23}$$

When $c_1 = 0.5$, the above equation reduces to

$$\bar{\mu} = -\frac{12K_0\mu_0}{3K_0 + 20\mu_0} < 0. \tag{7.7.24}$$

This is not a physically possible solution because the effective shear modulus should be positive. For $c_1 \neq 0.5$, the quadratic equation (7.7.23) yields two roots. Numerical analysis shows that both roots are real. One of them is always negative and other is given by

$$\frac{\bar{\mu}}{\mu_0} = \alpha + \frac{\beta}{1 - 2c_1} + \sqrt{\left(\alpha + \frac{\beta}{1 - 2c_1}\right)^2 + \frac{1 + \nu_0}{3(1 - 2\nu_0)(1 - 2c_1)}}, \tag{7.7.25}$$

where ν_0 is the Poisson ratio of the matrix material, and

$$\alpha = -\frac{3(1 + 3\nu_0)}{24(1 - 2\nu_0)}, \qquad \beta = \frac{11 - 19\nu_0}{24(1 - 2\nu_0)}. \tag{7.7.26}$$

In deriving (7.7.25), we have used the relationship

$$\frac{\mu_0}{K_0} = \frac{3(1 - 2\nu_0)}{2(1 + \nu_0)}. \tag{7.7.27}$$

Since $0 \leq \nu_0 \leq 0.5$, it can be shown that the right hand of (7.7.25) is positive only for $0 \leq c_1 < 0.5$. In other words, for spherical rigid particles dispersed in a matrix material, the self-consistent method yields a physically meaningful solution to the effective shear modulus only if the particle volume fraction is less than 50%.

Now, consider the differential method. Making use of (7.7.18) in (7.7.7) and (7.7.8) yields

$$\frac{d\bar{K}}{dc_1} - \frac{(3\bar{K} + 4\bar{\mu})}{3(1 - c_1)} = 0, \tag{7.7.28}$$

$$\frac{d\bar{\mu}}{dc_1} - \frac{5\bar{\mu}(3\bar{K} + 4\bar{\mu})}{6(1 - c_1)(\bar{K} + 2\bar{\mu})} = 0. \tag{7.7.29}$$

Solutions to this system of nonlinear differential equations are still fairly complicated.

To exam the high-concentration asymptotic behavior of the differential scheme predictions, let us expand the solution near $c_1 = 1$:

$$\overline{K} = \sum_{n=0}^{\infty} \frac{A_n}{(1 - c_1)^{m-n}}, \qquad \overline{\mu} = \sum_{n=0}^{\infty} \frac{B_n}{(1 - c_1)^{m-n}}, \tag{7.7.30}$$

where the asymptotic behavior is dictated by the constant m, which is to be determined from the solution procedure. Substituting (7.7.30) into (7.7.28) and (7.7.29) and keeping the leading order terms only, we obtain the following eigenvalue problem:

$$\begin{bmatrix} 3(m - 1) & -4 \\ 3(2m - 5) & 4(2m - 5) \end{bmatrix} \begin{bmatrix} A_0 \\ B_0 \end{bmatrix} = \begin{bmatrix} 0 \\ 0 \end{bmatrix}. \tag{7.7.31}$$

The corresponding eigenvalues are $m = 0$ and $m = 2$. The eigenvalue $m = 2$ should be chosen on physical ground. Therefore, the asymptotic behavior of the differential scheme for $c_1 \to 1$ is

$$\overline{K} \approx \frac{A_0}{(1 - c_1)^2}, \qquad \overline{\mu} \approx \frac{B_0}{(1 - c_1)^2}. \tag{7.7.32}$$

In summary, for the case of rigid particles, the Eshelby method yields finite values of the effective moduli in the high-concentration limit $c_1 \to 1$, while the self-consistent method yields no solution for $c_1 > 1$ in this case. Both the Mori–Tanaka and the differential methods predict unbounded effective moduli, but their asymptotic behavior is different. The Mori–Tanaka solution behaves as $1/(1 - c_1)$ while the differential solution behaves as $1/(1 - c_1)^2$.

Other interesting features of the differential scheme can also be explored. For example, in the case of the incompressible matrix, that is, $\overline{\mu}/\overline{K} \to 0$, (7.7.28) and (7.7.29) reduce to

$$\frac{d\overline{K}}{dc} - \frac{\overline{K}}{1 - c_1} = 0, \tag{7.7.33}$$

$$\frac{d\overline{\mu}}{dc} - \frac{5\overline{\mu}}{2(1 - c_1)} = 0. \tag{7.7.34}$$

The solutions that satisfy the initial conditions (7.7.9) are

$$\overline{K} = \frac{K_0}{1 - c}, \qquad \overline{\mu} = \frac{\mu_0}{(1 - c_1)^{5/2}}. \tag{7.7.35}$$

We see that the differential scheme does predict correctly

$$\overline{K} = K_0, \qquad \overline{\mu} = \mu_0 \tag{7.7.36}$$

in the low-concentration limit $c \to 0$. It also predicts that in the high-concentration limit $c \to 1$, shear modulus approaches to infinity with the order of $1/(1 - c_1)^{5/2}$.

Note that the solutions (7.7.32) and (7.7.35) are different in that the former is valid $c_1 = 1$, while the later is for the incompressible matrix, and is valid for all ranges of c_1 between 0 and 1.

Voids

Consider a matrix containing spherical voids. When the void content is zero ($c_1 = 0$), we should expect that the effective modulus is the same as that of the matrix. For high concentration of voids ($c_1 \to 1$), we should expect the effective modulus approaches to zero. Let us see how the various methods predict these limiting cases.

For voids, we can consider the limiting case of

$$\frac{\mu_1}{\mu_0} \to 0, \qquad \frac{K_1}{\mu_0} \to 0. \tag{7.7.37}$$

The Eshelby estimate reduces to

$$\overline{K} = K_0 - \frac{c_1 K_0 (3K_0 + 4\mu_0)}{4\mu_0}, \tag{7.7.38}$$

$$\overline{\mu} = \mu_0 - \frac{5c_1 \mu_0 (3K_0 + 4\mu_0)}{9K_0 + 8\mu_0}. \tag{7.7.39}$$

They correctly predict the low-concentration limiting value for $c_1 \to 0$. For $c_1 \to 1$, the Eshelby method yields

$$\overline{K} = -\frac{3K_0^3}{4\mu_0}, \qquad \overline{\mu} = -\frac{6\mu_0 (K_0 + 2\mu_0)}{9K_0 + 8\mu_0}. \tag{7.7.40}$$

These obviously are not physically admissible solutions.

The Mori–Tanaka method gives

$$\overline{K} = K_0 - \frac{c_1 K_0 (3K_0 + 4\mu_0)}{3K_0 + 4\mu_0 - 3(1 - c_1)K_0}, \tag{7.7.41}$$

$$\bar{\mu} = \mu_0 - \frac{5c_1\mu_0(3K_0 + 4\mu_0)}{5(3K_0 + 4\mu_0) - 6(1 - c_1)(K_0 + 2\mu_0)}. \tag{7.7.42}$$

For low concentration,

$$\bar{K} \approx K_0 - \frac{c_1K_0(3K_0 + 4\mu_0)}{4\mu_0}, \qquad \bar{\mu} \approx \mu_0 - \frac{5c_1\mu_0(3K_0 + 4\mu_0)}{9K_0 + 8\mu_0}. \tag{7.7.43}$$

For high concentration

$$\bar{K} \approx \frac{4\mu_0K_0}{3K_0 + 4\mu_0}(1 - c_1), \qquad \bar{\mu} \approx \frac{\mu_0(9K_0 + 8\mu_0)}{5(3K_0 + 4\mu_0)}(1 - c_1). \tag{7.7.44}$$

We see that the Mori–Tanaka method predicts both low- and high-concentration limits correctly, while the Eshelby method is only good for low concentration.

Similarly, the self-consistent estimate reduces

$$\bar{K} = K_0 - \frac{c_1K_0(3\bar{K} + 4\bar{\mu})}{4\bar{\mu}}, \qquad \bar{\mu} = \mu_0 - \frac{5c_1\mu_0(3\bar{K} + 4\bar{\mu})}{9\bar{K} + 8\bar{\mu}}. \tag{7.7.45}$$

Solving the first of (7.7.45) for $\bar{K}$ yields

$$\bar{K} = \frac{4(1 - c_1)K_0\bar{\mu}}{3c_1K_0 + 4\bar{\mu}}. \tag{7.7.46}$$

Making use of this in the second of (7.7.45) gives us the following quadratic equation for the effective shear modulus:

$$8\left(\frac{\bar{\mu}}{\mu_0}\right)^2 + \left[3(3 - c_1)\frac{K_0}{\mu_0} - 4(2 - 5c_1)\right]\frac{\bar{\mu}}{\mu_0} - 9(1 - 2c_1)\frac{K_0}{\mu_0} = 0. \tag{7.7.47}$$

When $c_1 = 0.5$, the two roots of this quadratic equation are, respectively, zero and

$$\overline{\mu} = -\left(\frac{15K_0}{16} + \frac{\mu_0}{4}\right) < 0.$$

Neither of them is physically meaningful. For $c_1 \neq 0.5$, the two roots are all real. Numerical analysis shows that one of these roots is always negative and the other root is given by

$$\frac{\overline{\mu}}{\mu_0} = \alpha(1 - 2c_1) - \beta + \sqrt{[\alpha(1 - 2c_1) - \beta]^2 + \frac{3(1 + \nu_0)(1 - 2c_1)}{4(1 - 2\nu_0)}}, \tag{7.7.48}$$

where

$$\alpha = \frac{3(3 - 7\nu_0)}{16(1 - 2\nu_0)}, \qquad \beta = \frac{7 + \nu_0}{16(1 - 2\nu_0)}. \tag{7.7.49}$$

Since $0 \leq \nu_0 \leq 0.5$, it can be shown that the right hand of (7.7.48) is positive only for $0 \leq c_1 < 0.5$. In other words, for spherical voids dispersed in an elastic matrix material, the self-consistent method yields a physically meaningful solution to the effective shear modulus only if the void volume fraction is less than 50%.

It is seen from the above cases that the self-consistent method has trouble when the particles are rigid or voids. In fact, it is generally true that the self-consistent method does not work well when the contrast between the inhomogeneities and matrix is too large.

Now, consider the differential scheme. Use of (7.7.37) leads to

$$\frac{d\overline{K}}{dc_1} + \frac{\overline{K}(3\overline{K} + 4\overline{\mu})}{4(1 - c_1)\overline{\mu}} = 0, \tag{7.7.50}$$

$$\frac{d\overline{\mu}}{dc_1} + \frac{5\overline{\mu}(3\overline{K} + 4\overline{\mu})}{(1 - c_1)[9\overline{K} + 8\overline{\mu}]} = 0. \tag{7.7.51}$$

Again, we first consider the asymptotic behavior near $c_1 = 1$:

$$\overline{K} = \sum_{n=0}^{\infty} A_n(1 - c_1)^{m+n}, \qquad \overline{\mu} = \sum_{n=0}^{\infty} B_n(1 - c_1)^{m+n}. \tag{7.7.52}$$

By substituting (7.7.52) into (7.7.50) and (7.7.51) and keeping the leading order terms only, we obtain the following eigenvalue problem:

$$\begin{bmatrix} 3 & 4(1 - m) \\ 3(3m - 5) & 4(2m - 5) \end{bmatrix} \begin{bmatrix} A_0 \\ B_0 \end{bmatrix} = \begin{bmatrix} 0 \\ 0 \end{bmatrix}. \tag{7.7.53}$$

The corresponding eigenvalues are $m = 0$ and $m = 2$. The eigenvalue $m = 2$ should be chosen on physical ground. Therefore, the asymptotic behavior of the differential scheme for $c_1 \to 1$ is

$$\overline{K} \approx A_0(1 - c_1)^2, \qquad \overline{\mu} \approx B_0(1 - c_1)^2. \tag{7.7.54}$$

Next, let us further assume that the matrix is incompressible, that is, $\overline{\mu}/\overline{K} \to 0$. In this case, Eq. (7.7.51) reduces to

$$\frac{d\overline{\mu}}{dc_1} + \frac{5\overline{\mu}}{3(1 - c_1)} = 0. \tag{7.7.55}$$

The solution that satisfies the initial condition (7.7.9) is

$$\overline{\mu} = \mu_0(1 - c_1)^{3/5}.$$

In summary, for the case of voids, the Eshelby method yields finite values of the effective moduli in the high-concentration limit $c_1 \to 1$, while the self-consistent method yields no physical solution for $c_1 > 0.5$ in this case. Both the Mori–Tanaka and the differential physical solution for methods yield vanishing effective moduli, but their asymptotic behavior is different. The Mori–Tanaka solution behaves as $1 - c_1$, while the differential solution behaves as $(1 - c_1)^2$.

PROBLEMS

7.1 Derive (7.2.17) from (7.2.16).

7.2 Show that the Eshelby estimate of the elastic constants for an isotropic spherical particulate-reinforced composite can also be written as

$$\overline{\mu} = \mu_0\left\{1 - \sum_{r=1}^{N} \frac{c_r(\mu_r - \mu_0)}{\mu_0 + 2\delta_0(\mu_r - \mu_0)}\right\}^{-1},$$

$$\overline{K} = K_0\left\{1 - \sum_{r=1}^{N} \frac{c_r(K_r - K_0)}{K_0 + 3\gamma_0(K_r - K_0)}\right\}^{-1},$$

where

$$\gamma_0 = \frac{K_0}{3K_0 + 4\mu_0} = \frac{1 + \upsilon_0}{9(1 - \upsilon_0)},$$

$$\delta_0 = \frac{3K_0 + 6\mu_0}{15K_0 + 20\mu_0} = \frac{4 - 5\upsilon_0}{15(1 - \upsilon_0)}.$$

7.3 Show, when all the inclusions are similar in shape, that alternative forms of the Mori–Tanaka effective stiffness and compliance tensors are, respectively, given by

$$\overline{\mathbf{L}} = \left\{\sum_{r=0}^{N} c_r(\mathbf{L}_0^* + \mathbf{L}_r)^{-1}\right\}^{-1} - \mathbf{L}_0^*,$$

$$\overline{\mathbf{M}} = \left\{\sum_{r=0}^{N} c_r(\mathbf{M}_r + \mathbf{M}_0^*)^{-1}\right\}^{-1} - \mathbf{M}_0^*,$$

where

$$\mathbf{L}_0^* = \mathbf{L}_0(\mathbf{S}_0^{-1} - \mathbf{I}), \qquad \mathbf{M}_0^* = [(\mathbf{I} - \mathbf{S}_0)^{-1} - \mathbf{I}]\mathbf{M}^{(0)}.$$

7.4 Prove (7.4.12).

7.5 Show that all three (Eshelby, Mori–Tanake, and self-consistent) estimates are the same for very low volume fractions of inhomogeneities (dilute concentration).

7.6 Assume spherical particles of equal size are dispersed in a matrix material. What is the highest possible volume fraction of the particles?

7.7 Assume circular cross-section fibers of equal radius are placed in a matrix. If all the fibers are aligned in the same direction, what is the highest possible fiber volume fraction?

7.8 Show that

$$[\mathbf{I} + \overline{\mathbf{S}}_r\overline{\mathbf{M}}(\mathbf{L}_r - \overline{\mathbf{L}})]^{-1} = (\mathbf{L}_r + \overline{\mathbf{H}}_r)^{-1}(\overline{\mathbf{L}} + \overline{\mathbf{H}}_r),$$

$$[\mathbf{M}_r - \overline{\mathbf{S}}_r(\mathbf{M}_r - \overline{\mathbf{M}})]^{-1}\,\overline{\mathbf{M}} = (\mathbf{M}_r + \overline{\mathbf{H}}_r^{-1})^{-1}(\overline{\mathbf{M}} + \overline{\mathbf{H}}_r^{-1}),$$

where $\overline{\mathbf{H}}_r = \overline{\mathbf{L}}(\overline{\mathbf{S}}_r^{-1} - \mathbf{I})$ is the Hill constraint tensor (4.5.14).

7.9 Prove

$$\mathbf{L}_r[\mathbf{I} + \bar{\mathbf{S}}_r\overline{\mathbf{M}}(\mathbf{L}_r - \bar{\mathbf{L}})]^{-1} = [\mathbf{M}_r - \bar{\mathbf{S}}_r(\mathbf{M}_r - \overline{\mathbf{M}})]^{-1},$$

$$\mathbf{M}_r[\mathbf{M}_r - \bar{\mathbf{S}}_r(\mathbf{M}_r - \overline{\mathbf{M}})]^{-1} = [\mathbf{I} + \bar{\mathbf{S}}_r\overline{\mathbf{M}}(\mathbf{L}_r - \bar{\mathbf{L}})]^{-1}.$$

SUGGESTED READINGS

Benveniste, Y. (1987). A New Approach to the Application of Mori–Tanaka's Theory in Composite Materials, *Mech. Mat.*, Vol. 6, pp. 147–157.

Christensen, R. M. (1990). A Critical Evaluation for a Class of Micromechanics Models, *J. Mech. Phys. Solids*, Vol. 38, pp. 379–404.

Christensen, R. M. (1991). *Mechanics of Composite Materials,* Krieger, Malabar, FL.

Hashin, Z. (1988). The Differential Scheme and Its Applications to Cracked Materials, *J. Mech. Phys. Solids,* Vol. 36, pp. 719–734.

Hill, R. (1963). Elastic Properties of Reinforced Solids: Some Theatrical Principles, *J. Mech. Phys. of Solids*, Vol. 11, pp. 357–372.

Hill, R. (1965). A Self-Consistent Mechanics of Composite Materials. *J. Mech. Phys. Solids*, Vol. 13, pp. 213–222.

McLaughlin, R. (1977). A Study of the Differential Scheme for Composite Materials, *Int. J. Engng. Sci.*, Vol. 15, pp. 237–244.

Mori, T. and K. Tanaka. (1973). Average Stress in Matrix and Average Elastic Energy of Materials with Misfitting Inclusions, *Acta Metall.*, Vol. 21, pp. 571–574.

Mura, T. (1987). *Micromechanics of Defects in Solids,* Martinus Nijhoff, Boston, Chapter 7, Section 45.

Nemat-Nasser, S. and M. Hori. (1993). *Micromechanics: Overall Properties of Heterogeneous Materials,* North-Holland, New York.

Norris, A. N. (1985). A Differential Scheme for the Effective Modulus of Composites, *Mech. Mat.*, Vol. 4, pp. 1–16.

Walpole, L. J. (1969). On the Overall Elastic Moduli of Composite Materials, *J. Mech. Phys. Solids*, Vol. 17, pp. 235–251.

Walpole, L. J. (1981). Elastic Behavior of Composite Materials: Theoretical Foundations, *Advances in Applied Mechanics,* Vol. 21, pp. 169–242.

8

DETERMINATION OF EFFECTIVE MODULI— MULTIINCLUSION APPROACHES

The methods discussed in the previous chapter are all based on a single-inclusion idea. In this chapter, we will introduce methods of determining the effective properties based on multiple inclusions. This includes coated inclusions.

8.1 COMPOSITE-SPHERE MODEL

The composite-sphere model was first introduced by Hashin (1962). It mainly applies to particulate-reinforced composites. This model postulates that a composite is made of an assembly of composite spheres of various sizes. Each composite sphere has a core of radius a comprised of the particle material with stiffness tensor $\mathbf{L}_1$ and a shell of thickness $b - a$ comprised of the matrix material with stiffness tensor $\mathbf{L}_0$. Regardless the size of the composite sphere, the ratio of a/b for all spheres in composite is such that the volume fraction of the particle in each composite sphere is that of the particle volume fraction c_1 of the entire composite, that is, $c_1 = (a/b)^3$. The size distribution of the composite spheres must be such that the entire space of the composite is fully occupied by the composite spheres. This requires the presence of composite spheres of infinitely small sizes; see Figure 8.1. Clearly, such a model may not be appropriate for a composite with uniform size particles. However, the microstructure so constructed enables us to find an exact solution for the effective bulk modulus.

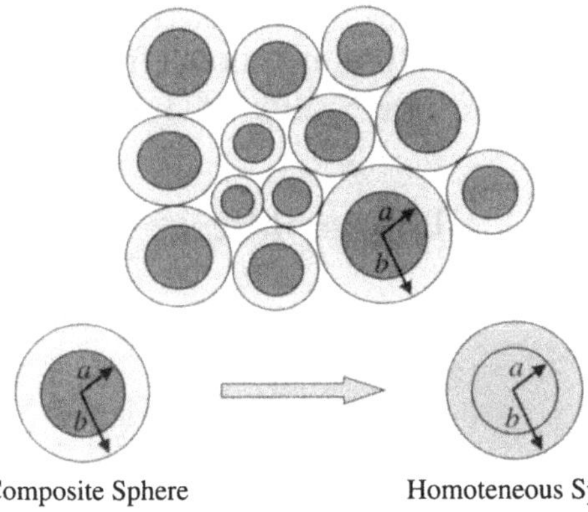

Figure 8.1 Composite-sphere model.

Let us consider first the effective bulk modulus of a typical sphere, for example, the rth composite sphere. This can be easily obtained by solving a spherically symmetric deformation problem; see Problem 8.1:

$$\overline{K}_r = K_0 + \frac{c_1(K_1 - K_0)(3K_0 + 4\mu_0)}{3K_0 + 4\mu_0 + 3(1 - c_1)(K_1 - K_0)}, \tag{8.1.1}$$

where $c_1 = (a/b)^3$ is the volume fraction of the particle material. It is interesting to note that this happens to be the Mori–Tanaka estimate of the effective bulk modulus of a particulate-reinforced composite; see Example 7.2.

Although (8.1.1) is for the rth composite sphere, it also applies to the entire representative volume element of the composite. To prove this, let the entire representative volume element be subjected to a hydrostatic pressure p, that is, the stress everywhere in the representative volume element of the composite is given by

$$\sigma_{ij} = p\delta_{ij}. \tag{8.1.2}$$

The corresponding complimentary energy density can then be written as

$$\Pi_c[\sigma_{ij}] = \frac{1}{2}\sigma_{ij}\varepsilon_{ij} = \frac{p}{2}\varepsilon_{ii} = \frac{p^2}{2\overline{K}}. \tag{8.1.3}$$

where $\overline{K}$ effective bulk modulus of the composite.

On the other hand, we also know that (8.1.2) is also a statically admissible stress field. Therefore, it follows from the minimum complimentary energy theorem (see Section 6.1) that

$$\frac{p^2}{2\overline{K}} = \Pi_c[\sigma_{ij}] \leq \Pi_c[\hat{\sigma}_{ij}] = \frac{p^2}{2\overline{K}_r}, \tag{8.1.4}$$

that is,

$$\frac{1}{\overline{K}} \leq \frac{1}{\overline{K}_r}. \tag{8.1.5}$$

Similarly, if we assume that the representative volume element of the composite is subjected to displacement boundary conditions so that the strain everywhere in the representative volume element is given by $\varepsilon_{ij} = \varepsilon_0\delta_{ij}$, then the use of the minimum potential energy theorem leads to

$$\overline{K} \leq \overline{K}_r. \tag{8.1.6}$$

Combining (8.1.5) and (8.1.6) yields the desired relationship:

$$\overline{K} = \overline{K}_r = K_0 + \frac{c_1(K_1 - K_0)(3K_0 + 4\mu_0)}{3K_0 + 4\mu_0 + (1 - c_1)(K_1 - K_0)}. \tag{8.1.7}$$

This equality can also be understood as follows. If the composite sphere is in a state of pure radial compression (dilatation), it can be substituted with a sphere of equivalent homogeneous material without perturbing the state of stress in its surroundings. Since the substitution can be performed on all composite spheres in the composite, and each sphere has the same $c_1 = (a/b)^3$, the effective bulk modulus of the entire composite should also be given by that of each individual composite sphere.

Unlike the effective bulk modulus, the effective shear modulus is rather difficult to estimate for the composite sphere model. This is due to the fact that a composite sphere subjected to shear deformation does not behave as a homogeneous sphere. Hence, the substitution scheme is not valid.

The composite sphere model was extended by Hashin and Rosen (1964) to the case of infinitely long cylinders imbedded in an isotropic matrix. Later, Hill (1964) and Hashin (1966) used the composite-sphere model to determine the effective modulus of fiber-reinforced composite materials; see Figure 8.2. They were able to find the exact solutions to four of the five elastic moduli (Christensen and Lo, 1979):

Longitudinal Young's modulus E_L:

$$E_L = c_f E_f + c_m E_m + \frac{4 c_f c_m (\upsilon_f - \upsilon_m)^2 \mu_m}{1 + 3c_f \mu_m/(3K_m + \mu_m) + 3c_m \mu_m/(3K_f + \mu_f)}, \quad (8.1.8)$$

Transverse bulk modulus K_T:

$$K_T = K_m + \frac{\mu_m}{3} + \frac{c_f}{3/(3K_f + \mu_f - 3K_m - \mu_m) + 3c_f/(3K_m + 4\mu_m)}, \quad (8.1.9)$$

Longitudinal shear modulus μ_L:

$$\mu_L = \frac{(c_f \mu_f + c_m \mu_m + \mu_f)\mu_m}{c_m \mu_f + c_f \mu_m + \mu_m}, \quad (8.1.10)$$

Longitudinal Poisson's ratio υ_{LT}:

$$\upsilon_{LT} = c_f \upsilon_f + c_m \upsilon_m + \frac{3c_f c_m(\upsilon_f - \upsilon_m)\mu_m[c_f \mu_m/(3K_m + \mu_m) + c_m \mu_m/(3K_f + \mu_f)]}{1 + 3c_f \mu_m/(3K_m + \mu_m) + 3c_m \mu_m/(3K_f + \mu_f)}, \quad (8.1.11)$$

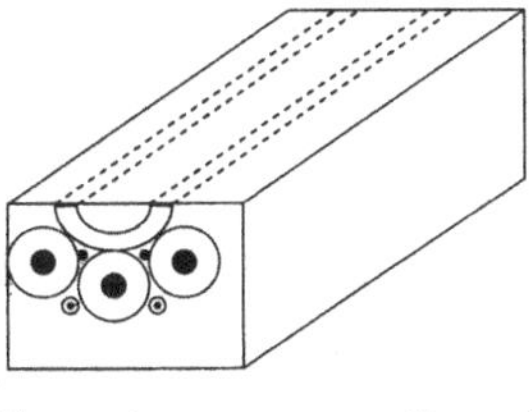

Figure 8.2 Two-phase composite-cylinder model.

where K_f, μ_f, and K_m, μ_m are the bulk and shear moduli of the fiber and the matrix, respectively.

The remaining modulus, the transverse shear modulus, cannot be obtained exactly using the two-phase composite-sphere model. It will be obtained in the next section using the three-phase model.

8.2 THREE-PHASE MODEL

The three-phase model or generalized self-consistent model (Christensen and Lo, 1979) is an extension of the composite-sphere model. As in the composite-sphere model, the three-phase model assumes that each particle of radius a is surrounded by a shell of matrix material with thickness $b - a$, so that the particle volume fraction is given by $c_1 = (a/b)^3$. Next, instead of using such composite spheres to fill in the entire representative volume element of the composite as in the composite-sphere model, the three-phase model assumes that the composite sphere is imbedded in an effective medium whose properties are yet to be determined; see Figure 8.3. This is very similar to the self-consistent method discussed in Sections 7.4 and 7.5; thus the name generalized self-consistent method. The reason it is called generalized is that in the three-phase model, each particle is in contact with or surrounded by the actual matrix material, while in the conventional self-consistent methods discussed in Sections 7.4 and 7.5, the particles are in indirect contact with the effective medium instead of the actual matrix.

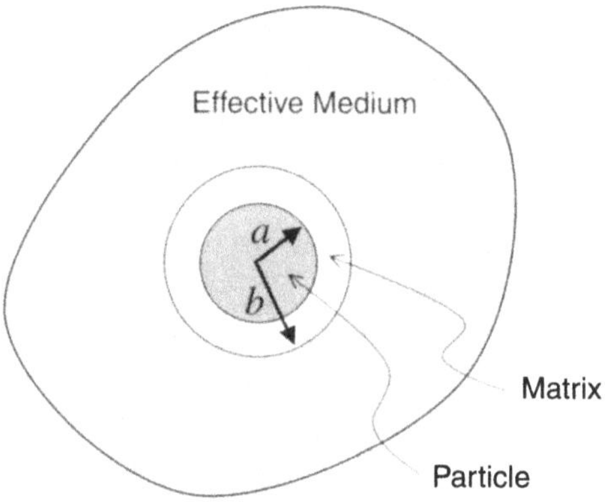

Figure 8.3 Three-phase model.

It turns out that the effective bulk modulus of a three-phase composite sphere is also given by (8.1.1); see Problem 8.2. Following the similar argument used to arrive at (8.1.7), one can also show that the effective bulk modulus of the representative volume element of a composite coincides with the effective bulk modulus of the three-phase composite sphere, that is,

$$\overline{K} = \overline{K}_r = K_0 + \frac{c_1(K_1 - K_0)(3K_0 + 4\mu_0)}{3K_0 + 4\mu_0 + 3(1 - c_1)(K_1 - K_0)}. \tag{8.2.1}$$

Figure 8.4 shows the comparison of effective bulk modulus obtained experimentally, and predicted by the three-phase model and the self-consistent model, for a polymer matrix with voids. The data are normalized by the matrix bulk modulus. The Poisson ratio of the matrix is 0.193. The experimental data are from Walsh et al. (1965). It is seen that the conventional self-consistent method underestimates the experimental results. As discussed in Section 7.7, the self-consistent method yields no physically admissible solution for void volume fraction above 0.5. The three-phase model, on the other hand, provides solutions for the entire range of void volume fraction and the results are in fairly good agreement with the experimental data, even when the volume fraction of voids is large.

Unlike the effective bulk modulus, obtaining the effective shear modulus is rather complicated. It involves solving the shear deformation of

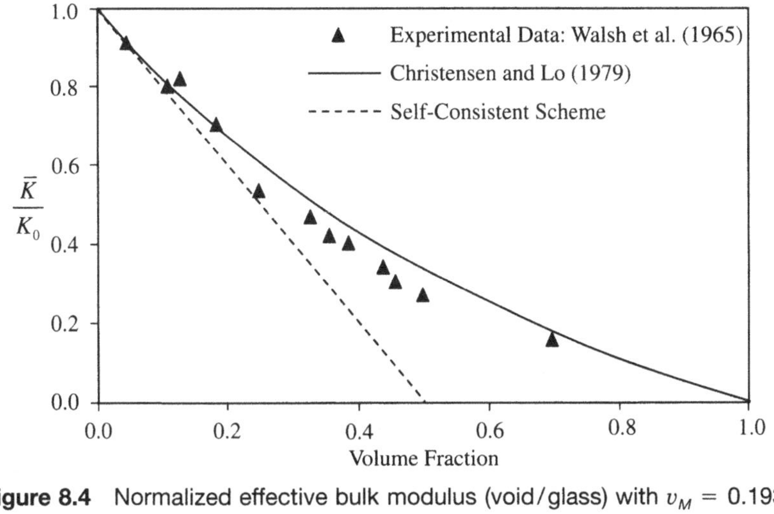

Figure 8.4 Normalized effective bulk modulus (void/glass) with $\upsilon_M = 0.193$.

a three-phase composite sphere. Detailed description of the solution procedure is given in Appendix 8.A. The solution can be written formally as

$$\frac{\bar{\mu}}{\mu_0} = \frac{-B \pm \sqrt{B^2 - AC}}{A}, \tag{8.2.2}$$

where the constants A, B, and C are given in Appendix 8.A.

Presented in Figure 8.5 are the effective modulus estimated by the three-phase model and the Hashin–Shtrikman bounds (Hashin and Shtrikman, 1963, Hashin, 1966; Walpole, 1966). The material parameters used in this comparison are $\mu_1/\mu_0 = 135.14$, $v_1 = 0.20$, and $v_0 = 0.35$. It is seen from Figure 8.5 that the effective shear modulus from the three-phase model is bounded by the Hashin–Shtrikman lower and upper bounds.

The three-phase model has also been applied to fiber-reinforced composite materials (Christensen and Lo, 1979) for obtaining the effective transverse shear modulus. Again, the solution procedure is rather complicated. The solution can be written formally as

$$\frac{\bar{\mu}_T}{\mu_m} = \frac{-B \pm \sqrt{B^2 - AC}}{A}, \tag{8.2.3}$$

where the constants A, B, and C are given in Appendix 8.B.

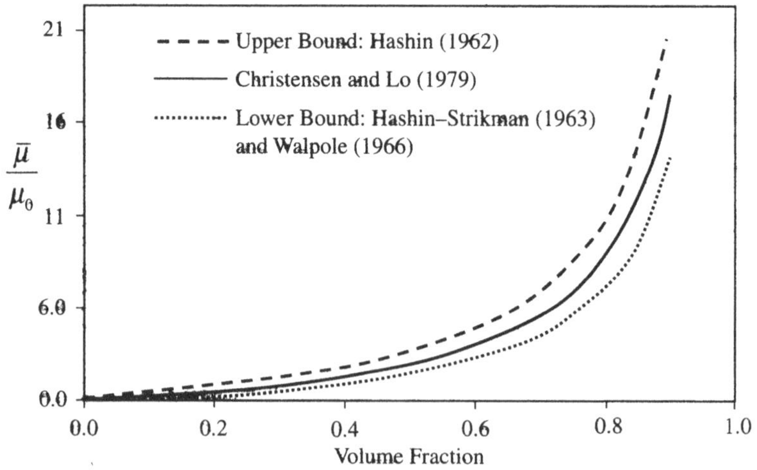

Figure 8.5 Normalized effective shear modulus for spherical model with $\mu_1/\mu_0 = 135.14$, $v_1 = 0.20$, and $v_0 = 0.35$ (boron/epoxy composite).

8.3 FOUR-PHASE MODEL

The four-phase model was developed in order to take into account more complex heterogeneities found in advanced composite materials. In the four-phase model, an interphase is introduced between the inhomogeneity and the surrounding matrix. This enables the modeling of composites with coated fibers or particles. This model is very commonly used in civil engineering applications (e.g., Li et al., 1999; Hashin and Monteiro 2003; Heukamp et al., 2004). Bardella et al. (2002) have also used the four-phase model to estimate the effective behavior of synthetic foam, which are materials composed of a polymer matrix reinforced with hollow glass inclusions and isolated porosities. The four-phase model can be applied in different ways. Some treat the four phases in a single formulation, while others split them into a three-phase model and a composite-sphere model. This later approach was taken by Li et al. (1999) to evaluate the effective elastic properties of an isotropic elastic medium. Their model was divided into two parts; one part corresponds to the three-phase model of Christensen and Lo (1979); the other part corresponds to the two-phase composite-sphere model of Hashin (1962); see Figure 8.6.

Note that the problem shown in Figure 8.6(*b*) is equivalent to the three-phase model. Therefore, the corresponding effective bulk modulus is given by (8.2.1):

$$\overline{K} = K_0 + \frac{(K_e - K_0)(3K_0 + 4\mu_0)(b^3/c^3)}{3K_0 + 4\mu_0 + 3(K_e - K_0)(1 - b^3/c^3)}, \tag{8.3.1}$$

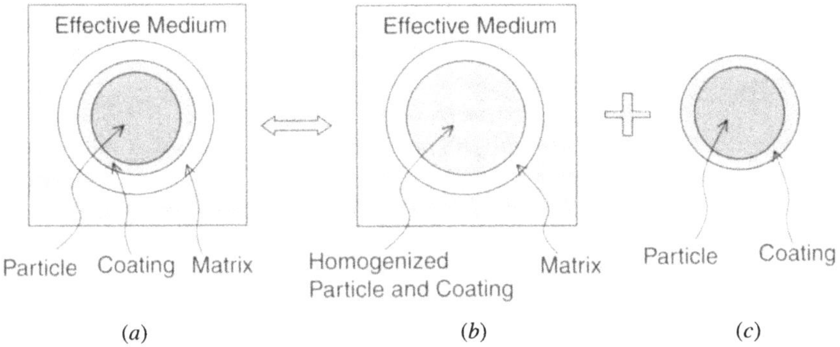

Figure 8.6 Decomposition of four-phase model: (*a*) four, (*b*) three, and (*c*) two phases.

where K_0 and μ_0 are, respectively, the bulk and shear moduli of the matrix, and K_e is the effective bulk modulus of the coated inhomogeneity, which can be obtained by solving the problem depicted in Figure 8.6(*a*):

$$K_e = K_c + \frac{(K_1 - K_c)(3K_c + 4\mu_c)(a^3/b^3)}{3K_c + 4\mu_c + 3(K_1 - K_c)(1 - a^3/b^3)}, \tag{8.3.2}$$

where K_c and μ_c are the bulk and shear moduli of the coating layer, respectively, and K_1 is the bulk modulus of the inhomogeneity. Substitution of (8.3.2) into (8.3.1) yields the effective bulk modulus of the representative volume element of the composite.

It is interesting to note that the effective bulk modulus depends not only on the elastic properties of the inhomogeneity and the matrix but also on the dimensions of the inhomogeneity and its coating material.

8.4 MULTICOATED INCLUSION PROBLEM

The ideas of the two-phase and three-phase models can be extended to develop the multiphase model. In the multiphase model, it is assumed that a spherical inhomogeneity comprised of isotropic material is coated with N layers of different isotropic materials. The interfaces between the inhomogeneity and the coating and between the coatings are all assumed perfect.

To obtain the effective bulk modulus, we subject the multicoated composite to pure dilatational deformation. It follows from the derivation in the previous sections that the effective modulus can be written as Herve and Zaoui (1993):

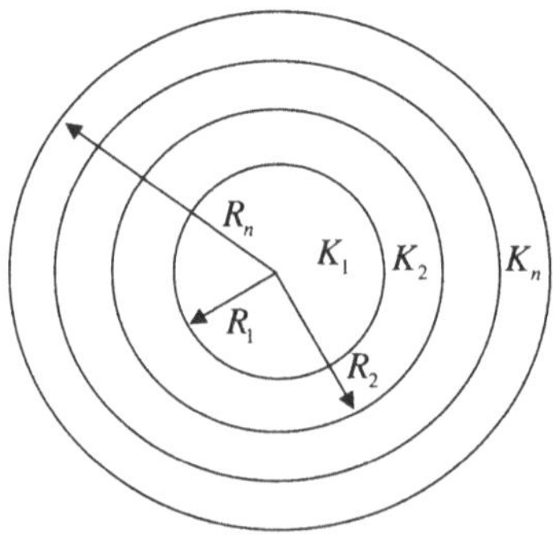

Figure 8.7 Multicoated inclusion schematic.

$$\overline{K}_n = K_n + \frac{(R_{n-1}^3/R_n^3)(\overline{K}_{n-1} - K_n)(3K_n + 4\mu_n)}{3K_n + 4\mu_n + 3(1 - R_{n-1}^3/R_n^3)(\overline{K}_{n-1} - K_n)}, \qquad n \geq 2, \tag{8.4.1}$$

where R_n and R_{n-1} are, respectively, the radii of phase n and $(n - 1)$, K_n and μ_n are, respectively, the bulk and shear moduli of phase n, and $\overline{K}_{n-1}$ is the effective bulk modulus of the $(n - 1)$-coated inclusion. If we define $\overline{K}_1 = K_1$ and $\overline{K}_N = \overline{K}$, (8.4.1) provides a recursive relationship to compute the effective bulk modulus.

For $n = 2$, the effective bulk modulus reduces to that of the two-phase composite sphere model:

$$\overline{K} = K_2 + \frac{c_2(K_1 - K_2)(3K_2 + 4\mu_2)}{3K_2 + 4\mu_2 + 3(1 - c_2)(K_1 - K_2)}, \tag{8.4.2}$$

where $c = R_1^3/R_2^3$ gives the volume fraction of the inhomogeneity, and the subscripts 1 and 2 refer to the inhomogeneity and the coating, respectively.

For $n = 3$, the effective bulk modulus can be expressed as

$$\overline{K} = K_3 + \frac{(R_2^3/R_3^3)(\overline{K}_2 - K_3)(3K_3 + 4\mu_3)}{3K_3 + 4\mu_3 + 3(1 - R_2^3/R_3^3)(\overline{K}_2 - K_3)}, \tag{8.4.3}$$

where

$$\overline{K}_2 = K_2 + \frac{(R_1^3/R_2^3)(K_1 - K_2)(3K_2 + 4\mu_2)}{3K_2 + 4\mu_2 + 3(1 - R_1^3/R_2^3)(K_1 - K_2)}. \tag{8.4.4}$$

Shown in Figure 8.8 are the prediction of the effective bulk modulus by the multicoated inclusion method. The data are normalized by the matrix's bulk modulus, K_3. The parameter $\beta = K_2/K_3$ is the ratio of the bulk moduli of the coating and the matrix. The Poisson ratios of the particle, matrix, and coating are all set to be 0.3 ($\upsilon_1 = \upsilon_2 = \upsilon_3 = 0.3$). The ratio of the bulk modulus of particle and of matrix is set to 6 ($K_1/K_3 = 6$), and the volume fraction of the particle with coating is $c_1 + c_2 = 0.2$.

It is seen that when $\beta > 1$ the interface is more rigid than the matrix. The case $\beta = 6$ corresponds to a simply coated inclusion (Christensen and Lo, 1979). The same remark stands for all values of β when the volume fraction of coating is in the neighborhood 0.2.

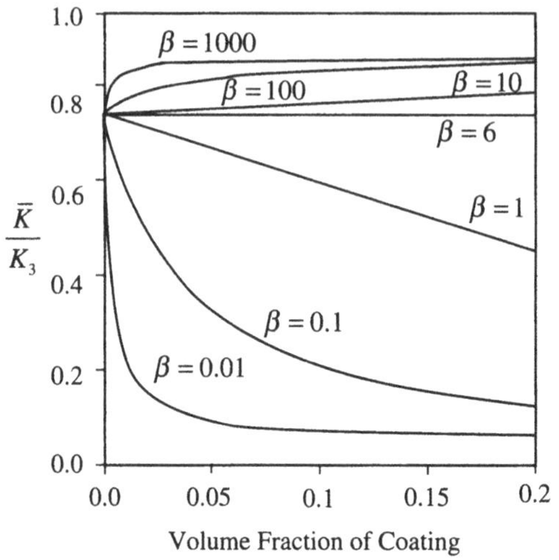

Figure 8.8 Normalized effective bulk modulus $\overline{K}/K_{N-1}$ (Hervé and Zaoui, 1993) of a composite material, made of coated particle (phase 2), for different values $\beta = \mu_1/\mu_M$ (1 denotes the inclusion) with $v_1 = v_2 = v_3 = 0.3$, $\mu_1/\mu_M = 6$, $c_1 + c_2 = 0.2$.

The effective shear modulus predicted by the multiphase model is given by

$$\frac{\overline{\mu}}{\mu_N} = \frac{-B \pm \sqrt{B^2 - 4AC}}{2A}, \tag{8.4.5}$$

where the constants A, B, and C are given in Appendix 8.C.

PROBLEMS

8.1 Consider a composite-sphere consists of a core of radius a and a shell of thickness $b - a$ surrounding the core; see Figure 8.9. Let both the core and shell be made of isotropic elastic material with

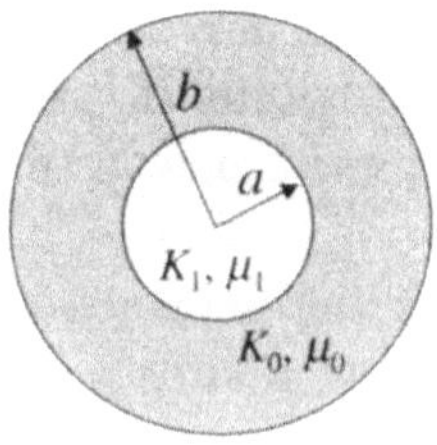

Figure 8.9

moduli K_0, μ_0 and K_1, μ_1, respectively. Please find the effective bulk modulus of the composite sphere.

8.2 Consider a three-phase concentric sphere as shown in Figure 8.10. Please find the effective bulk modulus of the composite sphere.

APPENDIX 8.A

To derive (8.2.2), the elasticity problem depicted in Figure 8.3 needs to be solved under simple shear deformation. This can be carried out by writing the displacement fields in a spherical coordinate system in each of the three regions: the inhomogeneity ($r \leq a$), the matrix ($a \leq r \leq b$), and the effective medium ($b \leq r$). The following equation can then be derived after making use of the equilibrium equations and the continuity conditions at the interfaces between the different regions:

$$A\left(\frac{\bar{\mu}}{\mu_0}\right)^2 + 2B\left(\frac{\bar{\mu}}{\mu_0}\right) + C = 0, \tag{8.A.1}$$

where

$$A = 8\left(\frac{\mu_1}{\mu_0} - 1\right)(4 - 5\upsilon_0)\eta_1 c^{10/3} - 2\left[63\left(\frac{\mu_1}{\mu_0} - 1\right)\eta_2 + 2\eta_1\eta_3\right]c^{7/3}$$
$$+ 252\left(\frac{\mu_1}{\mu_0} - 1\right)\eta_2 c^{5/3} - 50\left(\frac{\mu_1}{\mu_0} - 1\right)(7 - 12\upsilon_0 + 8\upsilon_0^2)\eta_2 c$$
$$+ 4(7 - 10\upsilon_0)\eta_2\eta_3,$$

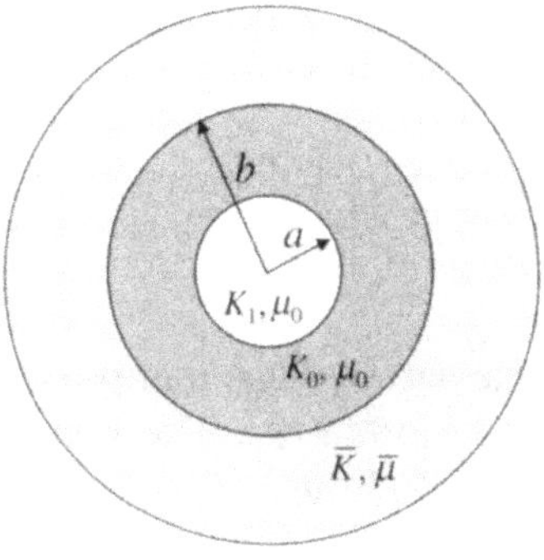

Figure 8.10

$$
\begin{aligned}
B = &-2\left(\frac{\mu_1}{\mu_0} - 1\right)(4 - 5v_0)\eta_1 c^{10/3} \\
&+ 2\left[63\left(\frac{\mu_1}{\mu_0} - 1\right)\eta_2 + 2\eta_1\eta_3\right]c^{7/3} - 252\left(\frac{\mu_1}{\mu_0} - 1\right)\eta_2 c^{5/3} \\
&+ 75\left(\frac{\mu_1}{\mu_0} - 1\right)(3 - v_0)v_0\eta_2 c + 3(15v_0 - 7)\eta_2\eta_3,
\end{aligned}
$$

$$
\begin{aligned}
C = &\,4\left(\frac{\mu_1}{\mu_0} - 1\right)(5v_0 - 7)\eta_1 c^{10/3} - 2\left[63\left(\frac{\mu_1}{\mu_0} - 1\right)\eta_2 + 2\eta_1\eta_3\right]c^{7/3} \\
&+ 252\left(\frac{\mu_1}{\mu_0} - 1\right)\eta_2 c^{5/3} + 25\left(\frac{\mu_1}{\mu_0} - 1\right)(v_0^2 - 7)\eta_2 c \\
&- (7 + 5v_0)\eta_2\eta_3,
\end{aligned}
$$

with

$$
\eta_1 = \left(\frac{\mu_1}{\mu_0} - 1\right)(49 - 50v_1v_0) + 35\left(\frac{\mu_1}{\mu_0}\right)(v_1 - 2v_0) + 35(2v_1 - v_0),
$$

$$
\eta_2 = 5v_1\left(\frac{\mu_1}{\mu_0} - 8\right) + 7\left(\frac{\mu_1}{\mu_0} + 4\right),
$$

$$
\eta_3 = \left(\frac{\mu_1}{\mu_0}\right)(8 - 10v_0) + (7 - 5v_0),
$$

where c is the volume fraction of the particles, that is, $c = (a/b)^3$.

APPENDIX 8.B

Detailed derivation of the three-phase composite cylinder model can be found in Christensen and Lo (1979). The results are list below for the effective transverse shear modulus.

$$A\left(\frac{\overline{\mu}_T}{\mu_m}\right)^2 + 2B\left(\frac{\overline{\mu}_T}{\mu_m}\right) + C = 0, \tag{8.B.1}$$

where

$$\begin{aligned} A = {} & 3c(1 - c)^2\left(\frac{\mu_f}{\mu_m} - 1\right)\left(\frac{\mu_f}{\mu_m} + \eta_f\right) \\ & + \left[\frac{\mu_f}{\mu_m}\,\eta_m + \eta_f\eta_m - \left(\frac{\mu_f}{\mu_m}\,\eta_m - \eta_f\right)c^3\right] \\ & \times \left[c\eta_m\left(\frac{\mu_f}{\mu_m} - 1\right) - \left(\frac{\mu_f}{\mu_m}\,\eta_m + 1\right)\right], \end{aligned} \tag{8.B.2}$$

$$\begin{aligned} B = {} & -3c(1 - c)^2\left(\frac{\mu_f}{\mu_m} - 1\right)\left(\frac{\mu_f}{\mu_m} + \eta_f\right) \\ & + \frac{1}{2}\left[\frac{\mu_f}{\mu_m}\,\eta_m + 1 + \left(\frac{\mu_f}{\mu_m} - 1\right)c\right] \\ & \times \left[(\eta_m - 1)\left(\frac{\mu_f}{\mu_m} + \eta_f\right) - 2\left(\frac{\mu_f}{\mu_m}\,\eta_m - \eta_f\right)c^3\right] \\ & + \frac{c}{2}\,(\eta_m - 1)\left(\frac{\mu_f}{\mu_m} - 1\right)\left[\frac{\mu_f}{\mu_m}\,\eta_m + \eta_f + \left(\frac{\mu_f}{\mu_m} - 1\right)c^3\right], \end{aligned} \tag{8.B.3}$$

$$\begin{aligned} C = {} & 3c(1 - c)^2\left(\frac{\mu_f}{\mu_m} - 1\right)\left(\frac{\mu_f}{\mu_m} + \eta_f\right) \\ & + \left[\frac{\mu_f}{\mu_m} + \eta_f - \left(\frac{\mu_f}{\mu_m}\,\eta_m - \eta_f\right)c^3\right] \\ & \times \left[\frac{\mu_f}{\mu_m}\,\eta_m + 1 + \left(\frac{\mu_f}{\mu_m} - 1\right)c\right], \end{aligned} \tag{8.B.4}$$

and

$$c = \left(\frac{a}{b}\right)^3, \qquad \eta_m = 3 - 4v_m, \qquad \eta_f = 3 - 4v_f. \qquad (8.B.5)$$

APPENDIX 8.C

The constants in (8.4.5) are given by

$$\begin{aligned} A = {} & 4R_n^{10}(1 - 2v_n)(7 - 10v_n)Z_{12} + 20R_n^7(7 - 12v_n + 8v_n^2)Z_{42} \\ & + 12R_n^5(1 - 2v_n)(Z_{14} - 7Z_{23}) + 20R_n^3(1 - 2v_n)^2 Z_{13} \\ & + 16(4 - 5v_n)(1 - 2v_n)Z_{43}, \end{aligned}$$

$$\begin{aligned} B = {} & 4R_n^{10}(1 - 2v_n)(15v_n - 7)Z_{12} + 60R_n^7(v_n - 3)v_n Z_{42} \\ & - 24R_n^5(1 - 2v_n)(Z_{14} - 7Z_{23}) - 40R_n^3(1 - 2v_n)^2 Z_{13} \\ & - 8(1 - 5v_n)(1 - 2v_n)Z_{43}, \end{aligned}$$

$$\begin{aligned} C = {} & -R_n^{10}(1 - 2v_n)(7 + 5v_n)Z_{12} + 10R_n^7(7 - v_n^2)Z_{42} \\ & + 12R_n^5(1 - 2v_n)(Z_{14} - 7Z_{23}) + 20R_n^3(1 - 2v_n)^2 Z_{13} \\ & - 8(1 - 5v_n)(1 - 2v_n)Z_{43}, \end{aligned}$$

where

$$Z_{\alpha\beta} = P_{\alpha 1}^{(n-1)} P_{\beta 2}^{(n-1)} - P_{\beta 1}^{(n-1)} P_{\alpha 2}^{(n-1)} \quad \text{with} \quad \alpha \in [1,4] \quad \text{and} \quad \beta \in [1,4],$$

$$P^{(n)} = \prod_{j=1}^{n} M^{(j)}.$$

$$M^{(k)} = \frac{1}{5(1 - v_{k+1})} \left[\begin{matrix} \dfrac{c_k}{3} & \dfrac{R_k^2(3b_k - 7c_k)}{5(1 - 2v_k)} \\ 0 & \dfrac{(1 - 2v_{k+1})b_k}{7(1 - 2v_k)} \\ \dfrac{R_k^5 \alpha_k}{2} & \dfrac{-R_k^7(2\alpha_k - 147\alpha_k)}{70(1 - 2v_k)} \\ \dfrac{-5}{6}(1 - 2v_{k+1})\alpha_k R_k^3 & \dfrac{7(1 - 2v_{k+1})\alpha_k R_k^5}{2(1 - 2v_k)} \end{matrix} \right.$$

$$\begin{bmatrix} \frac{-12\alpha_k}{R_k^5} & & \frac{4(f_k - 27\alpha_k)}{15(1-2v_k)R_k^3} \\ \frac{-20(1-2v_{k+1})\alpha_k}{7R_k^7} & & \frac{-12\alpha_k(1-2v_{k+1})}{7(1-2v)R} \\ \frac{d_k}{7} & \cdots & \frac{R_k^2(105(1-v_{k+1}) + 12\alpha_k(7-10v_k) - 7e_k)}{35(1-2v_k)} \\ & & \frac{e_k(1-2v_k)}{3(1-2v_k)} \end{bmatrix},$$

with

$$\alpha_k = \left(\frac{\mu_k}{\mu_{k+1}}\right)(7 + 5v_k)(7 - 10v_{k+1}) - (7 - 10v_k)(7 + 5v_{k+1}),$$

$$b_k = 4(7 - 10v_k) + \left(\frac{\mu_k}{\mu_{k+1}}\right)(7 + 5v_k),$$

$$c_k = (7 - 5v_{k+1}) + 2\left(\frac{\mu_k}{\mu_{k+1}}\right)(4 - 5v_{k+1}),$$

$$d_k = (7 + 5v_{k+1}) + 4\left(\frac{\mu_k}{\mu_{k+1}}\right)(7 - 10v_{k+1}),$$

$$e_k = 2(4 - 5v_k) + \left(\frac{\mu_k}{\mu_{k+1}}\right)(7 - 5v_k),$$

$$f_k = (4 - 5v_k)(7 - 5v_{k+1}) - \left(\frac{\mu_k}{\mu_{k+1}}\right)(4 - 5v_{k+1})(7 - 5v_k),$$

$$\alpha_k = \frac{\mu_k}{\mu_{k+1}} - 1.$$

REFERENCES

Christensen, R. M. and K. H. Lo. (1979). Solutions for Effective Shear Properties in Three Phase Sphere and Cylinder Models, *J. Mech. Phys. Solids*, Vol. 27, pp. 315–330.

Hashin, Z. (1962). The Elastic Moduli of Heterogeneous Materials, *J. Appl. Mech.,* Vol. 29, pp. 143–150.

Hashin, Z. (1966). Viscoelastic Fiber Reinforced Materials, *AIAA J.*, Vol. 4, pp. 1411–1417.

Hashin, Z. and B. W. Rosen. (1964). The Elastic Moduli of Reinforced-Reinforced Materials, *J. Appl. Mech.*, Vol. 31, pp. 223–228.

Hashin, Z. and S. Shtrikman. (1963). A Variational Approach to the Theory of the Elastic Behavior of Multiphase Materials, *J. Mech. Phys. Solids,* Vol. 11, pp. 127–140.

Hervé, E. and A. Zaoui. (1993). N-Layered Inclusion-Based Micromechanical Modeling, *Int. J. Eng. Sci.,* Vol. 31, pp. 1–10.

Hill, R. (1964). Theory of Mechanical Properties of Fibre-Strengthened Materials—I: Elastic Behavior," *J. Mech. Phys. Solids,* Vol. 12, pp. 199–212.

Li, G., Y. Zhao, and S. S. Pang. (1999). Four-Phase Sphere Modeling of Effective Bulk Modulus of Concrete, *Cement and Concrete Research,* Vol. 29, pp. 839–845.

Walpole, L. J. (1966). On the Bounds for Overall Elastic Moduli of Inhomogeneous Systems—I, *J. Mech. Phys. Solids,* Vol. 14, pp. 151–162.

Walpole, L. J. (1969). On the Overall Elastic Moduli of Composite Materials, *J. Mech. Phys. Solids,* Vol. 17, pp. 235–251.

Walsh, J. B., W. F. Brace, and A. W. England. (1965). The Effect of Porosity on Compressibility of Glass, *J. Am. Ceram. Soc.,* Vol. 48, pp. 605–608.

SUGGESTED READINGS

Bardella, L. and F. Genna. (2001). On the Elastic Behavior of Syntactic Foams, *Int. J. Solids Structures,* Vol. 38, pp. 7235–7260.

Benveniste, Y. (1987). A New Approach to the Application of Mori–Tanaka's Theory in Composite Materials, *Mech. Mater.,* Vol. 6, pp. 147–157.

Benveniste, Y., G. J. Dvorak, and T. Chen. (1989). Stress Field in Composite with Coated Inclusions, *Mech. Mater.,* Vol. 7, pp. 305–317.

Budiansky, B. and T. Wu. (1962). Theoretical Prediction of Plastic Strains in Polycrystals, In *Proc. 4th U.S. Nat. Cong. Theo. Appl. Mech.,* pp. 1175–1183.

Cherkaoui, M., H. Sabar, and M. Berveiller. (1994). Micromechanical Approach of the Coated Inclusion Problem and Applications to Composite Materials, *J. Eng. Mater. Tech.,* Vol. 116, pp. 274–278.

Cherkaoui, M., H. Sabar, and M. Berveiller. (1995). Elastic Composites with Coated Reinforcements: A Micromechanical Approach for Nonhomothetic Topology, *Int. J. Eng. Sci.,* Vol. 33, pp. 829–843.

Christensen, R. M. (1991). *Mechanics of Composite Materials,* Krieger, Malabar, FL.

El Mouden, M., M. Cherkaoui, A. Molinari, and M. Berveiller. (1998). The Overall Elastic Response of Materials Containing Coated Inclusions in a Periodic Array, *Int. J. Engng. Sci.,* Vol. 36, pp. 813–829.

Hashin, Z. and P. J. M. Monteiro. (2003). An Inverse Method to Determine the Elastic Properties of the Interphase between the Aggregate and the Cement Paste, *Cem. Conc. Res.,* Vol. 32, pp. 1291–1300.

Heukamp, F. H., E. Lemarchand, and F. J. Ulm. (2005). The Effect of Interfacial Properties on the Cohesion of Highly Filled Composite Materials, *Int. J, Solids Structures,* Vol. 42, 287–305.

Hill, R. (1963). Elastic Properties of Reinforced Solids: Some Theoretical Principles, *J. Mech. Phys. of Solids,* Vol. 11, pp. 357–372.

Hill, R. (1995). A Self-Consistent Mechanics of Composite Materials, *J. Mech. Phys. Solids,* Vol. 13, pp. 213–222.

Hori, M. and S. Nemat-Nasser. (1993). Double-Inclusion Model and Overall Moduli of Multi-Phase Composites, *Mech. Mater.,* Vol. 14, pp. 189–206.

Jasiuk, I. and M. W. Kouider. (1993). The Effect of an Inhomogeneous Interphase on the Elastic Constants of Transversely Isotropic Composites, *Mech. Mater.,* Vol. 15, pp. 53–63.

Mura, T. (1987). *Micromechanics of Defects in Solids,* Martinus Nijhoff, Boston.

Nemat-Nasser, S. and M. Hori. (1993). *Micromechanics: Overall Properties of Heterogeneous Materials,* North-Holland, New York.

Norris, A. N. (1985). A Differential Scheme for the Effective Modulus of Composites, *Mech. Mater.,* Vol. 4, pp. 1–16.

Qiu, Y. P. and G. J. Weng. (1991). Elastic Moduli of Thickly Coated Particle and Fiber-Reinforced Composites, *J. Appl. Mech.,* Vol. 58, pp. 388–398.

9

EFFECTIVE PROPERTIES OF FIBER-REINFORCED COMPOSITE LAMINATES

A fiber-reinforced composite laminate is a special kind of heterogeneous material. It typically consists of multiple layers of plies, each ply is thin sheet of matrix material reinforced by unidirectional fibers. Glass fiber- and graphite fiber-reinforced epoxy matrix composites are the most commonly used composite laminates. In this chapter, we will apply some of the theories developed in previous chapters to determine the effective properties for fiber-reinforced composite laminates.

9.1 UNIDIRECTIONAL FIBER-REINFORCED COMPOSITES

Consider a matrix material reinforced by circular cross-section fibers all aligned in the x_1 direction, as schematically shown in Figure 9.1. Let the volume fraction of the fiber be c_f and volume fraction of the matrix be c_m. Obviously, we have

$$c_m + c_f = 1. \tag{9.1.1}$$

We further assume that both the fiber and the matrix are isotropic, and their elastic properties are defined by the Young's modulus E_f, E_m and Poisson's ratio ν_f, ν_m, respectively.

Because of the cylindrical nature of the composite, the effective elastic properties are transversely isotropic. In the Voigt notation, this can be written as

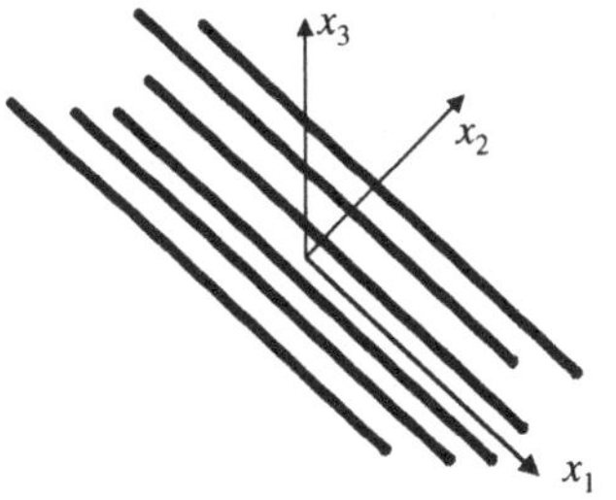

Figure 9.1 Unidirectional fiber-reinforced composite.

$$\begin{bmatrix} \sigma_{11} \\ \sigma_{22} \\ \sigma_{33} \\ \sigma_{23} \\ \sigma_{13} \\ \sigma_{12} \end{bmatrix} = \begin{bmatrix} C_{11} & C_{12} & C_{12} & 0 & 0 & 0 \\ C_{12} & C_{22} & C_{23} & 0 & 0 & 0 \\ C_{12} & C_{23} & C_{22} & 0 & 0 & 0 \\ 0 & 0 & 0 & \dfrac{C_{22} - C_{23}}{2} & 0 & 0 \\ 0 & 0 & 0 & 0 & C_{66} & 0 \\ 0 & 0 & 0 & 0 & 0 & C_{66} \end{bmatrix} \begin{bmatrix} \varepsilon_{11} \\ \varepsilon_{22} \\ \varepsilon_{33} \\ 2\varepsilon_{23} \\ 2\varepsilon_{13} \\ 2\varepsilon_{12} \end{bmatrix} \tag{9.1.2}$$

where C_{ij} are the effective Voigt elastic constants of the unidirectional fiber-reinforced composite. For convenience and without causing any confusion, we have omitted the overbar for the effective Voigt elastic constants, which are related to the effective elasticity tensor $\overline{L}_{ijkl}$ through the relationship described by Table 2.1.

Equation (9.1.2) is written in a coordinate system where the x_1 axis is in the direction of the fibers. In the coordinate system $\hat{x}_i$ shown in Figure 9.2, where the $\hat{x}_1$ axis forms an angle θ from the fiber direction, the expressions for the components $\hat{C}_{ij}$ of the Voigt elastic matrix are different. They are related to C_{ij} through a coordinate rotation; see (2.4.13):

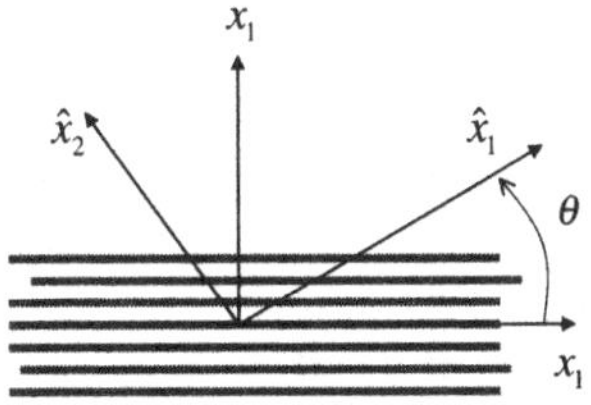

Figure 9.2 Fibers not aligned with coordinate axis.

$$
\begin{aligned}
\hat{C}_{11} &= Q_{11}m^4 + 2(C_{12} + 2C_{66})m^2n^2 + C_{22}n^4, \\
\hat{C}_{22} &= C_{11}n^4 + 2(C_{12} + 2C_{66})m^2n^2 + C_{22}m^4, \\
\hat{C}_{33} &= C_{33}, \\
\hat{C}_{12} &= (C_{11} + C_{22} - 4C_{66})m^2n^2 + C_{12}(m^4 + n^4), \\
\hat{C}_{13} &= C_{13}m^3 + C_{23}n^2, \\
\hat{C}_{23} &= C_{13}n^3 + C_{23}m^2, \\
\hat{C}_{36} &= (C_{23} - C_{13})mn, \qquad\qquad (9.1.3) \\
\hat{C}_{44} &= C_{44}m^3 + C_{55}n^2, \\
\hat{C}_{55} &= C_{55}m^3 + C_{44}n^2, \\
\hat{C}_{66} &= (C_{11} + C_{22} - 2C_{12})m^2n^2 + 2C_{22}mn(m^2 - n^2) + C_{66}(m^2 - n^2)^2, \\
\hat{C}_{45} &= (C_{44} - C_{55})mn, \\
\hat{C}_{16} &= (C_{22} - C_{12} - 2C_{66})mn^3 - (C_{11} - C_{12} - 2C_{66})m^3n, \\
\hat{C}_{26} &= (C_{22} - C_{12} - 2C_{66})m^3n - (C_{11} - C_{12} - 2C_{66})mn^3,
\end{aligned}
$$

in which $m = \cos\theta$, $n = \sin\theta$. Therefore, in a coordinate system not aligned with the fiber, the Voigt elasticity matrix has the following form:

$$
\hat{\mathbf{C}} = \begin{bmatrix} \hat{C}_{11} & \hat{C}_{12} & \hat{C}_{13} & 0 & 0 & \hat{C}_{16} \\ \hat{C}_{12} & \hat{C}_{22} & \hat{C}_{23} & 0 & 0 & \hat{C}_{26} \\ \hat{C}_{13} & \hat{C}_{23} & \hat{C}_{33} & 0 & 0 & \hat{C}_{36} \\ 0 & 0 & 0 & \hat{C}_{44} & \hat{C}_{45} & 0 \\ 0 & 0 & 0 & \hat{C}_{45} & \hat{C}_{55} & 0 \\ \hat{C}_{16} & \hat{C}_{26} & \hat{C}_{36} & 0 & 0 & \hat{C}_{66} \end{bmatrix}. \qquad (9.1.4)
$$

Now, let us go back to (9.1.2). Taking the inverse of the elasticity tensor yields

$$
\begin{bmatrix} \varepsilon_{11} \\ \varepsilon_{22} \\ \varepsilon_{33} \\ 2\varepsilon_{23} \\ 2\varepsilon_{13} \\ 2\varepsilon_{12} \end{bmatrix} = \begin{bmatrix} S_{11} & S_{12} & S_{12} & 0 & 0 & 0 \\ S_{12} & S_{22} & S_{23} & 0 & 0 & 0 \\ S_{12} & S_{23} & S_{22} & 0 & 0 & 0 \\ 0 & 0 & 0 & S_{44} & 0 & 0 \\ 0 & 0 & 0 & 0 & S_{66} & 0 \\ 0 & 0 & 0 & 0 & 0 & S_{66} \end{bmatrix} \begin{bmatrix} \sigma_{11} \\ \sigma_{22} \\ \sigma_{33} \\ \sigma_{23} \\ \sigma_{13} \\ \sigma_{12} \end{bmatrix}, \qquad (9.1.5)
$$

where

$$S_{11} = \frac{C_{22} + C_{23}}{-2C_{12}^2 + C_{11}(C_{22} + C_{23})}, \tag{9.1.6}$$

$$S_{12} = \frac{C_{12}}{2C_{12}^2 - C_{11}(C_{22} + C_{23})}, \tag{9.1.7}$$

$$S_{22} = \frac{C_{12}^2 - C_{11}C_{22}}{(C_{22} - C_{23})[2C_{12}^2 - C_{11}(C_{22} + C_{23})]}, \tag{9.1.8}$$

$$S_{23} = \frac{C_{12}^2 - C_{11}C_{23}}{(C_{22} - C_{23})[-2C_{12}^2 + C_{11}(C_{22} + C_{23})]}, \tag{9.1.9}$$

$$S_{44} = 2(S_{22} - S_{23}) = \frac{2}{C_{22} - C_{23}}, \qquad S_{66} = \frac{1}{C_{66}} \tag{9.1.10}$$

are the five elastic compliance constants. They can be related to the engineering constants easily as follows.

Under a uniaxial loading in the x_1 direction, we have

$$\varepsilon_{11} = S_{11}\sigma_{11}, \qquad \varepsilon_{22} = \varepsilon_{33} = S_{12}\sigma_{11}. \tag{9.1.11}$$

The first of (9.1.11) defines the longitudinal Young modulus in the fiber direction:

$$E_L = \frac{\sigma_{11}}{\varepsilon_{11}} = \frac{1}{S_{11}} = C_{11} - \frac{2C_{12}^2}{C_{22} + C_{23}}. \tag{9.1.12}$$

The contraction in the transverse direction that accompanies the uniaxial stress in the fiber direction defines the corresponding Poisson ratio through

$$\nu_{LT} = -\frac{\varepsilon_{22}}{\varepsilon_{11}} = -\frac{\varepsilon_{33}}{\varepsilon_{11}} = -\frac{S_{12}}{S_{11}} = \frac{C_{12}}{C_{22} + C_{23}}. \tag{9.1.13}$$

Similarly, when the uniaxial stress is in the x_2 direction (transverse to the fiber), we have

$$\varepsilon_{11} = S_{12}\sigma_{22}, \qquad \varepsilon_{22} = S_{22}\sigma_{22}, \qquad \varepsilon_{33} = S_{23}\sigma_{22}. \tag{9.1.14}$$

The second of (9.1.14) defines the Young modulus in the transverse direction:

$$E_T = \frac{\sigma_{22}}{\varepsilon_{22}} = \frac{1}{S_{22}} = C_{22} - \frac{C_{12}^2(C_{22} - 2C_{23}) + C_{11}C_{23}^2}{C_{11}C_{22} - C_{12}^2}. \tag{9.1.15}$$

In this case, the contraction in the fiber direction that accompanies the uniaxial stress in the transverse direction defines the corresponding Poisson ratio through

$$\nu_{TL} = -\frac{\varepsilon_{11}}{\varepsilon_{22}} = -\frac{S_{12}}{S_{22}} = \frac{C_{12}(C_{22} - C_{23})}{C_{11}C_{22} - C_{12}^2}. \tag{9.1.16}$$

Analogously, the contraction in the other transverse direction that accompanies the uniaxial stress in one transverse direction defines another Poisson ratio,

$$\nu_{TT} = -\frac{\varepsilon_{33}}{\varepsilon_{22}} = -\frac{S_{23}}{S_{22}} = \frac{C_{11}C_{23} - C_{12}^2}{C_{11}C_{22} - C_{12}^2}. \tag{9.1.17}$$

As a convention in labeling the Poisson ratios in this text, the first subscript indicates the direction of the uniaxial stress, and the second subscript indicates the direction normal to the loading direction in which the contraction takes place. It is clear from the above or simply from the material asymmetry that $\nu_{TL} \neq \nu_{LT}$. However, it can be easily verified by direction substitution of their definitions that

$$\frac{\nu_{LT}}{E_L} = \frac{\nu_{TL}}{E_T}. \tag{9.1.18}$$

If a simple shear $\sigma_{12} = \sigma_{21}$ is applied, we will have $2\varepsilon_{12} = S_{66}\sigma_{12}$. Thus the longitudinal shear modulus is defined as

$$\mu_L = \frac{\sigma_{12}}{2\varepsilon_{12}} = \frac{1}{S_{66}} = C_{66}. \tag{9.1.19}$$

Similarly, when the material is subjected to the simple shear σ_{23}, the corresponding shear strain is $2\varepsilon_{23} = S_{44}\sigma_{23}$. Thus, the corresponding shear modulus is defined by

$$\mu_T = \frac{\sigma_{23}}{2\varepsilon_{23}} = \frac{1}{S_{44}} = \frac{1}{2}(C_{22} - C_{23}). \tag{9.1.20}$$

It is clear from the above definitions that μ_L is the shear modulus when the fibers are shifted relative to each other in the longitudinal direction, while μ_T is the shear modulus when the fibers are shifted relative to each other in the transverse directions.

Finally, a plane strain bulk modulus can be introduced by applying biaxial stresses $\sigma_{22} = \sigma_{33} = \sigma$. Since the elastic properties are isotropic in the x_2x_3 plane, the corresponding strain field is given by $\varepsilon_{22} = \varepsilon_{33} = \varepsilon$. We thus have

$$\sigma = 2K_T\varepsilon, \tag{9.1.21}$$

where K_T is the plane strain bulk modulus within the isotropic plane given by

$$K_T = \tfrac{1}{2}(C_{22} + C_{23}). \tag{9.1.22}$$

So far, we have introduced two Young moduli, two shear moduli, one bulk modulus, and three Poisson ratios. Not all the eight engineering constants are independent. Since they are defined based on the five stiffness constants C_{ij} as shown in (9.1.2), it can be easily concluded that there are only five independent engineering constants. In addition to (9.1.18), the following relationships among these engineering constants can also be verified by direct substitution:

$$E_T = \frac{4\mu_T K_T}{K_T + \mu_T + 4\nu_{LT}^2\mu_T K_T / E_L}, \tag{9.1.23}$$

$$\nu_{TT} = \frac{K_T - \mu_T - 4\nu_{LT}^2\mu_T K_T / E_L}{K_T + \mu_T + 4\nu_{LT}^2\mu_T K_T / E_L}, \tag{9.1.24}$$

$$\nu_{LT}^2 = \left(-\nu_{TT} - \frac{1}{4}\frac{E_T}{K_T} + \frac{1}{4}\frac{E_T}{\mu_T}\right)\frac{E_L}{E_T}. \tag{9.1.25}$$

Making use of the definitions of the engineering elastic constants, the Voigt elasticity matrix for transversely isotropic solids with the x_1 direction being the axis of symmetry can be rewritten as

$$\mathbf{C} = \begin{bmatrix} E_L + 4\nu_{LT}^2 K_T & 2\nu_{LT}K_T & 2\nu_{LT}K_T & 0 & 0 & 0 \\ 2\nu_{LT}K_T & K_T + \mu_T & K_T - \mu_T & 0 & 0 & 0 \\ 2\nu_{LT}K_T & K_T + \mu_T & K_T + \mu_T & 0 & 0 & 0 \\ 0 & 0 & 0 & \mu_T & 0 & 0 \\ 0 & 0 & 0 & 0 & \mu_L & 0 \\ 0 & 0 & 0 & 0 & 0 & \mu_L \end{bmatrix}. \tag{9.1.26}$$

It is interesting to investigate the bounds on the various Poisson ratios. For example, If a material has a very large transverse shear modulus, that is, $\mu_T \rightarrow \infty$, then it follows from (9.1.24) that $\nu_{TT} \rightarrow -1$. On the other hand, if a material has very large plane strain bulk modulus and very large longitudinal Young's modulus, that is, $K_T \rightarrow \infty$ and $E_L \rightarrow \infty$, then it follows from (9.1.24) that $\nu_{TT} \rightarrow 1$. Therefore, we can conclude that

$$-1 < \nu_{TT} < 1. \tag{9.1.27}$$

Next, consider (9.1.23). Solving for ν_{LT}^2 gives

$$\nu_{LT}^2 = \frac{E_L}{E_T} - \frac{E_L}{4}\left(\frac{1}{K_T} + \frac{2}{\mu_T}\right). \tag{9.1.28}$$

Since the second term on the right-hand side of (9.1.28) is positive, we arrive at

$$|\nu_{LT}| \leq \sqrt{\frac{E_L}{E_T}}. \tag{9.1.29}$$

Making use of (9.1.18) also yields

$$|\nu_{TL}| \leq \sqrt{\frac{E_T}{E_L}}. \tag{9.1.30}$$

Let us now consider how the effective properties C_{ij} are related to the elastic properties of the fiber and the matrix. Several methods can be used for this purpose. We first consider a very simple approach.

Recall that in Chapter 6, we discussed the Reuss lower and Voigt upper bounds for the effective modulus of heterogeneous materials. In what follows, we will apply these bounds to unidirectional fiber-reinforced composite materials. Assume that both the fibers and the matrix are isotropic and linearly elastic materials. It then follows from (6.2.10) that the Voigt upper bound of the elasticity tensor is given by

$$\overline{\mathbf{L}}^V = c_m \mathbf{L}_m + c_f \mathbf{L}_f, \tag{9.1.31}$$

where $\mathbf{L}_f$ and $\mathbf{L}_m$ are the elasticity tensors of the fiber and the matrix, respectively. In terms of the Voigt notation, (9.1.31) implies that the nonzero components of the elasticity tensor can be written as

$$\overline{C}_{11}^V = \overline{C}_{22}^V = \overline{C}_{33}^V = \frac{c_f E_f (1 - \nu_f)}{(1 + \nu_f)(1 - 2\nu_f)} + \frac{c_m E_m (1 - \nu_m)}{(1 + \nu_m)(1 - 2\nu_m)}, \tag{9.1.32}$$

$$\overline{C}_{12}^V = \overline{C}_{21}^V = \overline{C}_{13}^V = \overline{C}_{31}^V = \overline{C}_{23}^V = \overline{C}_{32}^V = \frac{c_f E_f \nu_f}{(1 + \nu_f)(1 - 2\nu_f)} + \frac{c_m E_m \nu_m}{(1 + \nu_m)(1 - 2\nu_m)}, \tag{9.1.33}$$

$$\overline{C}_{44}^V = \overline{C}_{55}^V = \overline{C}_{66}^V = \frac{c_f E_f}{2(1 + \nu_f)} + \frac{c_m E_m}{2(1 + \nu_m)}, \tag{9.1.34}$$

where E_f, ν_f and E_m, ν_m are the Young moldulus and the Poisson ratio of the fiber and matrix, respectively. The corresponding upper bound for the effective Young's modulus is thus given by

$$E^V = C_{11}^V - \frac{2C_{12}^V C_{12}^V}{C_{22}^V + C_{12}^V} = c_f E_f + c_m E_m + \frac{2 c_f c_m E_f E_m (\nu_f - \nu_m)^2}{c_f E_f (1 + \nu_m)(1 - 2\nu_m) + c_m E_m (1 + \nu_f)(1 - 2\nu_f)}. \tag{9.1.35}$$

The corresponding upper bound for the effective Poisson's ratio is

$$\bar{\nu}^V = \frac{C_{12}^V(C_{11}^V - C_{12}^V)}{C_{11}^V C_{11}^V - C_{12}^V C_{12}^V}$$
$$= \frac{c_f \nu_f E_f(1 + \nu_m)(1 - 2\nu_m) + c_m \nu_m E_m(1 + \nu_f)(1 - 2\nu_f)}{c_f E_f(1 + \nu_m)(1 - 2\nu_m) + c_m E_m(1 + \nu_f)(1 - 2\nu_f)}. \quad (9.1.36)$$

If the Poisson ratios of the fiber and the matrix are the same, that is, $\nu_f = \nu_m = \nu$, we have

$$E^V = c_f E_f + c_m E_m, \qquad \bar{\nu}^V = \nu. \quad (9.1.37)$$

On the other hand, the Reuss lower bound (6.2.13) gives

$$\overline{\mathbf{L}}^R = [c_f \mathbf{M}_f + c_m \mathbf{M}_m]^{-1}. \quad (9.1.38)$$

In terms the Voigt notation, this means

$$\overline{C}_{11}^R = \overline{C}_{22}^R = \overline{C}_{33}^R = \frac{E_f E_m[(c_f E_m(1 - \nu_f) + c_m E_f(1 - \nu_m)]}{D}, \quad (9.1.39)$$

$$\overline{C}_{12}^R = \overline{C}_{21}^R = \overline{C}_{13}^R = \overline{C}_{31}^R = \overline{C}_{23}^R = \overline{C}_{32}^R = \frac{E_f E_m[(c_f E_m \nu_f + c_m E_f \nu_m)}{D}, \quad (9.1.40)$$

$$\overline{C}_{44}^R = \overline{C}_{55}^R = \overline{C}_{66}^R = \frac{E_f E_m[(c_f E_m(1 - 2\nu_f) + c_m E_f(1 - 2\nu_m)]}{2D}, \quad (9.1.41)$$

where

$$D = [(c_f E_m(1 - \nu_f) + c_m E_f(1 + \nu_m)] \times [(c_f E_m(1 - 2\nu_f) + c_m E_f(1 - 2\nu_m)]. \quad (9.1.42)$$

Following the procedures similar to (9.1.35) and (9.1.36), we obtain the lower bounds for the effective Young modulus and the effective Poisson ratio corresponding to (9.1.38),

$$\overline{E}^R = \frac{E_f E_m}{c_f E_m + c_m E_f}, \qquad \overline{\nu}^R = \frac{c_f \nu_f E_m + c_m \nu_m E_f}{c_m E_f + c_f E_m}. \tag{9.1.43}$$

Again, if the Poisson ratios of the fiber and the matrix are the same, that is, $\nu_f = \nu_m = \nu$, we have

$$\overline{E}^R = \frac{E_f E_m}{c_f E_m + c_m E_f}, \qquad \overline{\nu}^R = \nu. \tag{9.1.44}$$

It can be shown (see Problem 9.1) that the Voigt upper bound gives a better approximation in the fiber direction, while the Reuss lower bound gives a better approximation in the direction perpendicular (transverse) to the fiber. Therefore, by neglecting the difference in Poisson's ratio, one may take the Young modulus in the fiber direction to be

$$\overline{E}_L = \overline{E}^V = c_f E_f + c_m E_m, \tag{9.1.45}$$

and the Young modulus in the transverse direction to be

$$\overline{E}_T = \frac{E_f E_m}{c_f E_m + c_m E_f}. \tag{9.1.46}$$

Equations (9.1.45) and (9.1.46) are the well-known rule of mixture formulas used commonly in engineering practice. Based on the derivations above, it is clear that these are actually the upper and lower bounds when the Poisson ratios are the same for the matrix and the fibers.

More accurate estimate of the unidirectional fiber-reinforced composite can be obtained by using the techniques developed in Chapters 7 and 8. To this end, we first evaluate the Eshelby tensor for an infinitely long cylindrical inclusion of circular cross section. It follows from Appendix 4.B that in the limit of $a_2 = a_3 = a$ and $a_1 \to \infty$, we obtain the nonzero components of the Eshelby tensor:

$$S_{2222} = S_{3333} = \frac{5 - 4\nu}{8(1 - \nu)}, \qquad S_{2233} = S_{3322} = \frac{4\nu - 1}{8(1 - \nu)}, \tag{9.1.47}$$

$$S_{2211} = S_{3311} = \frac{\nu}{2(1 - \nu)}, \tag{9.1.48}$$

$$S_{2323} = S_{3223} = S_{2332} = S_{3232} = \frac{3 - 4\nu}{8(1 - \nu)}, \tag{9.1.49}$$

$$S_{3131} = S_{1331} = S_{3113} = S_{1313} = S_{1212} = S_{2112} = S_{1221} = S_{2121} = \tfrac{1}{4}. \tag{9.1.50}$$

Making use of the Eshelby tensor in, for example, the Mori–Tanaka estimate (7.3.8), we have the effective elasticity tensor of the unidirectional fiber-reinforced composite given by

$$\overline{\mathbf{L}} = (c_m\mathbf{L}_m + c_f\mathbf{L}_f\mathbf{T}_f)[c_m\mathbf{I} + c_f\mathbf{T}_f]^{-1}, \tag{9.1.51}$$

where

$$\mathbf{T}_f = [\mathbf{I} + \mathbf{S}(\mathbf{M}_m\mathbf{L}_f - \mathbf{I})]^{-1} \tag{9.1.52}$$

is the strain concentration tensor (7.2.4). In conjunction with the definitions of the engineering constants, (9.1.51) leads to

$$E_L = [\eta c + (1 - c)]E_m + \frac{2c(1 - c)(\nu_f - \nu_m)^2\eta E_m}{(1 - c)(1 + \nu_f)(1 - 2\nu_f) + \eta(1 + \nu_m)[1 + c(1 - 2\nu_m)]}, \tag{9.1.53}$$

$$K_T = \frac{\{(1 - c)(1 + \nu_f)(1 - 2\nu_f) + \eta(1 + \nu_m)[1 + c(1 - 2\nu_m)]\}E_m}{2(1 + \nu_m)[(1 + \nu_f)(1 - 2\nu_f)(c + 1 - 2\nu_m) + \eta(1 - c)(1 + \nu_m)(1 - 2\nu_m)]}, \tag{9.1.54}$$

$$\mu_T = \frac{[(1 - c)(1 + \nu_f) + \eta(1 + \nu_m)[c + (3 - 4\nu_m)]E_m}{2(1 + \nu_m)\{(1 - c)(1 + \nu_m)(3 - 4\nu_m)\eta + (1 + \nu_f)[1 + c(3 - 4\nu_m)]\}}, \tag{9.1.55}$$

$$\mu_L = \frac{[\eta(1 + c)(1 + \nu_m) + (1 - c)(1 + \nu_f)]E_m}{2(1 + \nu_m)[\eta(1 - c)(1 + \nu_m) + (1 + c)(1 + \nu_f)]}, \tag{9.1.56}$$

$$\nu_{LT} = \frac{2c\eta(1 - \nu_m^2)\nu_f + (1 - c)[(1 + \nu_f)(1 - 2\nu_f) + \eta(1 + \nu_m)]\nu_m}{2c\eta(1 - \nu_m^2) + (1 - c)[(1 + \nu_f)(1 - 2\nu_f) + \eta(1 + \nu_m)]}, \tag{9.1.57}$$

where $c = c_f$ is the fiber volume fraction, and

$$\eta = \frac{E_f}{E_m}.$$

The other engineering constants can be obtained through their relationships discussed earlier. For example, the transverse Young modulus E_T can be obtained from (9.1.23) based on the above equations. Unfortunately, the expression for E_T is extremely complex. We will not reproduce it here.

By neglecting the difference in Poisson's ratio between the fiber and the matrix, that is, $\nu_m = \nu_f = \nu$, we have

$$E_L = [\eta c + (1 - c)]E_m, \tag{9.1.58}$$

$$K_T = \frac{[1 + \eta - 2\nu - c(1 - 2\nu)(1 - \eta)]E_m}{2(1 + \nu)(1 - 2\nu)[1 - c(\eta - 1) + \eta - 2\nu]}, \tag{9.1.59}$$

$$\mu_T = \frac{[1 - c(1 - \eta) + \eta(3 - 4\nu)]E_m}{2(1 + \nu)[1 + (3 - 4\nu)(c + (1 - c)\eta)]}, \tag{9.1.60}$$

$$\mu_L = \frac{E_m[(1 - c) + \eta(1 + c)]}{2(1 + \nu)[(1 + c) + \eta(1 - c)]}, \tag{9.1.61}$$

$$\nu_{LT} = \nu. \tag{9.1.62}$$

Even in this case, the expression for E_T is still too long to be recorded here. However, thc following expression seems to give a pretty good approximation for most of the materials of practical interest:

$$E_T \approx \frac{2(1 - c) + \eta(1 + 2c)}{\eta(1 - c) + (2 + c)} E_m. \tag{9.1.63}$$

For the dilute case ($c << 1$), the following expression for E_T can be derived:

$$E_T = \frac{\eta E_m}{c + (1 - c)\eta} + \frac{\begin{array}{c} c(1 - \eta)^2[1 - 2\nu + \eta^2\nu^2(3 - 4\nu) \\ + 2\eta(1 - 3\nu + 5\nu^2 - 2\nu^3]E_m \end{array}}{\eta(1 + \eta - 2\nu)(1 + 3\eta - 4\eta\nu)}, \tag{9.1.64}$$

Generally speaking, the solutions from upper and lower bounds, as well as the ones from the Mori–Tanaka methods are approximate. Using the composite cylinder model, see Section 8.1, the exact relationships between the effective engineering constants and the moduli of the fibers and the matrix can be obtained (Christensen, 1979); see (8.1.8)–(8.1.11). Note that the composite cylinder model does not yield the exact solution for the transverse shear modulus μ_T.

In the event that $\nu_m = \nu_f = \nu$, (8.1.8)–(8.1.11) reduce to (9.1.58)–(9.1.62). In other words, the Mori–Tanaka estimate gives the exact solution when $\nu_m = \nu_f = \nu$.

9.2 EFFECTIVE PROPERTIES OF MULTILAYER COMPOSITES

In this section, we consider a fiber-reinforced composite material consisting of multiple layers. Each layer is assumed to be a unidirectional fiber-reinforced composite. The bonding between the layers is assumed perfect, that is, the displacement and tractions across the interfaces are continuous.

We further assume that the thickness of the layers is much greater than the fiber diameter. As discussed in Chapter 5, the effective properties of the layered composite can be obtained through a hierarchical approach. First, the effective properties of each layer are evaluated based on the method discussed in the previous section. Once this is done, the effective properties of the layered composite will be estimated by assuming that each layer is a homogeneous, transversely isotropic solid. In other words, the individual fibers are not explicitly accounted for in the second level of homogenization. Instead, only the collective effect of all the fibers in each layer is explicitly included in the effective properties of the laminate. Therefore, when we consider the effective properties of a multilayered composite, we may assume that each later is homogeneous.

To facilitate the discussions below, let us first define the following:

$$\boldsymbol{\sigma}_n = \begin{bmatrix} \sigma_{33} \\ \sigma_{23} \\ \sigma_{13} \end{bmatrix}, \quad \boldsymbol{\sigma}_t = \begin{bmatrix} \sigma_{11} \\ \sigma_{22} \\ \sigma_{12} \end{bmatrix}, \quad \boldsymbol{\varepsilon}_n = \begin{bmatrix} \varepsilon_{33} \\ 2\varepsilon_{23} \\ 2\varepsilon_{13} \end{bmatrix}, \quad \boldsymbol{\varepsilon}_t = \begin{bmatrix} \varepsilon_{11} \\ \varepsilon_{22} \\ 2\varepsilon_{12} \end{bmatrix}. \tag{9.2.1}$$

For the rest of this section, the summation convention is temporarily suspended. Sums will be explicitly indicated by the summation sign. This way, we can rewrite Hooke's law using the Voigt notation as

$$\boldsymbol{\sigma}_n = \mathbf{C}_{nn}\boldsymbol{\varepsilon}_n + \mathbf{C}_{nt}\boldsymbol{\varepsilon}_t, \qquad \boldsymbol{\sigma}_t = \mathbf{C}_{tn}\boldsymbol{\varepsilon}_n + \mathbf{C}_{tt}\boldsymbol{\varepsilon}_t, \tag{9.2.2}$$

where

$$C_{nn} = \begin{bmatrix} C_{33} & C_{34} & C_{35} \\ C_{34} & C_{44} & C_{45} \\ C_{35} & C_{45} & C_{55} \end{bmatrix}, \qquad \mathbf{C}_{nt} = \begin{bmatrix} C_{13} & C_{23} & C_{36} \\ C_{14} & C_{24} & C_{46} \\ C_{15} & C_{25} & C_{56} \end{bmatrix}, \tag{9.2.3}$$

$$C_{tn} = \begin{bmatrix} C_{13} & C_{14} & C_{15} \\ C_{23} & C_{24} & C_{25} \\ C_{36} & C_{46} & C_{56} \end{bmatrix}, \qquad \mathbf{C}_{tt} = \begin{bmatrix} C_{11} & C_{12} & C_{16} \\ C_{12} & C_{22} & C_{26} \\ C_{16} & C_{26} & C_{66} \end{bmatrix}. \tag{9.2.4}$$

Obviously,

$$\mathbf{C}_{nn} = (\mathbf{C}_{nn})^{\mathrm{T}}, \qquad \mathbf{C}_{tt} = (\mathbf{C}_{tt})^{\mathrm{T}}, \mathbf{C}_{tn} = (\mathbf{C}_{nt})^{\mathrm{T}}. \tag{9.2.5}$$

Also, because of the positive definiteness of the Voigt elasticity matrix, we know that $\mathbf{C}_{nn}$ and $\mathbf{C}_{tt}$ are positive definite matrices.

Now, consider a multilayer composite consisting of N layers of homogeneous materials as shown in Figure 9.3. We assume that each layer is homogeneous with its Voigt elastic matrix given by $\mathbf{C}^{(k)}$, $k = 1, 2, \ldots, N$. Therefore, for the kth layer, it follows from (9.2.2) that

$$\boldsymbol{\sigma}_n^{(k)} = \mathbf{C}_{nn}^{(k)}\boldsymbol{\varepsilon}_n^{(k)} + \mathbf{C}_{nt}^{(k)}\boldsymbol{\varepsilon}_t^{(k)}, \qquad \boldsymbol{\sigma}_t^{(k)} = \mathbf{C}_{tn}^{(k)}\boldsymbol{\varepsilon}_n^{(k)} + \mathbf{C}_{tt}^{(k)}\boldsymbol{\varepsilon}_t^{(k)}, \tag{9.2.6}$$

where the superscript (k) indicates that the base letter is associated with the kth layer. It is assumed that the representative volume element being considered is under a uniform state of deformation, even though the actual composite may be subjected to arbitrary loading. We further assume that the layers are perfectly bonded together, meaning that the

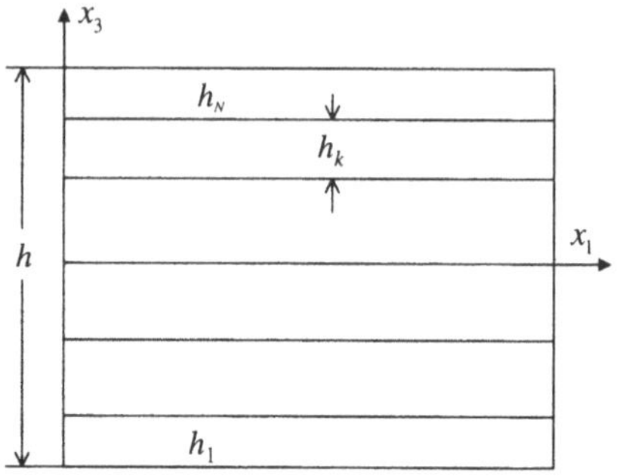

Figure 9.3 Multilayer composite.

traction and displacements are continuous across the layer interfaces. This immediately leads to

$$\boldsymbol{\sigma}_n^{(k)} = \boldsymbol{\sigma}_n, \qquad \boldsymbol{\varepsilon}_t^{(k)} = \boldsymbol{\varepsilon}_t, \qquad \text{for } k = 1, 2, \ldots, N \tag{9.2.7}$$

Making use of (9.2.7) in the first of (9.2.6) yields

$$\boldsymbol{\varepsilon}_n^{(k)} = (\mathbf{C}_{nn}^{(k)})^{-1}(\boldsymbol{\sigma}_n - \mathbf{C}_{nt}^{(k)}\boldsymbol{\varepsilon}_t). \tag{9.2.8}$$

Taking the average of $\boldsymbol{\varepsilon}_n^{(k)}$ over the representative volume element leads to

$$\boldsymbol{\varepsilon}_n = \sum_{k=1}^{N} c_k \boldsymbol{\varepsilon}_n^{(k)} = \sum_{k=1}^{N} c_k (\mathbf{C}_{nn}^{(k)})^{-1} \boldsymbol{\sigma}_n - \sum_{k=1}^{N} c_k (\mathbf{C}_{nn}^{(k)})^{-1} \mathbf{C}_{nt}^{(k)} \boldsymbol{\varepsilon}_t, \tag{9.2.9}$$

where $c_k = h_k/h$ is the volume fraction of the kth layer. By rearranging the terms in (9.2.9), we have

$$\boldsymbol{\sigma}_n = \overline{\mathbf{C}}_{nn}\boldsymbol{\varepsilon}_n + \overline{\mathbf{C}}_{nt}\boldsymbol{\varepsilon}_t, \tag{9.2.10}$$

where

$$\overline{\mathbf{C}}_{nn} = \left[\sum_{k=1}^{N} c_k (\mathbf{C}_{nn}^{(k)})^{-1}\right]^{-1}, \tag{9.2.11}$$

$$\overline{\mathbf{C}}_{nt} = \overline{\mathbf{C}}_{nn}\left[\sum_{k=1}^{N} c_k(\mathbf{C}_{nn}^{(k)})^{-1}\mathbf{C}_{n}^{(k)}\right]. \tag{9.2.12}$$

We next substitute (9.2.8) into the second of (9.2.6). This gives us

$$\begin{aligned}\boldsymbol{\sigma}_t^{(k)} &= \mathbf{C}_{tn}^{(k)}(\mathbf{C}_{nn}^{(k)})^{-1}(\boldsymbol{\sigma}_n - \mathbf{C}_{nt}^{(k)}\boldsymbol{\varepsilon}_t) + \mathbf{C}_{tt}^{(k)}\boldsymbol{\varepsilon}_t \\ &= \mathbf{C}_{tn}^{(k)}(\mathbf{C}_{nn}^{(k)})^{-1}\boldsymbol{\sigma}_n + [\mathbf{C}_{tt}^{(k)} - \mathbf{C}_{tn}^{(k)}(\mathbf{C}_{nn}^{(k)})^{-1}\mathbf{C2}_{nt}^{(k)}]\boldsymbol{\varepsilon}_t.\end{aligned} \tag{9.2.13}$$

Taking the average of $\boldsymbol{\sigma}_t^{(k)}$ over the representative volume element leads to

$$\begin{aligned}\boldsymbol{\sigma}_t = \sum_{k=1}^{N} c_k\boldsymbol{\sigma}_t^{(k)} &= \sum_{k=1}^{N} c_k\mathbf{C}_{tn}^{(k)}(\mathbf{C}_{nn}^{(k)})^{-1}\boldsymbol{\sigma}_n \\ &+ \sum_{k=1}^{N} c_k[\mathbf{C}_{tt}^{(k)} - \mathbf{C}_{tn}^{(k)}(\mathbf{C}_{nn}^{(k)})^{-1}\mathbf{C}_{nt}^{(k)}]\boldsymbol{\varepsilon}_t.\end{aligned} \tag{9.2.14}$$

Substitution of (9.2.10) into the above yields

$$\boldsymbol{\sigma}_t = \overline{\mathbf{C}}_{tn}\boldsymbol{\varepsilon}_n + \overline{\mathbf{C}}_{tt}\boldsymbol{\varepsilon}_t, \tag{9.2.15}$$

where

$$\overline{\mathbf{C}}_{tn} = \left[\sum_{k=1}^{N} c_k\mathbf{C}_{tn}^{(k)}(\mathbf{C}_{nn}^{(k)})^{-1}\right]\overline{\mathbf{C}}_{nn}, \tag{9.2.16}$$

$$\overline{\mathbf{C}}_{tt} = \sum_{k=1}^{N} c_k\mathbf{C}_{tt}^{(k)} + \sum_{k=1}^{N} c_k\mathbf{C}_{tn}^{(k)}(\mathbf{C}_{nn}^{(k)})^{-1}(\overline{\mathbf{C}}_{nt} - \mathbf{C}_{nt}^{(k)}). \tag{9.2.17}$$

Clearly, $\overline{\mathbf{C}}_{nn}$, $\overline{\mathbf{C}}_{nt}$, $\overline{\mathbf{C}}_{tn}$, and $\overline{\mathbf{C}}_{tt}$ provide the complete effective Voigt elastic matrix of the layered composite. Taking advantage of the symmetry properties (9.2.5), it can be shown that

$$\overline{\mathbf{C}}_{nn} = (\overline{\mathbf{C}}_{nn})^{\mathrm{T}}, \qquad \overline{\mathbf{C}}_{tt} = (\overline{\mathbf{C}}_{tt})^{\mathrm{T}}, \qquad \overline{\mathbf{C}}_{tn} = (\overline{\mathbf{C}}_{nt})^{\mathrm{T}}.$$

Next, we assume that each layer is a monoclinic material with its plane of symmetry in the x_1x_2 plane. In this case, the Voigt elastic matrix has the following form:

$$\mathbf{C}_{nn}^{(k)} = \begin{bmatrix} C_{33}^{(k)} & 0 & 0 \\ 0 & C_{44}^{(k)} & C_{45}^{(k)} \\ 0 & C_{45}^{(k)} & C_{55}^{(k)} \end{bmatrix}, \qquad \mathbf{C}_{nt}^{(k)} = \begin{bmatrix} C_{13}^{(k)} & C_{23}^{(k)} & C_{36}^{(k)} \\ 0 & 0 & 0 \\ 0 & 0 & 0 \end{bmatrix}, \tag{9.2.18}$$

$$\mathbf{C}_{tn}^{(k)} = (\mathbf{C}_{nt}^{(k)})^{\mathrm{T}}, \qquad \mathbf{C}_{tt}^{(k)} = \begin{bmatrix} C_{11}^{(k)} & C_{12}^{(k)} & C_{16}^{(k)} \\ C_{12}^{(k)} & C_{22}^{(k)} & C_{26}^{(k)} \\ C_{16}^{(k)} & C_{26}^{(k)} & C_{66}^{(k)} \end{bmatrix}. \tag{9.2.19}$$

Substituting these into (9.2.11) and (9.2.12) yields

$$\overline{\mathbf{C}}_{nn} = \begin{bmatrix} \displaystyle\sum_{k=1}^{N} \frac{c_k}{C_{33}^{(k)}} & 0 & 0 \\ 0 & \displaystyle\sum_{k=1}^{N} \frac{c_k C_{55}^{(k)}}{\Delta^{(k)}} & -\displaystyle\sum_{k=1}^{N} \frac{c_k C_{45}^{(k)}}{\Delta^{(k)}} \\ 0 & -\displaystyle\sum_{k=1}^{N} \frac{c_k C_{45}^{(k)}}{\Delta^{(k)}} & \displaystyle\sum_{k=1}^{N} \frac{c_k C_{44}^{(k)}}{\Delta^{(k)}} \end{bmatrix}^{-1}$$

$$= \begin{bmatrix} \left(\displaystyle\sum_{k=1}^{N} \frac{c_k}{C_{33}^{(k)}}\right) & 0 & 0 \\ 0 & \displaystyle\frac{1}{D}\sum_{k=1}^{N} \frac{c_k C_{44}^{(k)}}{\Delta^{(k)}} & \displaystyle\frac{1}{D}\sum_{k=1}^{N} \frac{c_k C_{45}^{(k)}}{\Delta^{(k)}} \\ 0 & \displaystyle\frac{1}{D}\sum_{k=1}^{N} \frac{c_k C_{45}^{(k)}}{\Delta^{(k)}} & \displaystyle\frac{1}{D}\sum_{k=1}^{N} \frac{c_k C_{55}^{(k)}}{\Delta^{(k)}} \end{bmatrix}, \tag{9.2.20}$$

where

$$D = \sum_{k=1}^{N} \frac{c_k C_{55}^{(k)}}{\Delta^{(k)}} \sum_{k=1}^{N} \frac{c_k C_{44}^{(k)}}{\Delta^{(k)}} - \left[\sum_{k=1}^{N} \frac{c_k C_{45}^{(k)}}{\Delta^{(k)}}\right]^2, \tag{9.2.21}$$

$$\Delta^{(k)} = C_{44}^{(k)} C_{55}^{(k)} - (C_{45}^{(k)})^2 > 0. \tag{9.2.22}$$

Next, making use of (9.2.20) in (9.2.16) yields

$$\overline{\mathbf{C}}_{tn} = \left(\sum_{k=1}^{N} \frac{c_k}{C_{33}^{(k)}}\right)^{-1} \sum_{k=1}^{N} \frac{c_k}{C_{33}^{(k)}} \begin{bmatrix} C_{13}^{(k)} & 0 & 0 \\ C_{23}^{(k)} & 0 & 0 \\ C_{36}^{(k)} & 0 & 0 \end{bmatrix}. \tag{9.2.23}$$

Finally, (9.2.17) leads to

$$\overline{\mathbf{C}}_{tt} = \sum_{k=1}^{N} c_k \begin{bmatrix} C_{11}^{(k)} & C_{12}^{(k)} & C_{16}^{(k)} \\ C_{12}^{(k)} & C_{22}^{(k)} & C_{26}^{(k)} \\ C_{16}^{(k)} & C_{26}^{(k)} & C_{66}^{(k)} \end{bmatrix}$$

$$- \sum_{k=1}^{N} \frac{c_k}{C_{33}^{(k)}} \begin{bmatrix} C_{13}^{(k)}C_{13}^{(k)} & C_{23}^{(k)}C_{13}^{(k)} & C_{36}^{(k)}C_{13}^{(k)} \\ C_{13}^{(k)}C_{23}^{(k)} & C_{23}^{(k)}C_{23}^{(k)} & C_{36}^{(k)}C_{23}^{(k)} \\ C_{13}^{(k)}C_{36}^{(k)} & C_{23}^{(k)}C_{36}^{(k)} & C_{36}^{(k)}C_{36}^{(k)} \end{bmatrix}$$

$$+ \left(\sum_{r=1}^{N} \frac{c_r}{C_{33}^{(r)}} \right)^{-1} \sum_{k=1}^{N} \sum_{m=1}^{N} \frac{c_m c_k}{C_{33}^{(m)} C_{33}^{(k)}}$$

$$\times \begin{bmatrix} C_{13}^{(m)}C_{13}^{(k)} & C_{23}^{(m)}C_{13}^{(k)} & C_{36}^{(m)}C_{13}^{(k)} \\ C_{13}^{(m)}C_{23}^{(k)} & C_{23}^{(m)}C_{23}^{(k)} & C_{36}^{(m)}C_{23}^{(k)} \\ C_{13}^{(m)}C_{36}^{(k)} & C_{23}^{(m)}C_{36}^{(k)} & C_{36}^{(m)}C_{36}^{(k)} \end{bmatrix}. \tag{9.2.24}$$

If we let $\overline{C}_{ij}$ be the elements of the 6×6 Voigt effective elastic matrix $\overline{\mathbf{C}}$, results from (9.2.20), (9.2.23), and (9.2.24) can be summarized as follows:

$$\overline{C}_{ij} = \sum_{k=1}^{N} c_k \left[C_{ij}^{(k)} - \frac{C_{i3}^{(k)} C_{j3}^{(k)}}{C_{33}^{(k)}} + \frac{C_{i3}^{(k)}}{C_{33}^{(k)}} \left(\sum_{r=1}^{N} \frac{c_r}{C_{33}^{(r)}} \right)^{-1} \left[\sum_{m=1}^{N} \frac{c_m C_{3j}^{(m)}}{C_{33}^{(m)}} \right] \right] \tag{9.2.25}$$

for $i, j = 1, 2, 3, 6$:

$$\overline{C}_{ij} = \frac{1}{D} \sum_{k=1}^{N} \frac{c_k C_{ij}^{(k)}}{\Delta^{(k)}} \qquad \text{for } i, j = 4, 5, \tag{9.2.26}$$

$$= 0 \qquad \text{for } i = 1, 2, 3, 6; \quad j = 4, 5. \tag{9.2.27}$$

Similarly, the compliance matrix can be partitioned as

$$\mathbf{S}_{nn} = \begin{bmatrix} S_{33} & S_{34} & S_{35} \\ S_{34} & S_{44} & S_{45} \\ S_{35} & S_{45} & S_{55} \end{bmatrix}, \qquad \mathbf{S}_{nt} = \begin{bmatrix} S_{13} & S_{23} & S_{36} \\ S_{14} & S_{24} & S_{46} \\ S_{15} & S_{25} & S_{56} \end{bmatrix}, \tag{9.2.28}$$

$$\mathbf{S}_{tn} = \begin{bmatrix} S_{13} & S_{14} & S_{15} \\ S_{23} & S_{24} & S_{25} \\ S_{36} & S_{46} & S_{56} \end{bmatrix}, \qquad \mathbf{S}_{tt} = \begin{bmatrix} S_{11} & S_{12} & S_{16} \\ S_{12} & S_{22} & S_{26} \\ S_{16} & S_{26} & S_{66} \end{bmatrix}, \tag{9.2.29}$$

Following the same procedure, elements of the effective compliance matrix are given by (see Appendix 9.A)

$$\overline{\mathbf{S}}_{tt} = \begin{bmatrix} \overline{S}_{11} & \overline{S}_{12} & \overline{S}_{16} \\ \overline{S}_{12} & \overline{S}_{22} & \overline{S}_{26} \\ \overline{S}_{16} & \overline{S}_{26} & \overline{S}_{66} \end{bmatrix} = \left[\sum_{k=1}^{N} c_k (\mathbf{S}_{tt}^{(k)})^{-1} \right]^{-1}, \tag{9.2.30}$$

$$\overline{\mathbf{S}}_{tn} = \begin{bmatrix} \overline{S}_{13} & \overline{S}_{14} & \overline{S}_{15} \\ \overline{S}_{23} & \overline{S}_{24} & \overline{S}_{25} \\ \overline{S}_{36} & \overline{S}_{46} & \overline{S}_{56} \end{bmatrix} = \overline{\mathbf{S}}_{tt} \left[\sum_{k=1}^{N} c_k (\mathbf{S}_{tt}^{(k)})^{-1} \mathbf{S}_{tn}^{(k)} \right], \tag{9.2.31}$$

$$\overline{\mathbf{S}}_{nt} = \begin{bmatrix} \overline{S}_{13} & \overline{S}_{23} & \overline{S}_{36} \\ \overline{S}_{14} & \overline{S}_{24} & \overline{S}_{46} \\ \overline{S}_{15} & \overline{S}_{25} & \overline{S}_{56} \end{bmatrix} = \left[\sum_{k=1}^{N} c_k \mathbf{S}_{nt}^{(k)} (\mathbf{S}_{tt}^{(k)})^{-1} \right] \overline{\mathbf{S}}_{tt}, \tag{9.2.32}$$

$$\overline{\mathbf{S}}_{nn} = \begin{bmatrix} \overline{S}_{33} & \overline{S}_{34} & \overline{S}_{35} \\ \overline{S}_{34} & \overline{S}_{44} & \overline{S}_{45} \\ \overline{S}_{35} & \overline{S}_{45} & \overline{S}_{55} \end{bmatrix} = \sum_{k=1}^{N} c_k \mathbf{S}_{nn}^{(k)} + \sum_{k=1}^{N} c_k \mathbf{S}_{nt}^{(k)} (\mathbf{S}_{tt}^{(k)})^{-1} (\overline{\mathbf{S}}_{tn} - \mathbf{S}_{tn}^{(k)}). \tag{9.2.33}$$

Example 9.1 Consider a layered material consisting of two alternating layers. The layers are identical, except that the fiber directions are perpendicular to each other between adjacent layers. We are to find the effective elastic constants for this layered composite.

To this end, let us denote the Voigt elasticity matrix for the layer whose fiber direction is in the x_1 direction by

$$\mathbf{C}^{(1)} = \begin{bmatrix} C_{11} & C_{12} & C_{12} & 0 & 0 & 0 \\ C_{12} & C_{22} & C_{23} & 0 & 0 & 0 \\ C_{12} & C_{23} & C_{22} & 0 & 0 & 0 \\ 0 & 0 & 0 & \dfrac{C_{22} - C_{23}}{2} & 0 & 0 \\ 0 & 0 & 0 & 0 & C_{66} & 0 \\ 0 & 0 & 0 & 0 & 0 & C_{66} \end{bmatrix}. \tag{9.2.34}$$

A 90° rotation about the x_3 axis yields the Voigt elasticity matrix for the adjacent layer:

$$\mathbf{C}^{(2)} = \begin{bmatrix} C_{22} & C_{12} & C_{23} & 0 & 0 & 0 \\ C_{12} & C_{11} & C_{12} & 0 & 0 & 0 \\ C_{23} & C_{23} & C_{22} & 0 & 0 & 0 \\ 0 & 0 & 0 & C_{66} & 0 & 0 \\ 0 & 0 & 0 & 0 & \dfrac{C_{22} - C_{23}}{2} & 0 \\ 0 & 0 & 0 & 0 & 0 & C_{66} \end{bmatrix}. \quad (9.2.35)$$

Making use of these in (9.2.25)–(9.2.27) with $N = 2$, $c_1 = c_2 = \frac{1}{2}$, we arrive at

$$\Delta^{(k)} = C_{44}^{(k)} C_{55}^{(k)} = \frac{(C_{22} - C_{23})C_{66}}{2}, \qquad D = \frac{1}{4}\left(\frac{2}{C_{22} - C_{23}} + \frac{1}{C_{66}}\right)^2. \quad (9.2.36)$$

Thus, the nonzero components of the effective Voigt elasticity matrix of the layered composite are

$$\overline{C}_{11} = \overline{C}_{22} = \frac{1}{2}\left[C_{11} + C_{22} - \frac{(C_{12} - C_{23})^2}{2C_{22}}\right], \quad (9.2.37)$$

$$\overline{C}_{33} = C_{22}, \qquad \overline{C}_{12} = C_{12} - \frac{(C_{12} - C_{23})^2}{4C_{22}},$$

$$\overline{C}_{13} = C_{23} = \frac{C_{12} + C_{23}}{2}, \quad (9.2.38)$$

$$\overline{C}_{44} = \overline{C}_{55} = \frac{2C_{66}(C_{22} - C_{23})}{2C_{66} + (C_{22} - C_{23})}, \qquad \overline{C}_{66} = C_{66}. \quad (9.2.39)$$

Making use of (9.2.30), the in-plane Young modulus and shear modulus can be written in terms of the engineering constants of the individual layer:

$$E_0 = \frac{(E_L + E_T)^2 - 4E_L^2 \nu_{TL}^2}{2(E_L + E_T)(1 - \nu_{LT}\nu_{TL})}, \qquad \nu_0 = \frac{2E_L}{E_L + E_T}\,\nu_{TL}. \quad (9.2.40)$$

9.3 EFFECTIVE PROPERTIES OF A LAMINA

In the composite material literature, a lamina is a thin layer of unidirectional fiber-reinforced composite material; see Figure 9.4. Because the thickness is much smaller relative to the in-plane dimensions, the deformation of a lamina can be approximated by a state of plane stress. This means

$$\sigma_{13} \approx \sigma_{23} \approx \sigma_{33} \approx 0. \tag{9.3.1}$$

It thus follows from Hooke's law that

$$L_{13kl}\varepsilon_{kl} = L_{23kl}\varepsilon_{kl} = L_{33kl}\varepsilon_{kl} = 0. \tag{9.3.2}$$

For a lamina, the effective elasticity tensor is transversely isotropic with the axis of symmetry being the x_1 axis, as shown in Figure 9.3, the above equation yields

$$\varepsilon_{13} = \varepsilon_{23} = 0 \quad \text{and} \quad \varepsilon_{33} = -\frac{L_{1311}}{L_{3333}}\varepsilon_{11} - \frac{l_{1322}}{L_{3333}}\varepsilon_{22}. \tag{9.3.3}$$

Substituting the above into Hooke's law again, we arrive at

$$\sigma_{11} = \left(L_{1111} - \frac{L_{1311}}{L_{3333}}\right)\varepsilon_{11} + \left(L_{1122} - \frac{L_{1322}}{L_{3333}}\right)\varepsilon_{22}, \tag{9.3.4}$$

$$\sigma_{22} = \left(L_{2211} - \frac{L_{1311}}{L_{3333}}\right)\varepsilon_{11} + \left(L_{2222} - \frac{L_{1322}}{L_{3333}}\right)\varepsilon_{22}. \tag{9.3.5}$$

Therefore, in the reduced form of the Voigt notation, the stress–strain relationship for a thin lamina can be written as

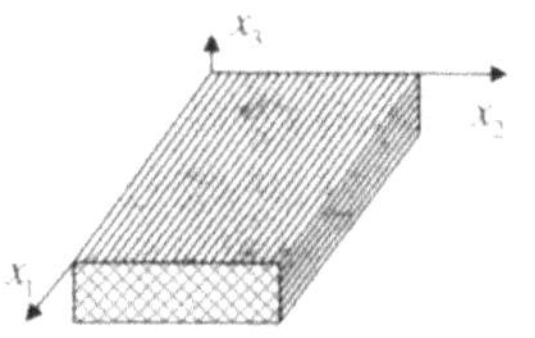

Figure 9.4 Thin lamina.

$$\begin{bmatrix} \sigma_1 \\ \sigma_2 \\ \sigma_6 \end{bmatrix} = \begin{bmatrix} Q_{11} & Q_{12} & 0 \\ Q_{12} & Q_{22} & 0 \\ 0 & 0 & Q_{66} \end{bmatrix} \begin{bmatrix} \varepsilon_1 \\ \varepsilon_2 \\ \varepsilon_6 \end{bmatrix}, \tag{9.3.6}$$

where

$$\begin{bmatrix} \sigma_1 \\ \sigma_2 \\ \sigma_6 \end{bmatrix} = \begin{bmatrix} \sigma_{11} \\ \sigma_{22} \\ \sigma_{12} \end{bmatrix}, \qquad \begin{bmatrix} \varepsilon_1 \\ \varepsilon_2 \\ \varepsilon_6 \end{bmatrix} = \begin{bmatrix} \varepsilon_{11} \\ \varepsilon_{22} \\ 2\varepsilon_{12} \end{bmatrix}, \tag{9.3.7}$$

and

$$Q_{ij} = C_{ij} - \frac{C_{i3}C_{j3}}{C_{33}}, \qquad i, j = 1, 2, 6 \tag{9.3.8}$$

with C_{ij} being the Voigt elastic constants of the lamina. Note that C_{ij} is transversely isotropic so that $C_{63} = 0$. Thus, $Q_{66} = C_{66}$. The matrix **Q** in (9.3.6) is called the *reduced* stiffness matrix of the lamina. The reduced compliance matrix **S** (not to be confused with the Eshelby tensor) for the lamina is obtained by inverting (9.3.6),

$$\begin{bmatrix} \varepsilon_1 \\ \varepsilon_2 \\ \varepsilon_6 \end{bmatrix} = \begin{bmatrix} S_{11} & S_{12} & 0 \\ S_{12} & S_{22} & 0 \\ 0 & 0 & S_{66} \end{bmatrix} \begin{bmatrix} \sigma_1 \\ \sigma_2 \\ \sigma_6 \end{bmatrix}, \tag{9.3.9}$$

where

$$S_{11} = \frac{Q_{22}}{Q_{11}Q_{22} - Q_{12}^2}, \qquad S_{22} = \frac{Q_{11}}{Q_{11}Q_{22} - Q_{12}^2}, \tag{9.3.10}$$

$$S_{12} = -\frac{Q_{12}}{Q_{11}Q_{22} - Q_{12}^2}, \qquad S_{66} = \frac{1}{Q_{66}}. \tag{9.3.11}$$

The Young moduli in the fiber and transverse to the fiber directions can thus be defined, respectively, as

$$E_L = \frac{1}{S_{11}} = \frac{Q_{11}Q_{22} - Q_{12}^2}{Q_{22}}, \qquad E_T = \frac{1}{S_{22}} = \frac{Q_{11}Q_{22} - Q_{12}^2}{Q_{11}}. \tag{9.3.12}$$

The shear modulus in the plane of the lamina is

$$\mu_L = Q_{66}. \qquad (9.3.13)$$

The corresponding Poisson ratios are

$$\nu_{LT} = -\frac{S_{12}}{S_{11}} = \frac{Q_{12}}{Q_{22}}, \qquad \nu_{TL} = -\frac{S_{12}}{S_{22}} = \frac{Q_{12}}{Q_{11}}. \qquad (9.3.14)$$

Note that ν_{LT} represents the contraction in the transverse direction when the lamina is stretched in the fiber direction, while ν_{TL} represents the contraction in the fiber direction when the lamina is stretched in the transverse direction. These are exactly the same engineering elastic constants as introduced in Section 9.1. As discussed in Section 9.1, only four of these engineering elastic constants are independent because

$$\nu_{LT}E_T = \frac{Q_{12}(Q_{11}Q_{22} - Q_{12}^2)}{Q_{11}Q_{22}} = \nu_{TL}E_L. \qquad (9.3.15)$$

Therefore, (9.3.12)–(9.3.14) can be inverted to obtain

$$Q_{11} = \frac{E_L}{1 - \nu_{LT}\nu_{TL}}, \qquad Q_{22} = \frac{E_T}{1 - \nu_{LT}\nu_{TL}}, \qquad Q_{66} = \mu_L, \quad (9.3.16)$$

$$Q_{12} = \frac{\nu_{LT}E_T}{1 - \nu_{LT}\nu_{TL}} = \frac{\nu_{TL}E_L}{1 - \nu_{LT}\nu_{TL}}. \qquad (9.3.17)$$

The corresponding components of the reduced compliance tensor are

$$S_{11} = \frac{1}{E_L}, \qquad S_{22} = \frac{1}{E_T}, \qquad S_{66} = \frac{1}{\mu_L}, \qquad (9.3.18)$$

$$S_{12} = -\frac{\nu_{LT}}{E_L} = -\frac{\nu_{TL}}{E_T}. \qquad (9.3.19)$$

Equations (9.3.6) and (9.3.9) are written in a coordinate system where the x_1 axis is in the direction of the fibers. In the coordinate system $\hat{x}_i$ shown in Figure 9.2, where the $\hat{x}_1$ axis forms an angle θ from the fiber direction, the stress–strain relationship becomes

$$\begin{bmatrix} \sigma_x \\ \sigma_y \\ \sigma_{xy} \end{bmatrix} = \begin{bmatrix} \hat{Q}_{11} & \hat{Q}_{12} & \hat{Q}_{13} \\ \hat{Q}_{12} & \hat{Q}_{22} & \hat{Q}_{23} \\ \hat{Q}_{13} & \hat{Q}_{23} & \hat{Q}_{66} \end{bmatrix} \begin{bmatrix} \varepsilon_x \\ \varepsilon_y \\ \gamma_{xy} \end{bmatrix}, \tag{9.3.20}$$

where

$$\hat{Q}_{11} = Q_{11}m^4 + 2(Q_{12} + 2Q_{66})m^2n^2 + Q_{22}n^4, \tag{9.3.21}$$

$$\hat{Q}_{22} = Q_{11}n^4 + 2(Q_{12} + 2Q_{66})m^2n^2 + Q_{22}m^4, \tag{9.3.22}$$

$$\hat{Q}_{12} = (Q_{11} + Q_{22} - 4Q_{66})m^2n^2 + Q_{12}(m^4 + n^4), \tag{9.2.23}$$

$$\hat{Q}_{16} = (Q_{11} - Q_{12} - 2Q_{66})m^3n + (Q_{12} - Q_{22} + 2Q_{66})mn^3, \tag{9.3.24}$$

$$\hat{Q}_{26} = (Q_{11} - Q_{12} - 2Q_{66})mn^3 + (Q_{12} - Q_{22} + 2Q_{66})m^3n, \tag{9.3.25}$$

$$\hat{Q}_{66} = (Q_{11} + Q_{22} - 2Q_{12} - 2Q_{66})m^2n^2 + Q_{66}(m^4 + n^4), \tag{9.3.26}$$

in which $m = \cos\theta$, $n = \sin\theta$.

Since the lamina is very thin, its deformation can be represented by the deformation of its midplane using the Kirchhoff assumption, that is, a cross-section plane perpendicular to the midplane prior to the deformation remains perpendicular to the midplane after deformation. Mathematically, this means

$$\varepsilon_x = \frac{\partial u_0}{\partial x} - z\frac{\partial^2 w_0}{\partial x^2}, \qquad \varepsilon_y = \frac{\partial \nu_0}{\partial y} - z\frac{\partial^2 w_0}{\partial y^2}, \tag{9.3.27}$$

$$\gamma_{xy} = 2\varepsilon_{xy} = \frac{\partial u_0}{\partial y} + \frac{\partial \nu_0}{\partial x} - 2z\frac{\partial^2 w_0}{\partial x \partial y}, \tag{9.3.28}$$

where u_0 and ν_0 are the in-plane displacement of the midplane and w_0 is the out-of-plane displacement (deflection) of the midplane of the lamina. The coordinate z indicates the location relative to the middle plane.

Substitution of (9.3.27) and (9.3.28) into (9.3.20) yields

$$\begin{bmatrix} \sigma_x \\ \sigma_y \\ \sigma_{xy} \end{bmatrix} = \begin{bmatrix} \hat{Q}_{11} & \hat{Q}_{12} & \hat{Q}_{13} \\ \hat{Q}_{12} & \hat{Q}_{22} & \hat{Q}_{23} \\ \hat{Q}_{13} & \hat{Q}_{23} & \hat{Q}_{66} \end{bmatrix} \begin{bmatrix} \varepsilon_x^0 \\ \varepsilon_y^0 \\ \gamma_{xy}^0 \end{bmatrix} + z \begin{bmatrix} \hat{Q}_{11} & \hat{Q}_{12} & \hat{Q}_{13} \\ \hat{Q}_{12} & \hat{Q}_{22} & \hat{Q}_{23} \\ \hat{Q}_{13} & \hat{Q}_{23} & \hat{Q}_{66} \end{bmatrix} \begin{bmatrix} k_x \\ k_y \\ k_{xy} \end{bmatrix}, \tag{9.3.29}$$

where

$$\begin{bmatrix} \varepsilon_x^0 \\ \varepsilon_y^0 \\ \gamma_{xy}^0 \end{bmatrix} = \begin{bmatrix} \dfrac{\partial u_0}{\partial x} \\ \dfrac{\partial v_0}{\partial y} \\ \dfrac{\partial u_0}{\partial y} + \dfrac{\partial v_0}{\partial x} \end{bmatrix} \tag{9.3.30}$$

is the midplane strain and

$$\begin{bmatrix} k_x \\ k_y \\ k_{xy} \end{bmatrix} = - \begin{bmatrix} \dfrac{\partial^2 w_0}{\partial y^2} \\ \dfrac{\partial^2 w_0}{\partial y^2} \\ 2\dfrac{\partial^2 w_0}{\partial x \partial y} \end{bmatrix} \tag{9.3.31}$$

is the midplane curvature.

9.4 EFFECTIVE PROPERTIES OF A LAMINATED COMPOSITE PLATE

Composite laminates in engineering applications typically consist of multiple unidirectional fiber-reinforced laminas (or plies) stacked up together. The stacking sequence of a laminate can be designated by a standard notation. For example, $[0_2/90_2/-45_3/45_3]_s$ is used to designate a laminate that starting from the bottom, that is, at $z = -h/2$ (h is the total thickness of the laminate), there are two plies at 0° orientation ($\theta = 0°$), two plies at 90° orientation, followed by three plies at −45°, and last three plies at +45° orientation. The subscript s indicates that the laminate is symmetrical with respect to the midplane ($z = 0$), that is, the top half of the laminate is a mirror image of the bottom half. For nonsymmetrical laminates, the total stacking sequence needs to be explicitly indicated. In this case, a subscript T will be used.

For most engineering laminates of practical interest, the thickness of each ply is in the order of millimeters. The diameter of the fiber is typically in the order of micrometers. To bridge these two length scales, we will take a hierarchical approach in order to estimate the effective properties of the laminate. First, the effective properties of the unidirectional ply are evaluated based on the method discussed in the previous chapter. Once this is done, the effective properties of the laminate will be estimated by assuming each ply as a homogeneous, transversely isotropic layer.

Now consider a generic laminate shown in Figure 9.3. We assume that the bonding between the plies is perfect, that is, displacements and tractions are continuous across the ply interfaces. Since the stresses in a laminate vary from layer to layer, it is convenient to introduce the resultant forces and moments:

$$\begin{bmatrix} N_x \\ N_y \\ N_{xy} \end{bmatrix} = \int_{-h/2}^{h/2} \begin{bmatrix} \sigma_x \\ \sigma_y \\ \sigma_{xy} \end{bmatrix} dz, \qquad \begin{bmatrix} M_x \\ M_y \\ M_{xy} \end{bmatrix} = \int_{-h/2}^{h/2} \begin{bmatrix} \sigma_x \\ \sigma_y \\ \sigma_{xy} \end{bmatrix} z\, dz. \tag{9.4.1}$$

In terms of the resultants, (9.3.29) can be rewritten as

$$\begin{bmatrix} N_x \\ N_y \\ N_{xy} \end{bmatrix} = \begin{bmatrix} A_{11} & A_{12} & A_{13} \\ A_{12} & A_{22} & A_{23} \\ A_{13} & A_{23} & A_{66} \end{bmatrix} \begin{bmatrix} \varepsilon_x^0 \\ \varepsilon_y^0 \\ \gamma_{xy}^0 \end{bmatrix} + z \begin{bmatrix} B_{11} & B_{12} & B_{13} \\ B_{12} & B_{22} & B_{23} \\ B_{13} & B_{23} & B_{66} \end{bmatrix} \begin{bmatrix} k_x \\ k_y \\ k_{xy} \end{bmatrix}, \tag{9.4.2}$$

$$\begin{bmatrix} M_x \\ M_y \\ M_{xy} \end{bmatrix} = \begin{bmatrix} B_{11} & B_{12} & B_{13} \\ B_{12} & B_{22} & B_{23} \\ B_{13} & B_{23} & B_{66} \end{bmatrix} \begin{bmatrix} \varepsilon_x^0 \\ \varepsilon_y^0 \\ \gamma_{xy}^0 \end{bmatrix} + z \begin{bmatrix} D_{11} & D_{12} & D_{13} \\ D_{12} & D_{22} & D_{23} \\ D_{13} & D_{23} & D_{66} \end{bmatrix} \begin{bmatrix} k_x \\ k_y \\ k_{xy} \end{bmatrix}, \tag{9.4.3}$$

where

$$A_{ij} = \sum_{k=1}^{n} (\hat{Q}_{ij})_k (h_k - h_{k-1}), \tag{9.4.4}$$

$$B_{ij} = \frac{1}{2} \sum_{k=1}^{n} (\hat{Q}_{ij})_k (h_k^2 - h_{k-1}^2), \tag{9.4.5}$$

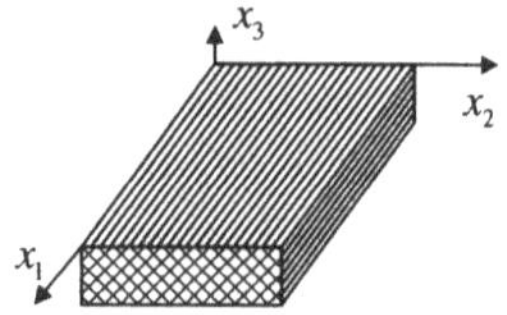

Figure 9.5

$$D_{ij} = \frac{1}{3} \sum_{k=1}^{n} (\hat{Q}_{ij})_k (h_k^3 - h_{k-1}^3) \tag{9.4.6}$$

are called the extensional stiffness matrix, coupling stiffness matrix, and bending stiffness matrix of the laminate, respectively.

PROBLEMS

9.1 Consider a unidirectional fiber-reinforced composite plate as shown in Figure 9.5 (fibers are in the x_1 direction). Find the lower and upper bounds for the effective Young modulus $\overline{E}_{11}$ and $\overline{E}_{22}$ by using the minimum potential and minimum complementary energy theorems, respectively. Compare your results with the rule of mixture predictions.

(Hint: Use $\sigma_{ij}^0 = \sigma_0 \delta_{i1} \delta_{j1}$ and $\varepsilon_{ij}^0 = \varepsilon_0 \delta_{i1} \delta_{j1}$ for $\overline{E}_{11}$. Use $\sigma_{ij}^0 = \sigma_0 \delta_{i2} \delta_{j2}$ and $\varepsilon_{ij}^0 = \varepsilon_0 \delta_{i2} \delta_{j2}$ for $\overline{E}_{22}$.)

9.2 Derive the extensional stiffness matrix, the coupling stiffness matrix, and the bending stiffness matrix of a generic laminate.

9.3 Prove (9.2.29)–(9.2.32).

9.4 Consider a fiber-reinforced cross-ply laminate as shown in Figure 9.4. Further, assume that all the layers have the same thickness and their elastic constants are given by E_L, E_T, μ_L, ν_{LT}, and ν_{TL}. They are related to the elastic properties of the fiber, the matrix, and the fiber volume fraction through the micromechanics schemes discussed in Section 9.1. Find the matrices **A**, **B**, and **D** as defined in (9.4.6)–(9.4.8).

APPENDIX 9.A

To facilitate the discussions below, let us first define the following:

$$\boldsymbol{\sigma}_n = \begin{bmatrix} \sigma_{33} \\ \sigma_{23} \\ \sigma_{13} \end{bmatrix}, \quad \boldsymbol{\sigma}_t = \begin{bmatrix} \sigma_{11} \\ \sigma_{22} \\ \sigma_{12} \end{bmatrix}, \quad \boldsymbol{\varepsilon}_n = \begin{bmatrix} \varepsilon_{33} \\ 2\varepsilon_{23} \\ 2\varepsilon_{13} \end{bmatrix}, \quad \boldsymbol{\varepsilon}_t = \begin{bmatrix} \varepsilon_{11} \\ \varepsilon_{22} \\ 2\varepsilon_{12} \end{bmatrix}. \tag{9.A.1}$$

Thus, we have

$$\boldsymbol{\varepsilon}_n = \mathbf{S}_{nt}\boldsymbol{\sigma}_t + \mathbf{S}_{nn}\boldsymbol{\sigma}_n, \qquad \boldsymbol{\varepsilon}_t = \mathbf{S}_{tt}\boldsymbol{\sigma}_t + \mathbf{S}_{tn}\boldsymbol{\sigma}_n, \tag{9.A.2}$$

where

$$\mathbf{S}_{nn} = \begin{bmatrix} S_{33} & S_{34} & S_{35} \\ S_{34} & S_{44} & S_{45} \\ S_{35} & S_{45} & S_{55} \end{bmatrix}, \qquad \mathbf{S}_{nt} = \begin{bmatrix} S_{13} & S_{23} & S_{36} \\ S_{14} & S_{24} & S_{46} \\ S_{15} & S_{25} & S_{56} \end{bmatrix}, \tag{9.A.3}$$

$$\mathbf{S}_{tn} = \begin{bmatrix} S_{13} & S_{14} & S_{15} \\ S_{23} & S_{24} & S_{25} \\ S_{36} & S_{46} & S_{56} \end{bmatrix}, \qquad \mathbf{S}_{tt} = \begin{bmatrix} S_{11} & S_{12} & S_{16} \\ S_{12} & S_{22} & S_{26} \\ S_{16} & S_{26} & S_{66} \end{bmatrix}. \tag{9.A.4}$$

Obviously,

$$\mathbf{S}_{nn} = (\mathbf{S}_{nn})^{\mathrm{T}}, \qquad \mathbf{S}_{tt} = (\mathbf{S}_{tt})^{\mathrm{T}}, \ \mathbf{S}_{tn} = (\mathbf{S}_{nt})^{\mathrm{T}}. \tag{9.A.5}$$

Also, because of the positive definiteness of the Voigt compliance matrix, we know that $\mathbf{S}_{nn}$ and $\mathbf{S}_{tt}$ are positive definite matrices.

Now, consider a multilayered composite consisting of N layers of homogeneous materials as shown in Figure 9.3. We assume that each layer is homogeneous with its Voigt elastic matrix given by $\mathbf{S}^{(k)}$, $k = 1, 2, \ldots, N$. Therefore, for the kth layer, it follows from (9.A.2) that

$$\boldsymbol{\varepsilon}_n^{(k)} = \mathbf{S}_{nn}^{(k)}\boldsymbol{\sigma}_n^{(k)} + \mathbf{S}_{nt}^{(k)}\boldsymbol{\sigma}_t^{(k)}, \qquad \boldsymbol{\varepsilon}_t^{(k)} = \mathbf{S}_{tn}^{(k)}\boldsymbol{\sigma}_n^{(k)} + \mathbf{S}_{tt}^{(k)}\boldsymbol{\sigma}_t^{(k)}, \tag{9.A.6}$$

where the superscript (k) indicates the quantity associated with the kth layer. It is assumed that the representative element being considered is under a uniform state of deformation, even though the actual composite may be subjected to arbitrary loading. We further assume that the layers are perfectly bonded together, meaning that the traction and displacements are continuous across the layer interfaces. This immediately leads to

$$\boldsymbol{\sigma}_n^{(k)} = \boldsymbol{\sigma}_n, \qquad \boldsymbol{\varepsilon}_t^{(k)} = \boldsymbol{\varepsilon}_t, \qquad \text{for } k = 1, 2, \ldots, N. \tag{9.A.7}$$

Making use of the above in the second of (9.A.6) yields

$$\boldsymbol{\sigma}_t^{(k)} = (\mathbf{S}_{tt}^{(k)})^{-1}(\boldsymbol{\varepsilon}_n - \mathbf{S}_{tn}^{(k)}\boldsymbol{\sigma}_n). \tag{9.A.8}$$

Taking the average of $\boldsymbol{\sigma}_t^{(k)}$ over the representative volume element leads to

$$\boldsymbol{\sigma}_n = \sum_{k=1}^{N} c_k \boldsymbol{\sigma}_t^{(k)} = \sum_{k=1}^{N} c_k (\mathbf{S}_{tt}^{(k)})^{-1} \boldsymbol{\varepsilon}_t - \sum_{k=1}^{N} c_k (\mathbf{S}_{tt}^{(k)})^{-1} \mathbf{S}_{tn}^{(k)} \boldsymbol{\sigma}_n, \tag{9.A.9}$$

where $c_k = h_k/h$ is the volume fraction of the kth layer. By rearranging the terms in (9.A.9), we have

$$\boldsymbol{\varepsilon}_t = \overline{\mathbf{S}}_{tt}\boldsymbol{\sigma}_t + \overline{\mathbf{S}}_{tn}\boldsymbol{\sigma}_n, \tag{9.A.10}$$

where

$$\overline{\mathbf{S}}_{tt} = \begin{bmatrix} \overline{S}_{11} & \overline{S}_{12} & \overline{S}_{16} \\ \overline{S}_{12} & \overline{S}_{22} & \overline{S}_{26} \\ \overline{S}_{16} & \overline{S}_{26} & \overline{S}_{66} \end{bmatrix} = \left[\sum_{k=1}^{N} c_k (\mathbf{S}_{tt}^{(k)})^{-1} \right]^{-1}, \tag{9.A.11}$$

$$\overline{\mathbf{S}}_{tn} = \begin{bmatrix} \overline{S}_{13} & \overline{S}_{14} & \overline{S}_{15} \\ \overline{S}_{23} & \overline{S}_{24} & \overline{S}_{25} \\ \overline{S}_{36} & \overline{S}_{46} & \overline{S}_{56} \end{bmatrix} = \overline{\mathbf{S}}_{tt} \left[\sum_{k=1}^{N} c_k (\mathbf{S}_{tt}^{(k)})^{-1} \mathbf{S}_{tn}^{(k)} \right]. \tag{9.A.12}$$

We next substitute (9.A.8) into the first of (9.A.6). This gives us

$$\boldsymbol{\varepsilon}_n^{(k)} = [\mathbf{S}_{nn}^{(k)} - \mathbf{S}_{nt}^{(k)}(\mathbf{S}_{tt}^{(k)})^{-1}\mathbf{S}_{tn}^{(k)}]\boldsymbol{\sigma}_n + \mathbf{S}_{nt}^{(k)}(\mathbf{S}_{tt}^{(k)})^{-1}\boldsymbol{\varepsilon}_t. \tag{9.A.13}$$

Taking the average of $\boldsymbol{\sigma}_t^{(k)}$ over the representative volume element leads to

$$\boldsymbol{\varepsilon}_n = \sum_{k=1}^{N} c_k \boldsymbol{\varepsilon}_n^{(k)} = \sum_{k=1}^{N} c_k [\mathbf{S}_{nn}^{(k)} - \mathbf{S}_{nt}^{(k)}(\mathbf{S}_{tt}^{(k)})^{-1}\mathbf{S}_{tn}^{(k)}]\boldsymbol{\sigma}_n + \sum_{k=1}^{N} c_k \mathbf{S}_{nt}^{(k)}(\mathbf{S}_{tt}^{(k)})^{-1}\boldsymbol{\varepsilon}_t.$$

Substitution of (9.A.10) into the above yields

$$\boldsymbol{\varepsilon}_n = \bar{\mathbf{S}}_{nn}\boldsymbol{\sigma}_n + \bar{\mathbf{S}}_{nt}\boldsymbol{\sigma}_t, \tag{9.A.14}$$

where

$$\bar{\mathbf{S}}_{nt} = \begin{bmatrix} \bar{S}_{13} & \bar{S}_{23} & \bar{S}_{36} \\ \bar{S}_{14} & \bar{S}_{24} & \bar{S}_{46} \\ \bar{S}_{15} & \bar{S}_{25} & \bar{S}_{56} \end{bmatrix} = \left[\sum_{k=1}^{N} c_k \mathbf{S}_{nt}^{(k)} (\mathbf{S}_{tt}^{(k)})^{-1} \right] \bar{\mathbf{S}}_{tt}, \tag{9.A.15}$$

$$\bar{\mathbf{S}}_{nn} = \begin{bmatrix} \bar{S}_{33} & \bar{S}_{34} & \bar{S}_{35} \\ \bar{S}_{34} & \bar{S}_{44} & \bar{S}_{45} \\ \bar{S}_{35} & \bar{S}_{45} & \bar{S}_{55} \end{bmatrix} = \sum_{k=1}^{N} c_k \mathbf{S}_{nn}^{(k)} + \sum_{k=1}^{N} c_k \mathbf{S}_{nt}^{(k)} (\mathbf{S}_{tt}^{(k)})^{-1} (\bar{\mathbf{S}}_{tn} - \mathbf{S}_{tn}^{(k)}). \tag{9.A.16}$$

Clearly, $\bar{\mathbf{S}}_{nn}$, $\bar{\mathbf{S}}_{nt}$, $\bar{\mathbf{S}}_{tn}$, and $\bar{\mathbf{S}}_{tt}$ provide the complete effective Voigt compliant matrix of the layered composite. Taking advantage of the symmetry properties (9.A.5), it can be shown that

$$\bar{\mathbf{S}}_{nn} = (\bar{\mathbf{S}}_{nn})^{\mathrm{T}}, \qquad \bar{\mathbf{S}}_{tt} = (\bar{\mathbf{S}}_{tt})^{\mathrm{T}}, \qquad \bar{\mathbf{S}}_{tn} = (\bar{\mathbf{S}}_{nt})^{\mathrm{T}}.$$

Next, we assume that each layer is a monoclinic material with its plane of symmetry in the x_1x_2 plane. In this case, the Voigt elastic matrix has the following forms:

$$\mathbf{S}_{nn}^{(k)} = \begin{bmatrix} S_{33}^{(k)} & 0 & 0 \\ 0 & S_{44}^{(k)} & S_{45}^{(k)} \\ 0 & S_{45}^{(k)} & S_{55}^{(k)} \end{bmatrix}, \qquad \mathbf{S}_{nt}^{(k)} = \begin{bmatrix} S_{13}^{(k)} & S_{23}^{(k)} & S_{36}^{(k)} \\ 0 & 0 & 0 \\ 0 & 0 & 0 \end{bmatrix}, \tag{9.A.17}$$

$$\mathbf{S}_{tn}^{(k)} = (\mathbf{S}_{nt}^{(k)})^{\mathrm{T}}, \qquad \mathbf{S}_{tt}^{(k)} = \begin{bmatrix} S_{11}^{(k)} & S_{12}^{(k)} & S_{16}^{(k)} \\ S_{12}^{(k)} & S_{22}^{(k)} & S_{26}^{(k)} \\ S_{16}^{(k)} & S_{26}^{(k)} & S_{66}^{(k)} \end{bmatrix}. \tag{9.A.18}$$

REFERENCE

Christensen, R. M. (1979). *Mechanics of Composite Materials,* Krieger, Malabar, FL.

SUGGESTED READINGS

Agarwal, B. D. and L. J. Broutman. (1990). *Analysis and Performance of Fiber Composites,* Wiley, New York.

Chawla, K. K. (1987). *Composite Materials—Science and Engineering,* Springer, New York.

Hyer, M. W. (1998). *Stress Analysis of Fiber-Reinforced Composite Materials,* McGraw-Hill, Boston.

10

BRITTLE DAMAGE AND FAILURE OF ENGINEERING COMPOSITES

Micromechanics has been used extensively in the study of damage and failure of engineering materials. There is a large body of literature in this area. It is certainly not possible, nor the intent of this chapter, to provide an extensive summary of existing results. The objective here is to take a few examples that are illustrative of how the micromechanics methods and techniques are being used to understand and model certain commonly seen damage and failure modes in engineering composite materials. We will discuss one example in each of the three areas; interfaces, fiber, and matrix.

10.1 IMPERFECT INTERFACES

Let the interface between two media be denoted by S. As discussed in Section 4.2, we call an interface S a perfect interface if the two media are mechanically coherent, that is, the displacement and traction fields are continuous across this interface. The continuity of displacement and traction across a perfect interface can be written as

$$\Delta u_i \equiv u_i(S^+) - u_i(S^-) = 0, \tag{10.1.1}$$

$$\Delta\sigma_{ij}n_j \equiv [\sigma_{ij}(S^+) - \sigma_{ij}(S^-)]n_j = 0, \tag{10.1.2}$$

where n_j is the unit normal vector of the interface S, and $u_i(S^+)$ and $\sigma_{ij}(S^+)$ are the values of $u_i(\mathbf{x})$ and $\sigma_{ij}(\mathbf{x})$ evaluated at the positive side of S, while $u_i(S^-)$ and $\sigma_{ij}(S^-)$ are the values of $u_i(\mathbf{x})$ and $\sigma_{ij}(\mathbf{x})$ evaluated at the negative side of S. It is assumed that n_j points to the positive side of S; see Figure 4.3.

In many engineering applications, the interfaces in a composite material may not be perfect. In this section, we will discuss one of the simplest models used to model the imperfect interfaces. For quasi-static problems, the imperfection of interfaces is reflected mainly by the discontinuity of displacement field across the interface. The interfacial spring model therefore postulates that the traction is continuous across the interface

$$\Delta\sigma_{ij}n_j \equiv [\sigma_{ij}(S^+) - \sigma_{ij}(S^-)]n_j = 0, \tag{10.1.3}$$

but the displacement field is discontinuous. The jump in the displacement field across the interface is proportional to the traction vector on the interface, that is,

$$\Delta u_i \equiv u_i(S^+) - u_i(S^-) = \eta_{ij}\sigma_{ik}n_k, \tag{10.1.4}$$

where the second-order tensor η_{ij} represents the compliance of the interface. For simplicity, we assume that η_{ij} is symmetric and positive definite. It is clear from (10.1.4) that $\eta_{ij} = 0$ corresponds to a perfect interface, while $\eta_{ij} \to \infty$ represents complete debond ($\sigma_{ik}n_k = 0$). From this point of view, a slight deviation from the perfect interface case can be modeled by the limit of $\eta_{ij} \to 0$.

A special form of η_{ij} that has some physical significance is given by

$$\eta_{ij} = \alpha\delta_{ij} + (\beta - \alpha)n_in_j. \tag{10.1.5}$$

It can be easily shown that α and β represent the compliance in the tangential and normal directions of the interfaces, respectively, that is,

$$\Delta u_i(\delta_{ik} - n_in_k) = \alpha\sigma_{ij}n_j(\delta_{ik} - n_in_k), \tag{10.1.6}$$

$$\Delta u_in_i = \beta\sigma_{ij}n_jn_i. \tag{10.1.7}$$

When $\beta = 0$, such constitutive characterization of the interfaces allows for relative sliding between the two surfaces, but not separation. Furthermore, the free-sliding case studied by Mura (1987) can be achieved by setting $\alpha \to \infty$ with $\beta = 0$. Therefore, solutions to the case of small α with $\beta = 0$ provide the approximations in contrast to the sliding interface. It also should be mentioned that material interpenetration may take place when $\beta \neq 0$, which is a violation of the compatibility requirement. Further discussions on this may be found in Achenbach and Zhu (1989).

For the remainder of this section, we focus on the effective elastic properties of composite materials with imperfect interfaces. First, we will solve the Eshelby inclusion problem with imperfect interface. Then, we will use the Mori–Tanaka approach to estimate the effective elastic properties. Finally, the bounds of the effective properties for such composite will be developed based on the classical variational principles.

Consider an elastic solid of infinite extent containing an ellipsoidal inclusion Ω with eigenstrain distribution ε_{ij}^*. Assume that the interface between the inclusion and the surrounding medium is imperfect and the displacement jump across the interface is governed by (10.1.6) and (10.1.7). We will call this eigenstrain problem the modified Eshelby inclusion problem. In what follows, we would like to develop a solution to this modified Eshelby inclusion problem.

It follows from (3.3.5) that the displacement field should satisfy the following equations of equilibrium:

$$L_{ijkl}u_{k,lj}(\mathbf{x}) - L_{ijkl}\varepsilon_{kl,j}^*(\mathbf{x}) = 0 \quad \text{in } \Omega. \tag{10.1.8}$$

$$u_i(\mathbf{x}) \to 0 \quad \text{as } |\mathbf{x}| \to \infty. \tag{10.1.9}$$

Furthermore, it follows from (2.6.23) and (2.6.24) that the Green function $G_{ij}^\infty(\mathbf{x}, \mathbf{y})$ satisfies

$$L_{ijkl}\frac{\partial^2 G_{km}^\infty(\mathbf{x}, \mathbf{y})}{\partial x_l\, \partial x_j} + \delta_{im}\delta(\mathbf{x} - \mathbf{y}) = 0, \tag{10.1.10}$$

$$G_{ij}^\infty(\mathbf{x}, \mathbf{y}) \to 0 \quad \text{as } |\mathbf{x}| \to \infty, \tag{10.1.11}$$

where $\delta(\mathbf{x})$ is the three-dimensional Dirac delta function.

Now, let us multiply (10.1.10) by $u_i(\mathbf{x})$ and integrate the resulting equation over a volume V. This leads to

$$\int_V u_i(\mathbf{x})L_{ijkl}\frac{\partial^2 G_{km}^\infty(\mathbf{x}, \mathbf{y})}{\partial x_l\, \partial x_j}\, dV(\mathbf{x}) = \begin{cases} -u_m(\mathbf{y}) & \mathbf{y} \in V \\ 0 & \mathbf{y} \notin V \end{cases}. \tag{10.1.12}$$

By using the divergence theorem, we can rewrite the volume integral in (10.1.12) as

$$\int_V u_i(\mathbf{x}) L_{ijkl} \frac{\partial^2 G_{km}(\mathbf{x}, \mathbf{y})}{\partial x_l \, \partial x_j} \, dV(\mathbf{x})$$
$$= \int_S u_i(\mathbf{x}) L_{ijkl} \frac{\partial G_{km}(\mathbf{x}, \mathbf{y})}{\partial x_l} n_j \, dS(\mathbf{x}) - \int_V u_{i,j}(\mathbf{x}) L_{ijkl} \frac{\partial G_{km}(\mathbf{x}, \mathbf{y})}{\partial x_l} \, dV(\mathbf{x}), \tag{10.1.13}$$

where S is the surface of V, and n_i is the unit outward normal of S.

Next, multiplying (10.1.8) by $G_{im}(\mathbf{x}, \mathbf{y})$ and integrating the result over the volume V lead to

$$\int_V G_{im}^{\infty}(\mathbf{x}, \mathbf{y}) L_{ijkl} u_{k,lj}(\mathbf{x}) \, dV(\mathbf{x}) - \int_V G_{im}^{\infty}(\mathbf{x}, \mathbf{y}) L_{ijkl} \varepsilon_{kl,j}^{*}(\mathbf{x}) \, dV(\mathbf{x}) = 0. \tag{10.1.14}$$

Following the procedures that led to (10.1.13), the first volume integral in the above equation can be written as

$$\int_V G_{im}^{\infty}(\mathbf{x}, \mathbf{y}) L_{ijkl} u_{k,lj}(\mathbf{x}) \, dV(\mathbf{x})$$
$$= \int_S G_{im}^{\infty}(\mathbf{x}, \mathbf{y}) L_{ijkl} u_{k,l}(\mathbf{x}) n_j \, dS(\mathbf{x}) - \int_V \frac{\partial G_{im}^{\infty}(\mathbf{x}, \mathbf{y})}{\partial x_j} L_{ijkl} u_{k,l}(\mathbf{x}) \, dV(\mathbf{x}). \tag{10.1.15}$$

Note that the last volume integral on the right-hand side of (10.1.13) is the same as that of (10.1.15). Therefore, by subtracting (10.1.12) from (10.1.14), we arrive at

$$\int_S L_{ijkl} \left[G_{im}^{\infty}(\mathbf{x}, \mathbf{y}) u_{k,l}(\mathbf{x}) - u_i(\mathbf{x}) \frac{\partial G_{im}^{\infty}(\mathbf{x}, \mathbf{y})}{\partial x_l} \right] n_j \, dS(\mathbf{x})$$
$$- \int_V G_{im}^{\infty}(\mathbf{x}, \mathbf{y}) L_{ijkl} \varepsilon_{kl,j}^{*}(\mathbf{x}) \, dV(\mathbf{x}) = \begin{cases} u_m(\mathbf{y}) & \mathbf{y} \in V \\ 0 & \mathbf{y} \notin V \end{cases}. \tag{10.1.16}$$

Finally, making use of the divergence theorem on the volume integral, we arrive at

$$\int_S L_{ijkl}\left\{G^{\infty}_{im}(\mathbf{x}, \mathbf{y})[u_{k,l}(\mathbf{x}) - \varepsilon^*_{kl}(\mathbf{x})] - u_i(\mathbf{x}) \frac{\partial G^{\infty}_{im}(\mathbf{x}, \mathbf{y})}{\partial x_l}\right\} n_j \, dS(\mathbf{x})$$

$$+ \int_V \frac{\partial G^{\infty}_{im}(\mathbf{x}, \mathbf{y})}{\partial x_l} L_{ijkl}\varepsilon^*_{kl}(\mathbf{x}) \, dV(\mathbf{x}) = \begin{cases} u_m(\mathbf{y}) & \mathbf{y} \in V \\ 0 & \mathbf{y} \notin V \end{cases}. \qquad (10.1.17)$$

Now, taking the region V to be the inclusion Ω, and letting $\mathbf{y} \in \Omega$ where the eigenstrain is given, the above integral representation yields

$$u_m(\mathbf{y}) = \int_{S^-} L_{ijkl}\left\{G^{\infty}_{im}(\mathbf{x}, \mathbf{y})[u_{k,l}(\mathbf{x}) - \varepsilon^*_{kl}(\mathbf{x})] - u_i(\mathbf{x}) \frac{\partial G^{\infty}_{im}(\mathbf{x}, \mathbf{y})}{\partial x_l}\right\} n_j \, dS(\mathbf{x})$$

$$+ \int_V \frac{\partial G^{\infty}_{im}(\mathbf{x}, \mathbf{y})}{\partial x_l} L_{ijkl}\varepsilon^*_{kl}(\mathbf{x}) \, dV(\mathbf{x}). \qquad (10.1.18)$$

Note that for $\mathbf{y} \in S^-$, Hooke's law (3.3.2) yields

$$L_{ijkl}[u_{k,l}(\mathbf{x}) - \varepsilon^*_{kl}(\mathbf{x})] = \sigma_{ij}(\mathbf{x}). \qquad (10.1.19)$$

Therefore, (10.1.18) can be rewritten as

$$u_m(\mathbf{y}) = \int_{S^-} \left[G^{\infty}_{im}(\mathbf{x}, \mathbf{y})\sigma_{kl}(\mathbf{x}) - u_i(\mathbf{x})L_{ijkl} \frac{\partial G^{\infty}_{im}(\mathbf{x}, \mathbf{y})}{\partial x_l}\right] n_j \, dS(\mathbf{x})$$

$$+ \int_V \frac{\partial G^{\infty}_{im}(\mathbf{x}, \mathbf{y})}{\partial x_l} L_{ijkl}\varepsilon^*_{kl}(\mathbf{x}) \, dV(\mathbf{x}). \qquad (10.1.20)$$

Next, take V to be the region exterior to Ω. Then, for $\mathbf{y}$ inside Ω, or equivalently, for $\mathbf{y}$ outside V where the eigenstrain is zero, (10.1.17) yields

$$\int_{S^+} L_{ijkl}\left[G^{\infty}_{im}(\mathbf{x}, \mathbf{y})u_{k,l}(\mathbf{x}) - u_i(\mathbf{x}) \frac{\partial G^{\infty}_{im}(\mathbf{x}, \mathbf{y})}{\partial x_l}\right] n_j \, dS(\mathbf{x}) = 0, \qquad (10.1.21)$$

or, make use of Hooke's law,

$$\int_{S^+} \left[G^{\infty}_{im}(\mathbf{x}, \mathbf{y})\sigma_{kl}(\mathbf{x}) - u_i(\mathbf{x})L_{ijkl} \frac{\partial G^{\infty}_{im}(\mathbf{x}, \mathbf{y})}{\partial x_l} \right] n_j \, dS(\mathbf{x}) = 0. \quad (10.1.22)$$

Note that both (10.1.20) and (10.1.22) are valid for $\mathbf{y}$ inside Ω. Thus, it follows from subtracting (10.1.21) from (10.1.18) that

$$u_m(\mathbf{x}) = \int_V \frac{\partial G^{\infty}_{im}(\mathbf{x}, \mathbf{y})}{\partial y_l} L_{ijkl}\varepsilon^*_{kl}(\mathbf{y}) \, dV(\mathbf{y}) + \int_S \Delta u_i(\mathbf{y})L_{ijkl} \frac{\partial G^{\infty}_{im}(\mathbf{x}, \mathbf{y})}{\partial y_l} n_j \, dS(\mathbf{y}), \quad (10.1.23)$$

where we have used the traction continuity condition (10.1.3). For convenience, we have also switched the variables $\mathbf{x}$ and $\mathbf{y}$ and made use of the symmetry properties of the infinite domain Green's function (2.6.26). The above equation is an integral representation of the displacement field within the inclusion. An integral equation can be derived by letting the point $\mathbf{x}$ approach the boundary S. Solving such integral equations typically requires numerical methods. In what follows, we will discuss some special cases of (10.1.23) and develop approximate and asymptotic solutions.

First, we assume that the eigenstrain is uniform on Ω. Then, (10.1.23) leads to

$$u_{m,n}(\mathbf{x}) = \int_V \frac{\partial G^{\infty}_{im}(\mathbf{x}, \mathbf{y})}{\partial y_l \, \partial x_n} L_{ijkl}\varepsilon^*_{kl} \, dV(\mathbf{y}) + \int_S \Delta u_i(\mathbf{y})L_{ijkl} \frac{\partial G^{\infty}_{im}(\mathbf{x}, \mathbf{y})}{\partial y_l \, \partial x_n} n_j \, dS(\mathbf{y}), \quad (10.1.24)$$

or

$$\varepsilon_{ij}(\mathbf{x}) = L_{klmn}\varepsilon^*_{kl} \int_\Omega \Gamma^{\infty}_{ijmn}(\mathbf{y}, \mathbf{x}) \, dV(\mathbf{y}) + \int_S L_{klmn} \, \Delta u_k(\mathbf{y})\Gamma^{\infty}_{ijmn}(\mathbf{y}, \mathbf{x})n_l \, dS(\mathbf{y}), \quad (10.1.25)$$

where $\Gamma^{\infty}_{ijmn}(\mathbf{y}, \mathbf{x})$ is defined by (4.3.6). It then follows from (4.3.8) that

$$\varepsilon_{ij}(\mathbf{x}) = \varepsilon^*_{kl}S_{ijkl} + \int_S L_{klmn} \, \Delta u_k(\mathbf{y})\Gamma^{\infty}_{ijmn}(\mathbf{y}, \mathbf{x}) \, n_l \, dS(\mathbf{y}), \quad (10.1.26)$$

where S_{ijkl} is the Eshelby inclusion tensor. Making use of (10.1.4) in (10.1.26) gives

$$\varepsilon_{ij}(\mathbf{x}) = S_{ijkl}\varepsilon_{kl}^* + \int_S L_{klmn}\eta_{kp}\sigma_{pq}(\mathbf{y})\Gamma_{ijmn}^{\infty}(\mathbf{y}, \mathbf{x})n_q n_l \, dS(\mathbf{y}). \quad (10.1.27)$$

Clearly, the first term in (10.1.27) is identical to the original Eshelby solution. The effect of imperfect interface comes through the surface integral involving the displacement jump at the interface. Since perfect interface corresponds to $\eta_{ij} = 0$, a weakly imperfect interface may be modeled by a very small η_{ij}. Thus, (10.1.27) provides a convenient basis for constructing asymptotic solutions for the imperfect interface problem.

To this end, we make use of Hooke's law (3.3.2) in (10.1.27) to obtain

$$\begin{aligned}\varepsilon_{ij}(\mathbf{x}) &= \varepsilon_{kl}^* S_{ijkl} - \int_S L_{klmn}L_{pqst}\eta_{kp}\varepsilon_{st}^*\Gamma_{ijmn}^{\infty}(\mathbf{y}, \mathbf{x})n_q n_i \, dS(\mathbf{y}) \\ &\quad + \int_S L_{klmn}L_{pqst}\eta_{kp}\varepsilon_{st}(\mathbf{y})\Gamma_{ijmn}^{\infty}(\mathbf{y}, \mathbf{x})n_q n_l \, dS(\mathbf{y}). \quad (10.1.28)\end{aligned}$$

An iteration procedure can now be formulated,

$$\begin{aligned}\varepsilon_{ij}^{(n)}(\mathbf{x}) &= \varepsilon_{kl}^* S_{ijkl} - \int_S L_{klmn}L_{pqst}\eta_{kp}\varepsilon_{st}^*\Gamma_{ijmn}^{\infty}(\mathbf{y}, \mathbf{x})n_q n_l \, dS(\mathbf{y}) \\ &\quad + \int_S L_{klmn}L_{pqst}\eta_{kp}\varepsilon_{st}^{(n-1)}(\mathbf{y})\Gamma_{ijmn}^{\infty}(\mathbf{y}, \mathbf{x})n_q n_l \, dS(\mathbf{y})\end{aligned}$$

for $n = 1, 2, \ldots$. The initial value can be set to

$$\varepsilon_{ij}^{(0)} = S_{ijkl}\varepsilon_{ij}^*. \quad (10.1.29)$$

The leading order solution for small η_{ij} can now be obtained as

$$\varepsilon_{ij}(\mathbf{x}) \approx \varepsilon_{ij}^{(1)}(\mathbf{x}) = S_{ijkl}\varepsilon_{kl}^* + Z_{ijst}(\mathbf{x})(I_{stkl} - S_{stkl})\varepsilon_{kl}^*, \quad (10.1.30)$$

where

$$Z_{ijst}(\mathbf{x}) = L_{klmn}L_{stpq}\int_S \eta_{kp}\Gamma_{ijmn}^{\infty}(\mathbf{y}, \mathbf{x})n_q n_l \, dS(\mathbf{y}). \quad (10.1.31)$$

Alternatively, (10.1.30) can also be written as

$$\varepsilon_{ij}(\mathbf{x}) = S^M_{ijkl}(\mathbf{x})\varepsilon^*_{kl}, \tag{10.1.32}$$

where

$$S^M_{ijkl}(\mathbf{x}) = S_{ijkl} + Z_{ijst}(\mathbf{x})(I_{stkl} - S_{stkl}) \tag{10.1.33}$$

can be viewed as the modified Eshelby inclusion tensor for an ellipsoidal inclusion with weakly imperfect interface. One significant difference of the modified Eshelby inclusion problem is that the strain within the ellipsoidal inclusion is no longer uniform even though the eigenstrain is. A quantity that will be needed later is the average of the modified Eshelby tensor

$$\bar{S}^M_{ijkl}(\mathbf{x}) = \frac{1}{\Omega}\int_\Omega S^M_{ijkl}(\mathbf{x})\, dV. \tag{10.1.34}$$

To carry out the integral, consider

$$\begin{aligned}\int_\Omega Z_{ijst}(\mathbf{x})\, dV &= L_{klmn}L_{stpq}\int_S \eta_{kp}\left[\int_\Omega \Gamma^\infty_{ijmn}(\mathbf{y}, \mathbf{x})\, dV(\mathbf{x})\right] n_q n_l\, dS(\mathbf{y})\\ &= L_{klmn}L_{stpq}\int_S \eta_{kp}P^\infty_{ijmn}n_q n_l\, dS(\mathbf{y})\\ &= S_{ijmn}L_{stpq}\int_S \eta_{mp}n_q n_n\, dS(\mathbf{y}),\end{aligned} \tag{10.1.35}$$

where (4.3.7) has been used. Making use of the above, the integral in (10.1.34) can be carried out to yield

$$\bar{S}^M_{ijkl} = S_{ijkl} + S_{ijmn}R_{mnpq}L_{pqst}(I_{stkl} - S_{stkl}), \tag{10.1.36}$$

where

$$R_{mnpq} = \frac{1}{4\Omega}\int_S (\eta_{mp}n_q n_n + \eta_{mq}n_p n_n + \eta_{np}n_q n_m + \eta_{nq}n_p n_m)\, dS(\mathbf{y}). \tag{10.1.37}$$

Clearly, we have $R_{mnpq} = R_{nmpq} = R_{nmqp} = R_{mnqp}$. Note that R_{mnpq} depends on the interface properties through η_{ij} and the geometry of the inclusion. Expressions of R_{mnpq} for ellipsoids, cylinders, and spheres are given in Appendix 10.A.

Now, we are ready to consider a composite material consisting of randomly oriented and shaped inhomogeneities embedded in a matrix with stiffness tensor $\mathbf{L}_0$. Let the stiffness tenors of the inhomogeneities be $\mathbf{L}_1, \mathbf{L}_2, \mathbf{L}_3, \ldots, \mathbf{L}_N$, as shown in Figure 7.1. We further assume that interfaces between the inhomogeneities and the matrix are not perfect. The interfacial jump condition is given by (10.1.4).

We first consider the average stress and average strain fields in the composite. It follows from (5.4.1) and (5.4.2) that the average stress over the entire representative volume element of the composite is

$$\overline{\boldsymbol{\sigma}} = \frac{1}{D}\int_D \boldsymbol{\sigma}\, dV = \sum_{r=0}^{N} c_r \overline{\boldsymbol{\sigma}}_r, \tag{10.1.38}$$

where

$$\overline{\boldsymbol{\sigma}}_r = \frac{1}{\Omega_r}\int_{\Omega_r} \boldsymbol{\sigma}\, dV$$

is the average stress on the rth inhomogeneity, while the average strain over the entire representative volume element of the composite is given by

$$\overline{\boldsymbol{\varepsilon}} = \frac{1}{D}\int_D \boldsymbol{\varepsilon}\, dV = \sum_{r=0}^{N} c_r \overline{\boldsymbol{\varepsilon}}_r + \frac{1}{2D}\sum_{r=1}^{N}\int_{S_r} (\Delta\mathbf{u} \otimes \mathbf{n} + \mathbf{n} \otimes \Delta\mathbf{u})\, dS, \tag{10.1.39}$$

where S_r is the surface of Ω_r, $\overline{\boldsymbol{\varepsilon}}_r$ is the average stain in the rth inhomogeneity

$$\overline{\boldsymbol{\varepsilon}}_r = \frac{1}{\Omega_r}\int_{\Omega_r} \boldsymbol{\varepsilon}\, dV, \tag{10.1.40}$$

and the symbol $\otimes$ indicates a dyad, see Section 1.4. By using the displacement jump condition (10.1.4), the surface integral in (10.1.39) can be written as

$$\int_{S_r} (\Delta\mathbf{u} \otimes \mathbf{n} + \mathbf{n} \otimes \Delta\mathbf{u})\, dS = \int_{S_r} [(\boldsymbol{\eta}\cdot\boldsymbol{\sigma}\cdot\mathbf{n}) \otimes \mathbf{n} + \mathbf{n} \otimes (\boldsymbol{\eta}\cdot\boldsymbol{\sigma}\cdot\mathbf{n})]\, dS. \tag{10.1.41}$$

The integral is rather difficult to evaluate. So, we will make an ad hoc approximation (Qu, 1993a),

$$\int_{S_r} [(\boldsymbol{\eta} \cdot \boldsymbol{\sigma} \cdot \mathbf{n}) \otimes \mathbf{n} + \mathbf{n} \otimes (\boldsymbol{\eta} \cdot \boldsymbol{\sigma} \cdot \mathbf{n})]\, dS$$
$$\approx \int_{S_r} [(\boldsymbol{\eta} \cdot \overline{\boldsymbol{\sigma}}_r \cdot \mathbf{n}) \otimes \mathbf{n} + \mathbf{n} \otimes (\boldsymbol{\eta} \cdot \overline{\boldsymbol{\sigma}}_r \cdot \mathbf{n})]\, dS, \qquad (10.1.42)$$

that is, replacing the stress distribution along S_r by the average stress on Ω_r. Clearly, (10.1.42) would be exact if the stress is uniform on Ω_r. The right-hand side of (10.1.42) can be further written as

$$\int_{S_r} [(\boldsymbol{\eta} \cdot \overline{\boldsymbol{\sigma}}_r \cdot \mathbf{n}) \otimes \mathbf{n} + \mathbf{n} \otimes (\boldsymbol{\eta} \cdot \overline{\boldsymbol{\sigma}}_r \cdot \mathbf{n})]\, dS = 2\Omega_r \mathbf{R}_r \overline{\boldsymbol{\sigma}}_r, \qquad (10.1.43)$$

where the fourth-order tensor $\mathbf{R}_r$ is defined by (10.1.37). Finally, we can write the average strain on the composite as

$$\overline{\boldsymbol{\varepsilon}} = \sum_{r=0}^{N} c_r \overline{\boldsymbol{\varepsilon}}_r + \sum_{r=1}^{N} c_r \mathbf{R}_r \overline{\boldsymbol{\sigma}}_r. \qquad (10.1.44)$$

Let us now consider the effective elastic properties of the composite. To this end, let the representative volume element of the composite be subjected to the following traction boundary condition:

$$\boldsymbol{\sigma} \cdot \mathbf{n}|_S = \overline{\boldsymbol{\sigma}} \cdot \mathbf{n}, \qquad (10.1.45)$$

where $\overline{\boldsymbol{\sigma}}$ is the average stress over the representative volume element of the composite, according to the average stress theorem introduced in Section 5.4 (It can be easily shown that the average stress theorem is still valid in this case). The Mori–Tanaka method assumes that the average strain in the rth inhomogeneity can be written as [see (7.3.5)]

$$\overline{\boldsymbol{\varepsilon}}_r = \overline{\boldsymbol{\varepsilon}}_0 + \overline{\boldsymbol{\varepsilon}}_r^{pt}, \qquad r = 1, 2, \ldots, N, \qquad (10.1.46)$$

where $\overline{\boldsymbol{\varepsilon}}_r^{pt}$ is the average $\overline{\boldsymbol{\varepsilon}}_r^{pt}$ perturbance of the strain field in the rth inhomogeneity. The corresponding average stress is

$$\overline{\boldsymbol{\sigma}}_r = \mathbf{L}_r \overline{\boldsymbol{\varepsilon}}_r = \mathbf{L}_r(\overline{\boldsymbol{\varepsilon}}_0 + \overline{\boldsymbol{\varepsilon}}_r^{pt}), \qquad r = 1, 2, \ldots, N. \qquad (10.1.47)$$

The Mori–Tanaka method further assumes that each inhomogeneity Ω_r $(r > 0)$ is embedded, in the absence of other inhomogeneities, in a uniform material of stiffness $\mathbf{L}_0$, which was prestressed by $\overline{\boldsymbol{\sigma}}_0$. When the inhomogeneity is simulated by an inclusion with eigenstrain $\boldsymbol{\varepsilon}_r^*$, the equivalent inclusion equation becomes (see Section 4.4)

$$\overline{\boldsymbol{\sigma}}_r = \mathbf{L}_r(\overline{\boldsymbol{\varepsilon}}_0 + \overline{\boldsymbol{\varepsilon}}_r^{pt}) = \mathbf{L}_0(\overline{\boldsymbol{\varepsilon}}_0 + \overline{\boldsymbol{\varepsilon}}_r^{pt} - \boldsymbol{\varepsilon}_r^*). \tag{10.1.48}$$

It is further assumed that

$$\overline{\boldsymbol{\varepsilon}}_r^{pt} = \overline{\mathbf{S}}_r^M \boldsymbol{\varepsilon}_r^*, \tag{10.1.49}$$

where $\overline{\mathbf{S}}_r^M$ is the modified Eshelby tensor averaged over Ω_r as given by (10.1.36). Substituting (10.1.49) into (10.1.48) yields an equation for $\boldsymbol{\varepsilon}_r^*$, which can be solved to obtain

$$\boldsymbol{\varepsilon}_r^* = [(\mathbf{L}_r - \mathbf{L}_0)\overline{\mathbf{S}}_r^M + \mathbf{L}_0]^{-1}(\mathbf{L}_r - \mathbf{L}_0)\overline{\boldsymbol{\varepsilon}}_0. \tag{10.1.50}$$

The total strain in the rth inhomogeneity can then be written as

$$\overline{\boldsymbol{\varepsilon}}_r = \overline{\boldsymbol{\varepsilon}}_0 + \overline{\boldsymbol{\varepsilon}}_r^{pt} = \overline{\boldsymbol{\varepsilon}}_0 + \overline{\mathbf{S}}_r^M \boldsymbol{\varepsilon}_r^* = \mathbf{T}_r^M \overline{\boldsymbol{\varepsilon}}_0, \tag{10.1.51}$$

where

$$\mathbf{T}_r^M = [\mathbf{I} + \overline{\mathbf{S}}_r^M \mathbf{L}_0^{-1}(\mathbf{L}_r - \mathbf{L}_0)]^{-1}. \tag{10.1.52}$$

The above derivation is very similar to the steps leading to (7.1.6). The average stress on the rth inhomogeneity then follows from substituting (10.1.51) into (10.1.47),

$$\overline{\boldsymbol{\sigma}}_r = \mathbf{L}_r \mathbf{T}_r^M \overline{\boldsymbol{\varepsilon}}_0 = \mathbf{H}_r^M \overline{\boldsymbol{\sigma}}_0, \tag{10.1.53}$$

where

$$\mathbf{H}_r^M = \mathbf{L}_r \mathbf{T}_r^M \mathbf{M}_0$$

is the local stress concentration tensor; see (7.3.11). Note that, although (10.1.53) was derived for $r > 0$, it can be easily see that it is also valid for $r = 0$. Making use of (10.1.53) in the average stress leads to

$$\bar{\boldsymbol{\sigma}} = \sum_{r=0}^{N} c_r \bar{\boldsymbol{\sigma}}_r = \sum_{r=0}^{N} c_r \mathbf{H}_r^M \bar{\boldsymbol{\sigma}}_0. \tag{10.1.54}$$

Consequently,

$$\bar{\boldsymbol{\sigma}}_0 = \left(\sum_{r=0}^{N} c_r \mathbf{H}_r^M \right)^{-1} \bar{\boldsymbol{\sigma}}. \tag{10.1.55}$$

Introducing (10.1.55) back to (10.1.53) yields

$$\bar{\boldsymbol{\sigma}}_r = \mathbf{B}_r^M \bar{\boldsymbol{\sigma}}, \tag{10.1.56}$$

where

$$\mathbf{B}_r^M = \mathbf{H}_r^M \left(\sum_{r=0}^{N} c_r \mathbf{H}_r^M \right)^{-1} \tag{10.1.57}$$

is the global stress concentration tensor; see (7.3.12). It then follows from (10.1.56) that the average strain in the rth inhomogeneity is

$$\bar{\boldsymbol{\varepsilon}}_r = \mathbf{M}_r \mathbf{B}_r^M \bar{\boldsymbol{\sigma}}. \tag{10.1.58}$$

Making use of (10.1.56) and (10.1.58) in (10.1.44) yields

$$\bar{\boldsymbol{\varepsilon}} = \sum_{r=0}^{N} c_r \mathbf{M}_r \mathbf{B}_r^M \bar{\boldsymbol{\sigma}} + \sum_{r=1}^{N} c_r \mathbf{R}_r \mathbf{B}_r^M \bar{\boldsymbol{\sigma}}.$$

This immediately leads to

$$\begin{aligned}
\overline{\mathbf{M}} &= \sum_{r=0}^{N} c_r \mathbf{M}_r \mathbf{B}_r^M + \sum_{r=1}^{N} c_r \mathbf{R}_r \mathbf{B}_r^M \\
&= c_0 \mathbf{M}_0 \mathbf{B}_0^M + \sum_{r=1}^{N} c_r (\mathbf{M}_r + \mathbf{R}_r) \mathbf{B}_r^M \\
&= \left[c_0 \mathbf{M}_0 + \sum_{r=1}^{N} c_r (\mathbf{M}_r + \mathbf{R}_r) \mathbf{L}_r \mathbf{T}_r^M \mathbf{M}_0 \right] \left(\sum_{r=0}^{N} c_r \mathbf{L}_r \mathbf{T}_r^M \mathbf{M}_0 \right)^{-1} \\
&= \left[c_0 \mathbf{I} + \sum_{r=1}^{N} c_r (\mathbf{M}_r + \mathbf{R}_r) \mathbf{L}_r \mathbf{T}_r^M \right] \left(\sum_{r=0}^{N} c_r \mathbf{L}_r \mathbf{T}_r^M \right)^{-1}.
\end{aligned} \tag{10.1.59}$$

This gives the Mori–Tanaka estimate of the effective compliance of the composite material with imperfect interfaces:

$$\overline{\mathbf{M}} = \left[c_0\mathbf{I} + \sum_{r=1}^{N} c_r(\mathbf{M}_r + \mathbf{R}_r)\mathbf{L}_r\mathbf{T}_r^M\right]\left(\sum_{r=0}^{N} c_r\mathbf{L}_r\mathbf{T}_r^M\right)^{-1}. \qquad (10.1.60)$$

By specifying displacement boundary conditions, one can obtain the effective stiffness tensor; see Problem 10.1:

$$\overline{\mathbf{L}} = \left(\sum_{r=0}^{N} c_r\mathbf{L}_r\mathbf{T}_r^M\right)\left[c_0\mathbf{I} + \sum_{r=1}^{N} c_r(\mathbf{M}_r + \mathbf{R}_r)\mathbf{L}_r\mathbf{T}_r^M\right]^{-1}. \qquad (10.1.61)$$

Obviously, we have

$$\overline{\mathbf{L}}\overline{\mathbf{M}} = \overline{\mathbf{M}}\overline{\mathbf{L}} = \mathbf{I}. \qquad (10.1.62)$$

Furthermore, it can be easily verified that when $\boldsymbol{\eta} \to 0$, we have $\mathbf{R}_r \to 0$ and $\mathbf{T}_r^M \to \mathbf{T}_r$. Therefore, the effective elastic properties (10.1.59) and (10.1.61) reduce to those of the perfect interface cases (7.3.14) and (7.3.8), respectively.

Example 10.1 Consider a composite consisting of an isotropic matrix with $\mathbf{L}_0$ and isotropic, spherical particles of $\mathbf{L}_1$. The interface between the particles and the matrix is described by the displacement jump condition (10.1.4) where

$$\eta_{ij} = \alpha(\delta_{ij} - n_i n_j). \qquad (10.1.63)$$

We are to find the effective elastic properties of the composite.

Since the materials are isotropic, we may represent them as

$$\mathbf{L}_0 = (3K_0, 2\mu_0), \qquad \mathbf{L}_1 = (3K_1, 2\mu_1). \qquad (10.1.64)$$

The composite should also be isotropic, that is, $\overline{\mathbf{L}} = (3\overline{K}, 2\overline{\mu})$. It follows from Appendix 10.A that

$$\mathbf{R}_1 = \frac{2\alpha}{5d}(0, 3), \qquad (10.1.65)$$

where d is the diameter of the particles. The modified Eshelby tensor is thus given by

$$\bar{\mathbf{S}}_1^M = \mathbf{S}_1 + \mathbf{S}_1\mathbf{R}_1\mathbf{L}_0(\mathbf{I} - \mathbf{S}_1) = (3\gamma_0^M, 2\delta_0^M), \qquad (10.1.66)$$

where

$$(3\gamma_0^M, 2\delta_0^M) = \left(3\gamma_0, 2\delta_0 + \frac{24\alpha\mu_0\delta_0}{5d}(1 - 2\delta_0)\right). \qquad (10.1.67)$$

Thus,

$$\mathbf{T}_1^M = [\mathbf{I} + \bar{\mathbf{S}}_1^M\mathbf{L}_0^{-1}(\mathbf{L}_1 - \mathbf{L}_0)]^{-1}$$

$$= \left(\frac{K_0}{K_0 + 3\gamma_0^M(K_1 - K_0)}, \frac{\mu_0}{\mu_0 + 2\delta_0^M(\mu_1 - \mu_0)}\right). \qquad (10.1.68)$$

The effective stiffness tensor is obtained from

$$\bar{\mathbf{L}} = (c_0\mathbf{L}_0 + c_1\mathbf{L}_1\mathbf{T}_1^M)(c_0\mathbf{I} + c_1(\mathbf{M}_1 + \mathbf{R}_1)\mathbf{L}_1\mathbf{T}_1^M)^{-1}. \qquad (10.1.69)$$

Carrying out the matrix algebra, we finally arrive at

$$\bar{K} = K_0 + \frac{c_1K_0(K_1 - K_0)}{K_0 + 3\gamma_0(1 - c_1)(K_1 - K_0)}, \qquad (10.1.70)$$

$$\bar{\mu} = \mu_0 + \frac{c_1\mu_0[5d(\mu_1 - \mu_0) - 12\alpha\mu_0\mu_1]}{5d\mu_0 + 10d\delta_0^M(1 - c_1)(\mu_1 - \mu_0) + 12c_1\alpha\mu_0\mu_1}. \qquad (10.1.71)$$

It is seen that the effective bulk modulus is the same as that of the perfect interface case. Only the effective shear modulus is affected by the imperfect interface. It is also noticed that the effective shear modulus depends not only on the volume fraction of the particles but also on the particle size. This is very different from the perfect interface case where only the volume fraction of the particles matters.

To close this section, we state two variation principles for composites with imperfect interfaces (Qu, 1993b). The proof of them is straightforward; see Problem 10.2.

Principle of Minimum Potential Energy Among all kinematically admissible displacement fields, the true one makes the following functional minimum:

$$\Pi(\mathbf{u}) = \frac{1}{2}\int_D \boldsymbol{\varepsilon}:\mathbf{L}:\boldsymbol{\varepsilon}\, dV - \int_{S_\sigma} \mathbf{u}\cdot\boldsymbol{\sigma}\cdot\mathbf{n}\, dS + \frac{1}{2}\int_\Gamma \Delta\mathbf{u}\cdot\boldsymbol{\eta}^{-1}\cdot\Delta\mathbf{u}\, dS, \tag{10.1.72}$$

where S_σ is the portion of the external boundary where traction is prescribed, and Γ is the union of all internal interfaces.

Principle of Minimum Complimentary Energy Among all statically admissible stress fields, the true one makes the following functional minimum:

$$\Pi_c(\boldsymbol{\sigma}) = \frac{1}{2}\int_D \boldsymbol{\sigma}:\mathbf{M}:\boldsymbol{\sigma}\, dV - \int_{S_u} \mathbf{u}\cdot\boldsymbol{\sigma}\cdot\mathbf{n}\, dS + \frac{1}{2}\int_\Gamma \mathbf{n}\cdot\boldsymbol{\sigma}\cdot\boldsymbol{\eta}\cdot\boldsymbol{\sigma}\cdot\mathbf{n}\, dS, \tag{10.1.73}$$

where S_u is the portion of the external boundary where displacement is prescribed, and Γ is the union of all internal interfaces.

10.2 FIBER BRIDGING

Fibers in engineering composites typically have higher elastic modulus than the matrix. Because of this, the effective elastic modulus of the composite is higher than that of the matrix material. In addition to the enhancement of elastic stiffness, fibers can also enhance the fracture toughness of the matrix material. In this section, we will consider a very simple example to illustrate how the fibers in a composite affect the composite's fracture behavior.

Let us consider a unidirectional fiber-reinforced composite material. Assume that the matrix has a crack perpendicular to the fiber direction, as shown in Figure 10.1. When the composite is subjected to a uniform tensile load σ_∞ in the fiber direction at infinity, a mode I stress intensity is created at the crack tips, which is characterized by the stress intensity factor K_I. According to linear elastic fracture mechanics, the crack may start growing leading to the fracture of the material if the stress intensity factor K_I becomes greater than the fracture toughness of the matrix material K_{Ic}, that is,

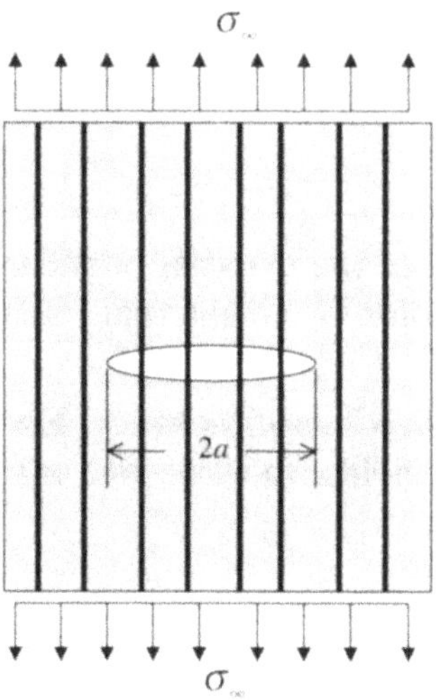

Figure 10.1 Unidirectional fiber-reinforced composite with Griffith crack.

$$K_I \geq K_{Ic}. \tag{10.2.1}$$

If, under the same given load, K_I can be reduced, it effectively improves the fracture toughness of the material. We will see from the following analysis that fibers that bridge the crack surfaces can effectively reduce K_I, thus improving the fracture strength of the composite.

For simplicity, we will limit ourselves to plane strain deformation, so the crack can be considered as a two-dimensional slit, which is referred to as the Griffith crack in fracture mechanics literature. We further assume that the size of the existing crack in the composite is much greater than the fiber diameter. This assumption is valid for most engineering applications where the fiber diameter is in the order of micrometers, while the flaw size is typically several hundred micrometers, or millimeters. Under this assumption, the unidirectional fiber-reinforced composite can be viewed as a homogeneous, transversely isotropic solid as discussed in Section 9.1.

To proceed, we first transfer the externally applied load σ_∞ to the crack surface. A simple linear superposition argument would show that loading on the crack surfaces is equivalent to loading at infinity in that they produce the same stress intensity factors. So, in what follows, we will consider σ_∞ being applied on the cracks surface instead of at infinity.

If there were no fibers bridging the crack faces, the stress intensity factor of a Griffith crack of length $2a$ in a homogeneous, transversely isotropic elastic solid subject to normal loading σ_∞ on the crack surfaces is given by (see Tada et al., 1985; Qu and Bassani, 1993)

$$K_I = 2\sqrt{\frac{a}{\pi}} \int_0^a \frac{\sigma_\infty \, dx}{\sqrt{a^2 - x^2}}. \tag{10.2.2}$$

However, the bridging fibers limit the opening of the crack faces, so the actual (or effective) stress intensity factor at the crack tip is rather different from (10.2.2). In other words, not all the applied stress σ_∞ is working to open the crack surfaces. Instead, a portion of the applied stress σ_∞ has to stretch the bridging fibers. Or, one may view this in a different way. The bridging fibers exert a stress $p(x)$ on the crack surfaces, which works against the applied stress σ_∞ by trying to close the crack. Thus, effectively, the problem at hand is equivalent to a crack without bridging fibers, but the externally applied load is $\sigma_\infty - p(x)$. Therefore, the actual stress intensity factor should be, following (10.2.2),

$$K_I = 2\sqrt{\frac{a}{\pi}} \int_0^a \frac{[\sigma_\infty - p(x)] \, dx}{\sqrt{a^2 - x^2}}. \tag{10.2.3}$$

What we need to do next is to find $p(x)$.

Since the total load is shared between the fibers and the matrix, we have

$$\sigma_\infty = c_f \sigma_f^\infty + c_m \sigma_m^\infty, \tag{10.2.4}$$

where, as defined before, c_f and c_m are the volume fractions of the fiber and the matrix, respectively. Away from the crack, the axial stresses in the fiber and in the matrix are denoted by σ_f^∞ and σ_m^∞, respectively. Since the axial strain is the same for both fiber and matrix (no gross fiber pull-out), one must have

$$\frac{\sigma_f^\infty}{E_f} = \frac{\sigma_m^\infty}{E_m} = \frac{\sigma_\infty}{E_L}, \tag{10.2.5}$$

where, again, E_f and E_m are the Young moduli of the fiber and matrix, respectively, while E_L is the effective Young modulus of the unidirectional fiber-reinforced composite in the fiber direction. Combining (10.2.4) and (10.2.5) yields the stress in the fiber

$$\sigma_f^\infty = \frac{\sigma_\infty E_f}{c_m E_m (1 + \eta)}, \tag{10.2.6}$$

where

$$\eta = \frac{E_f c_f}{E_m c_m}. \tag{10.2.7}$$

This is the axial stress in the fiber far away from the crack (or when the crack is absent). We now consider the axial stress in the fiber near the crack faces. Shown in Figure 10.2 is a representative element near the crack faces containing one fiber. The center of the fiber is x distance away from the center of the crack. Due to geometric singularities, the stresses near the intersection of the fiber and crack surfaces are very high. It is perceivable that certain portions of the fiber–matrix interface had failed and sliding between the fiber and matrix had occurred. We assume that the length of the sliding region is l and the shear stress on the interface due to friction is denoted by τ (assume it is a constant). Outside the sliding area, the axial stress in the fiber is given by (10.2.6). Within the sliding area, the axial stress in the fiber can be obtained by balancing the axial stress in the fiber and the shear stress on the fiber surface due to friction,

$$\sigma_f(z) = \sigma_f^\infty + \frac{2\pi b \tau}{\pi b^2} z = \sigma_f^\infty + \frac{2\tau}{b} z, \tag{10.2.8}$$

where b is the fiber radius. According to Hooke's law, the strain in the fiber follows from (10.2.8),

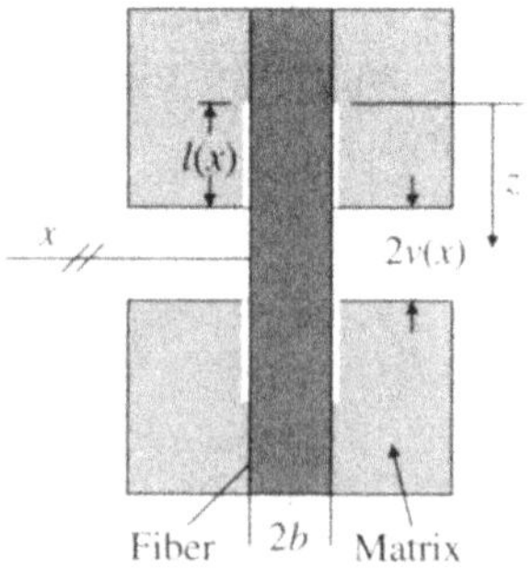

Figure 10.2 Representative element of crack surfaces r distance from crack center.

$$\varepsilon_f(z) = \frac{\sigma_f^\infty}{E_f} + \frac{2\tau}{bE_f} z. \tag{10.2.9}$$

Integration of (10.2.9) with respect to z leads to the total elongation of the fiber within the sliding area

$$2\Delta_f(x) = 2\int_0^{l(x)} \varepsilon_f(z)\, dz = 2\left(\frac{\sigma_f^\infty}{E_f} + \frac{\tau l(x)}{bE_f}\right) l(x). \tag{10.2.10}$$

The factor 2 in the above equation comes from the fact that the sliding area is symmetric with respect to the crack plane.

We now consider the matrix surrounding the fiber. Analogous to (10.2.8), we assume that the stress in the matrix within the sliding area is also a linear function of z. Further, the axial stress in the matrix must be zero on the crack surfaces. Thus, we write

$$\sigma_m(z) = \left[1 - \frac{z}{l(x)}\right] \sigma_m^\infty. \tag{10.2.11}$$

The corresponding strain thus follows:

$$\varepsilon_m(z) = \frac{\sigma_m^\infty}{E_m}\left[1 - \frac{z}{l(x)}\right]. \tag{10.2.12}$$

Consequently, integration of (10.2.12) gives the axial elongation of the matrix within the sliding area

$$2\Delta_m(x) = 2\int_0^{l(x)} \varepsilon_m(z)\, dz = \frac{\sigma_m^\infty l(x)}{E_m}. \tag{10.2.13}$$

Comparing (10.2.10) and (10.2.13), we have the total crack opening displacement

$$2v(x) = 2\Delta_f(x) - 2\Delta_m(x) = 2\left[\frac{\sigma_f^\infty}{E_f} + \frac{\tau l(x)}{bE_f}\right] l(x) - \frac{\sigma_m^\infty l(x)}{E_m}. \tag{10.2.14}$$

Substituting (10.2.6) into (10.2.14) yields the following crack opening displacement:

$$v(x) = \left[\frac{\tau l(x)}{bE_f} + \frac{\sigma_\infty}{2(E_f c_f + E_m c_m)} \right] l(x). \tag{10.2.15}$$

This can also be solved to obtain the length of the sliding zone,

$$l(x) = \frac{\eta \sigma_\infty b}{4(1 + \eta) c_f \tau} \left[\sqrt{1 + \frac{16(1 + \eta)^2 E_f c_f^2 \tau v(x)}{\eta^2 \sigma_\infty^2 b}} - 1 \right]. \tag{10.2.16}$$

Clearly, for the fiber at the crack tip, the crack opening displacement is zero, that is, $v(a) = 0$. It thus follows from (10.2.16) that $l(a) = 0$. That means that the fibers outside the cracked area do no show any pull-out, which is consistent with our intuition.

If we now substitute (10.2.16) back to (10.2.8), we can calculate the stress exerted on the crack surface by the bridging fibers:

$$p(x) = c_f \sigma_f(l) = \frac{\eta \sigma_\infty}{2(1 + \eta)} \left[\sqrt{1 + \frac{16(1 + \eta)^2 E_f c_f^2 \tau v(x)}{\eta^2 \sigma_\infty^2 b}} + 1 \right]. \tag{10.2.17}$$

Note that $p(x)$ is positive. It represents the stress carried by the bridging fibers over the cracked area. One may also view $p(x)$ as the stress acting on the crack surface trying to close the crack opening by working against the externally applied load σ_∞. Equation (10.2.17) is a relationship between $p(x)$ and $v(x)$.

For convenience, let us recast (10.2.17) into a nondimensional form:

$$\frac{p(x)}{\sigma_\infty} = \frac{\eta}{2(1 + \eta)} \left[\sqrt{1 + \frac{4(1 + \eta) \nu(x)}{\eta^2 \nu_0}} + 1 \right], \tag{10.2.18}$$

where

$$\nu_0 = \frac{\sigma_\infty^2 b}{4(1 + \eta) E_f c_f^2 \tau}. \tag{10.2.19}$$

Since $p(x) \leq \sigma_\infty$, it can be easily shown from (10.2.18) that $\nu(x) \leq \nu_0$. Shown in Figure 10.3 is the relationship between the bridging force $p(x)$ and the parameter $\eta = E_f c_f / E_m c_m$ for various crack opening displacement values. It is seen that as η increases (either higher fiber

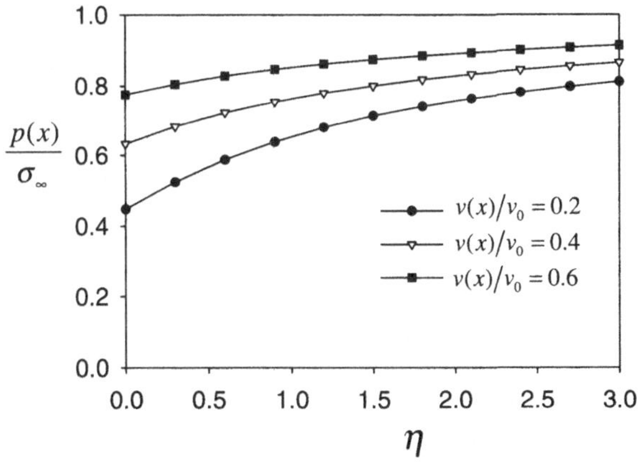

Figure 10.3 Relationship between bridging force and $\eta = E_f c_f / E_m c_m$.

modulus, or more fibers, or both), the bridge forces increase as well. Higher crack opening displacement also causes higher bridging forces. For a given crack, the crack opening displacement is typically highest near the center of the crack. Therefore, that is where the bridging force is the highest. Higher bridging means higher stress in the bridging fiber. So, conceivably, some of the fibers near the center of the crack may break. Such a situation has been studied by a number of investigators (e.g., Cui, 1992; Xia et al., 1994; Budiansky and Cui, 1995).

As discussed earlier, the problem at hand is equivalent to a crack in a homogeneous, transversely isotropic solid subject to normal pressure $\sigma_\infty - p(x)$. The crack opening displacement in this case is given by (see, e.g., Qu and Bassani, 1993)

$$\begin{aligned} v(r) &= \frac{2D}{\pi} \int_0^a [\sigma_\infty - p(t)] \log \left| \frac{\sqrt{a^2 - x^2} + \sqrt{a^2 - t^2}}{\sqrt{a^2 - x^2} - \sqrt{a^2 - t^2}} \right| dt \\ &= 2D\sigma_\infty \sqrt{a^2 - x^2} - \frac{2D}{\pi} \int_0^a p(t) \log \left| \frac{\sqrt{a^2 - x^2} + \sqrt{a^2 - t^2}}{\sqrt{a^2 - x^2} - \sqrt{a^2 - t^2}} \right| dt, \end{aligned} \tag{10.2.20}$$

where

$$D = \sqrt{\left(\frac{A_{11}A_{22}}{2}\right)\left(\frac{2A_{12} + A_{66}}{2A_{11}} + \sqrt{\frac{A_{22}}{A_{11}}}\right)}, \tag{10.2.21}$$

$$A_{11} = \frac{1 - \nu^2}{E_T}, \qquad A_{22} = \frac{E_L - \nu^2 E_T}{E_L^2},$$

$$A_{12} = -\frac{\nu_{TT}(1 + \nu)}{E_L}, \qquad A_{66} = \frac{1}{\mu_L}. \tag{10.2.22}$$

In deriving the above expressions, it has been assumed that $\nu_f = \nu_m = \nu$. The other engineering constants E_L, E_T, and μ_L, and ν_{TT} are defined in Section 9.1. They are related to the elastic properties of the fiber and matrix through the formulas discussed in the same section.

Substituting (10.2.17) into (10.2.20) yields an integral equation for the crack opening displacement. Numerical methods are needed to obtain the solution to the integral equation. Once the crack opening displacement $v(x)$ is solved, the pressure $p(x)$ can be evaluated from (10.2.17). Once $p(x)$ is known, linear elastic fracture mechanics dictates that fracture may occur when

$$K_I = 2\sqrt{\frac{a}{\pi}} \int_0^a \frac{[\sigma_\infty - p(x)]\, dx}{\sqrt{a^2 - x^2}} = K_{Ic}. \tag{10.2.23}$$

Since $p(x)$ is positive, the stress intensity factor calculated from (10.2.23) is lower than the case without fiber bridging. Therefore, the bridge fiber effectively increases the fracture toughness of the composite.

10.3 TRANSVERSE MATRIX CRACKS

Transverse matrix cracking is one of the most common damage mode in cross-ply laminates. It has been observed in both ceramic matrix and polymer matrix composites that the evolution of transverse matrix cracking shows two distinct stages. When tensile stress is applied in the fiber direction of the 0° ply, cracks are formed in the 90° ply at a load level much lower than the ultimate strength of the laminate. These transverse matrix cracks are tunneling cracks in that they almost instantaneously run through the entire specimen width and the thickness of the 90° ply but arrest at the 0°/90° ply interfaces. As the load further increases, more transverse cracks are initiated. At some critical load, the number of cracks ceases to increase; instead, the existing cracks may penetrate into the adjacent 0° layers and eventually fracture the 0° fibers (for ceramic matrix composite CMCs), or divert into the 0°/90° ply interfaces (for polymer matrix composite PMCs). Catastrophic failure becomes imminent when the 0° fibers are broken.

In what follows, we will develop a micromechanics model for the onset and subsequent multiplication of transverse cracks in the 90° layers at the first stage of damage evolution based on the Griffith fracture criterion. The cross section of a typical fiber-reinforced cross-ply laminate is schematically shown in Figure 10.4. For convenience, the following convention is adopted here. The layers with vertical fibers in the y direction are called 0° layers (or plies) and the layers with fibers perpendicular to the page (in the z direction) are called 90° layers (or plies). It is assumed that the cross-ply laminates have equal distribution of 0° and 90° plies.

The effective modulus of the fiber-reinforced cross-ply laminate shown in Figure 10.4 has been studied in Section 9.2. What we will need here is the plane-strain Young modulus in the y direction of the uncracked cross-ply laminate; see (9.2.40):

$$E_0 = \frac{(E_L + E_T)^2 - 4E_L^2\nu_{TL}^2}{2(E_L + E_T)(1 - \nu_{LT}\nu_{TL})}, \qquad \nu_0 = \frac{2E_L}{E_L + E_T}\nu_{TL}, \tag{10.3.1}$$

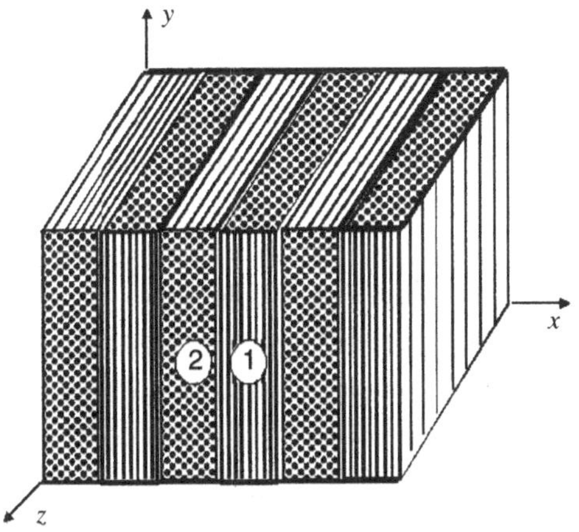

Figure 10.4 Cross section of cross-ply laminate.

where ν_{TL} ν_{LT}, E_L, and E_T are the effective engineering elastic constants of the unidirectional fiber-reinforced composite. These engineering elastic constants can be related to the Young moduli of the fiber and the matrix materials as discussed in Chapter 9. If we assume further that the Poisson ratios for the fiber and the matrix are the same, that is, $\nu_m = \nu_f = \nu$, the above equations reduce to

$$E_0 = \frac{(E_L/E_T + 1)^2 - 4\nu^2(E_L/E_T)^2}{2(E_L/E_T + 1)(1 - \nu^2)} E_T, \qquad \nu_0 = \nu. \quad (10.3.2)$$

It is obvious from (9.1.58) to (9.1.64) that the ratio E_L/E_T depends only on the fiber volume fraction c and the ratio $\eta = E_f/E_m$. Therefore, one can conclude that, besides a common factor E_m, for a cross-ply laminate with equal number of 0° and 90° plies of equal thickness, the effective elastic properties depend only on ν, η, and c.

As the matrix cracks multiply, the effective stiffness of the composite decreases. We will use a differential scheme to estimate the effective modulus as a function of the crack density. The idea is based on the notion of incremental construction of the strain energy by introducing one crack at a time in the 90° layer. Suppose that at a given crack density f in the 90° layer, the composite is homogenized with effective Young's modulus $\overline{E}_c(f)$, where the dependence of $\overline{E}_c$ on the crack density is explicitly indicated. The fundamental assumption of the differential method is that when an additional crack is introduced in the 90° layer, the change of strain energy due to this addition is the energy released from the formation of the new crack.

To this end, consider a pair of 0° and 90° layers in the cross-ply laminate as shown in Figure 10.5. Assume that there are N cracks in the 90° layer (material 1) within the height L. Let the representative element be subjected to an applied strain $\bar{\varepsilon}$ in the y direction. Then, at this given crack density, the total strain energy in this pair of layers can be written as

$$U(f) = \tfrac{1}{2}\overline{E}_c(f)\bar{\varepsilon}^2(4t)L, \quad (10.3.3)$$

where

$$h = \frac{L}{2N}. \quad (10.3.4)$$

When an additional crack is introduced, the crack density becomes

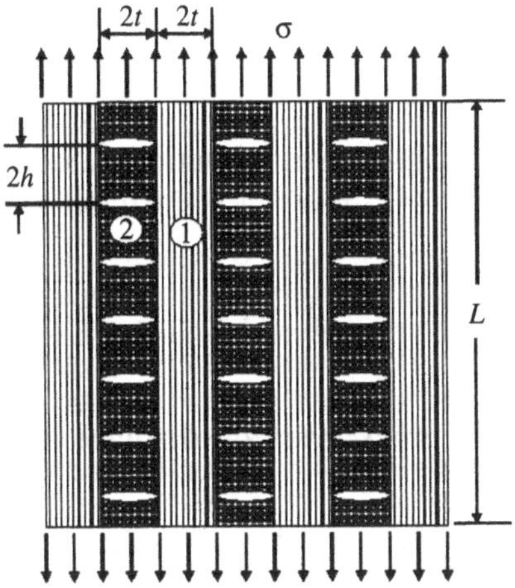

Figure 10.5 Transverse matrix cracks in cross-ply laminate under uniaxial loading.

$$f_1 = \frac{t}{h_1}, \qquad h_1 = \frac{L}{2(N+1)}. \tag{10.3.5}$$

The total strain energy at this crack density becomes

$$U(f_1) = \tfrac{1}{2}\overline{E}_c(f_1)\overline{\varepsilon}^2(4t)L. \tag{10.3.6}$$

Based on the differential scheme, the difference between (10.3.3) and (10.3.6) is the energy released due to one transverse crack given by

$$U(f_1) - U(f) = -\frac{1}{2}\left\{2\int_{-t}^{t} \sigma(x)\delta(x)\,dx\right\}, \tag{10.3.7}$$

where $\delta(x)$ is the crack opening displacement defined by

$$\Delta(x) = u_y(x, 0^+) - u_y(x, 0^-) \qquad \text{for } -t \le x \le t. \tag{10.3.8}$$

Note that the right-hand side of (10.3.7) is the energy released when a crack is formed in the laminate with effective Young's modulus $\overline{E}_c(f)$. Therefore, the stress $\sigma(x)$ in the 90° layer (prior to the formation of this crack) should be uniform and given by

$$\sigma(x) = \left[2\overline{E}_c(f) - \frac{E_L}{1 - \nu^2 E_T/E_L} \right] \bar{\varepsilon}, \tag{10.3.9}$$

where E_L is the longitudinal (fiber direction) modulus of the individual layers. Substituting (10.3.9) into the right-hand side of (10.3.7) yields

$$U(f_1) - U(f) = -\left[2\overline{E}_c(f) - \frac{E_L}{1 - \nu^2 E_T/E_L} \right] \bar{\varepsilon}^2 t \overline{\Delta}, \tag{10.3.10}$$

where $\overline{\Delta}$ is the average crack opening displacement defined by

$$\overline{\Delta} = \frac{1}{2t} \int_{-t}^{t} \Delta_0(x)\, dx, \tag{10.3.11}$$

and $\Delta_0(x)$ is the crack opening displacement due to a unit applied strain field, that is,

$$\Delta_0(x) = \frac{\Delta(x)}{\bar{\varepsilon}}. \tag{10.3.12}$$

Since $\Delta(x)$ can be calculated by the finite-element method for given $\bar{\varepsilon}$, $\Delta_0(x)$ can then be obtained through linear superposition from (10.3.12).

Next, consider the left-hand side of (10.3.10). It follows from (10.3.3) and (10.3.6) that

$$U(f_1) - U(f) = 2\bar{\varepsilon}^2 Lt[\overline{E}_c(f_1) - \overline{E}_c(f)]. \tag{10.3.13}$$

On the other hand, making use of (10.3.4) and (10.3.5) yields

$$L = \frac{2t}{f_1 - f}. \tag{10.3.14}$$

Substitution of (10.3.14) into (10.3.13) gives

$$U(f_1) - U(f) = 4\bar{\varepsilon}^2 t^2 \frac{[\overline{E}_c(f_1) - \overline{E}_c(f)]}{f_1 - f}. \tag{10.3.15}$$

Obviously, in the limiting of $f_1 \rightarrow f$, (10.3.15) can be written as a differential equation:

$$U(f_1) - U(f) = 4\bar{\varepsilon}^2 t^2 \frac{d\overline{E}_c(f)}{df}. \tag{10.3.16}$$

Combining (10.3.10)–(10.3.16) yields

$$\frac{d\overline{E}_c}{df} = -\frac{\overline{\Delta}}{4t}\left[2\overline{E}_c(f) - \frac{E_L}{1 - \nu^2 E_T/E_L}\right]. \tag{10.3.17}$$

In addition to (10.3.17), $\overline{E}_c(f)$ should also satisfy the initial condition that when there is no crack ($f = 0$),

$$\overline{E}_c(0) = E_0. \tag{10.3.18}$$

Equations (10.3.17) and (10.3.18) form an initial value problem for the effective Young modulus of the composite. The solution to the initial value problem (10.3.17) and (10.3.18) is given by (Qu and Hoiseth, 1998)

$$\overline{E}_c(f) = \overline{E}_0\left\{\exp\left(-\frac{f\overline{\Delta}}{2t}\right) + \frac{E_L}{E_L + E_T}\left[1 - \exp\left(-\frac{f\overline{\Delta}}{2t}\right)\right]\right\}. \tag{10.3.19}$$

It is seen that this solution approaches the theoretical limit

$$\frac{\overline{E}_c(f)}{E_0} \to \frac{E_L}{E_T + E_L}, \tag{10.3.20}$$

as $f \to \infty$ in which only the 0° layers carry the load.

It is seen from (10.3.19) that once $\overline{\Delta}$, the crack opening displacement of a single crack in an infinitely long ($h/t \to \infty$) specimen, is known, the effective elastic modulus of the laminate with any crack density can be predicted. Shown in Figure 10.6 is the predicted Young modulus $\overline{E}_c$ as a function of crack density f for $c = 0.55$ and various values of η. The solid lines are the analytical solution from (10.3.19), and the symbols are the exact solutions from the finite-element method (Qu and Hoiseth, 1998). It is found that for the wide range of materials considered, the maximum difference between the finite element and the analytical solutions is less than 5 percent.

The effective Young modulus given by (10.3.19) can only give the stiffness as a function of crack density. To predict the effective (overall) stress–strain behavior, damage (cracking) evolution must be modeled.

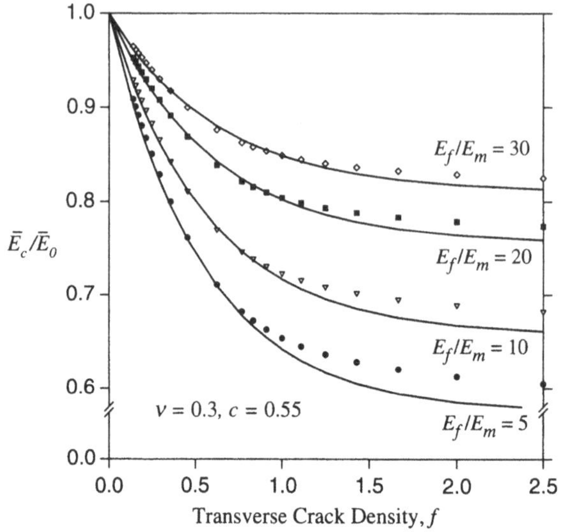

Figure 10.6 Effective Young's modulus. Lines are from the analytical solution and symbols are from the finite-element solutions.

To this end, a fracture criterion is required. Once a fracture criterion is established, solutions of crack density can be obtained as a function of the applied strain $\bar{\varepsilon}$.

As the cross-ply laminate is loaded in the y direction, transverse cracks develop in the 90° layers. Assume the transverse cracks initiate as tunneling cracks, namely, they span the entire 90° ply and propagate in the width direction of the laminate (the z direction). Then, based on linear elastic fracture mechanics, the energy released due to the formation of a new tunneling crack can be written as (Hutchinson and Suo, 1991),

$$G_{ss}(f) = \frac{1}{4t}\int_{-t}^{t} \sigma(x)\Delta(x)\,dx, \tag{10.3.21}$$

where $\Delta(x)$ is the crack opening displacement given in (10.3.11) and $\sigma(x) = \sigma_y(x, 0)$, that is, the normal stress on the crack line prior to cracking. Note that both $\Delta(x)$ and $\sigma(x)$ depend on the crack density f. Therefore, G_{ss} is a function of f.

Through simple dimensional analysis, the energy release rate can be written as

$$G_{ss}(f) = \bar{\varepsilon}^2 t\bar{E}_0 g(\eta, \nu, c, f), \qquad (10.3.22)$$

where $\bar{\varepsilon}$ is the applied strain, $\bar{E}_0$ is the overall plane strain Young modulus for the uncracked composite defined in (10.3.2), and $g(\eta, \nu, c, f)$ is the nondimensional energy release rate.

The energy release rate $G_{ss}(f)$ can also be related to the effective stiffness through the principle of conservation of energy. To develop such a relationship, consider a specimen containing N cracks in the 90° layer. It follows from the conservation of strain energy that (Qu and Hoiseth, 1998)

$$\frac{1}{2}\bar{E}_c(f)\bar{\varepsilon}^2(2hNt) = \frac{1}{2}\bar{E}_0\bar{\varepsilon}^2(2hNt) - \frac{Nt}{2}G_{ss}(f), \qquad (10.3.23)$$

where $2hN$ is the height of the specimen (thickness is unit). Substituting (10.3.22) into (10.3.23) yields

$$g(\eta, \nu, c, f) = \left[1 - \frac{\bar{E}_c(f)}{\bar{E}_0}\right]\frac{2}{f}, \qquad (10.3.24)$$

where $\bar{E}_c$ is the effective Young modulus determined in the previous section. Making use of (10.3.24) gives the steady-state energy release rate for tunneling cracks

$$\frac{G_{ss}(f)}{\bar{\varepsilon}^2 t\bar{E}_0} = g(\eta, \nu, c, f) = \frac{2E_T}{(E_L + E_T)f}\left[1 - \exp\left(-\frac{\bar{\delta}f}{2t}\right)\right]. \qquad (10.3.25)$$

An interesting limit of (10.3.25) as $f \to 0$ is

$$G_{ss}(0) = \frac{E_T}{E_T + E_L}\bar{\varepsilon}^2\bar{\delta}\bar{E}_0. \qquad (10.3.26)$$

This can also be obtained directly from (10.3.21) by using (10.3.9) and (10.3.11).

Next, consider a 90° layer with an existing set of transverse matrix cracks of density f. Assume that a new set of cracks bisecting the existing set is in the process of tunneling across the layer; see Figure 10.7. The energy released by the cracks in the process of tunneling is (Hutchinson and Suo, 1991)

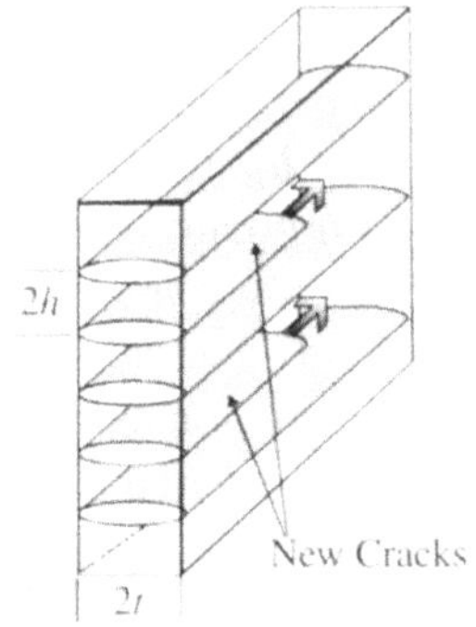

Figure 10.7 Multiplication of transverse matrix cracking in cross-ply laminate.

$$2G_{ss}(2f) - G_{ss}(f). \tag{10.3.27}$$

Then, based on the Griffith energy criterion in linear elastic fracture mechanics, a new set of cracks will be initiated if the following equation is satisfied:

$$G_c^{(2)} = 2G_{ss}(2f) - G_{ss}(f), \tag{10.3.28}$$

where $G_c^{(2)}$ is the plane strain mode I toughness of the 90° layer (material 2).

Substituting (10.3.26) into (10.3.28) yields

$$G_c^{(2)} = \frac{2\bar{\varepsilon}^2 t\bar{E}_0 E_T}{(E_L + E_T)f}\left[\exp\left(-\frac{\bar{\delta}f}{2t}\right) - \exp\left(-\frac{\bar{\delta}f}{t}\right)\right]. \tag{10.3.29}$$

Equation (10.3.29) is the desired evolution equation for the crack density f as a function of the applied strain $\bar{\varepsilon}$. For $\nu = 0.3$, $c = 0.55$, the crack density f as a function of $\bar{\varepsilon}\sqrt{tE_0/G_c^{(2)}}$ is plotted in Figure 10.8 for $\eta = 1, 5, 10, 20, 30$.

Figure 10.9 presents a comparison between the theoretical predictions and the experimental data for crack density versus applied load. The experimental data is from Varna and Berglund (1991). The tested material is an AS/3501-06 carbon fiber–epoxy $[0_2/90_2]_s$ laminate. In the numerical calculations, no adjustment parameters are used. All material constants are taken directly from Varna and Berglund (1991). It is seen that the predicted crack density versus load curve (dotted line) has the same trend as the experimental one but is slightly shifted to the right. This may be attributed to many factors. Chief among them

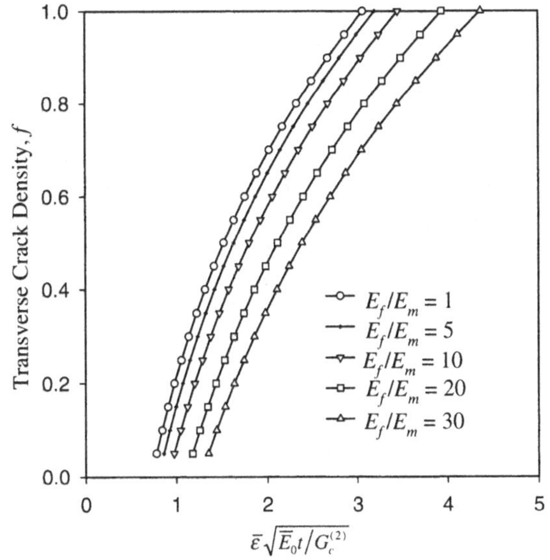

Figure 10.8 Crack density vs. applied strain.

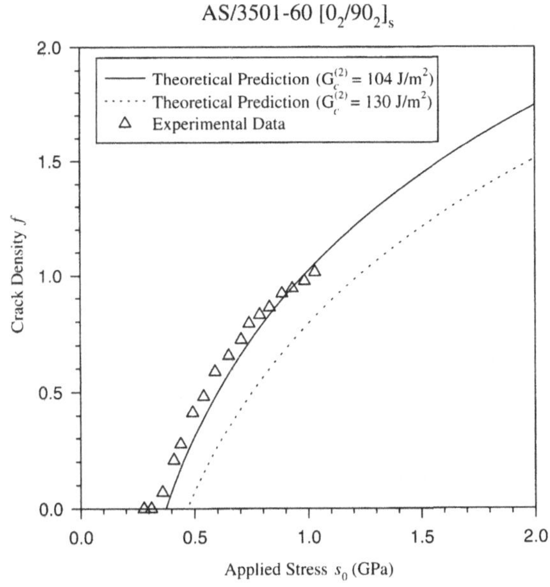

Figure 10.9 Comparison between experimental and analytical results for crack density vs. applied load.

is the fact that the model is based on the assumption that the laminate contains an infinite number of layers, while the experimental data were obtained for a $[0_2/90_2]_s$ laminate. Since the model predicts the trend correctly, a simple parameter adjustment can be done in practical applications to correlate the prediction with the measurement. For example, in the case considered here, by reducing the toughness of the 90° layer in the direction of the fiber from $G_c^{(2)} = 130\ \mathrm{J/m^2}$ as reported in Varna and Berglund (1991) to $G_c^{(2)} = 104\ \mathrm{J/m^2}$, the model predicts excellent agreement with the experiments as shown by the solid line in Figure 10.10.

Note that, in the limit of $f \to 0$, Eq. (10.3.29) becomes

$$G_c^{(2)} = \frac{\bar{\varepsilon}^2 \bar{\delta} \bar{E}_0 E_T}{E_T + E_L}. \tag{10.3.30}$$

Thus, the threshold strain at which transverse matrix cracking starts to occur is

$$\bar{\varepsilon}_{\text{th}} = \sqrt{\frac{G_c^{(2)}(E_T + E_L)}{\bar{\delta} \bar{E}_0 E_T}}. \tag{10.3.31}$$

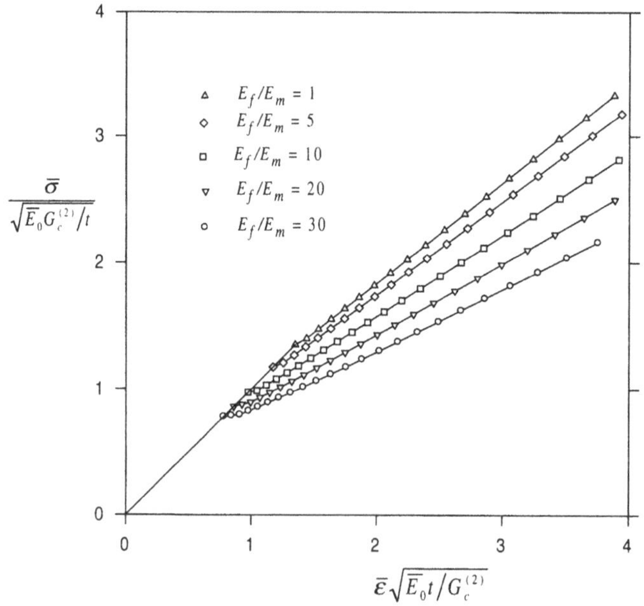

Figure 10.10 Effective stress–strain curves.

Now, consider the effect of matrix cracking on the overall stress–strain behavior. To begin the calculation, the threshold strain $\bar{\varepsilon}_{th}$ is obtained from (10.3.31) for the given composite system. Then, the strain $\bar{\varepsilon}$ is increased gradually. For each value of $\bar{\varepsilon} > \bar{\varepsilon}_{th}$, the corresponding crack density f is computed from (10.3.29). Substituting this value of f into (10.3.19) yields the corresponding effective Young modulus at this given strain. Finally, the stress $\bar{\sigma}(f)$ corresponding to this strain is evaluated from

$$\bar{\sigma}(f) = \bar{E}_c(f)\bar{\varepsilon}(f). \tag{10.3.32}$$

For $\nu = 0.3$, $c = 0.55$, the overall stress–strain curves for various values of η are presented in Figure 10.10. Note that these curves are invariant to $G_c^{(2)}$ because of the normalization.

PROBLEMS

10.1 Consider a composite material consisting of randomly oriented and shaped inhomogeneities embedded in a matrix with stiffness tensor $\mathbf{L}_0$. Let the stiffness tenors of the inhomogeneities be $\mathbf{L}_1$, $\mathbf{L}_2$, $\mathbf{L}_3$, . . . , $\mathbf{L}_N$, as shown in Figure 7.1. We further assume that interfaces between the inhomogeneities and the matrix are not perfect. The interfacial jump condition is given by (10.1.4). Find the effective stiffness tensor of this composite.

10.2 Prove the two variational principles stated in Section 10.1.

APPENDIX 10.A

By substituting (10.1.5) into (10.1.36), we obtain the following:

$$R_{ijkl} = \alpha P_{ijkl} + (\beta - \alpha)Q_{ijkl} \tag{10.A.1}$$

with

$$P_{ijkl} = \frac{3}{16\pi}\int_0^{\pi}\left[\int_0^{2\pi}(\delta_{ik}\hat{n}_j\hat{n}_l + \delta_{jk}\hat{n}_i\hat{n}_l + \delta_{il}\hat{n}_j\hat{n}_k + \delta_{jl}\hat{n}_i\hat{n}_k)\frac{d\theta}{n}\right] \times \sin\phi\, d\phi, \tag{10.A.2}$$

$$Q_{ijkl} = \frac{3}{4\pi}\int_0^{\pi}\left[\int_0^{2\pi}\frac{\hat{n}_i\hat{n}_j\hat{n}_k\hat{n}_l}{n}\,d\theta\right]\sin\phi\, d\phi, \tag{10.A.3}$$

where

$$n = \sqrt{\hat{n}_i \hat{n}_i}, \tag{10.A.4}$$

$$\hat{\mathbf{n}} = \left(\frac{\sin\phi\cos\theta}{a_1}, \frac{\sin\phi\sin\theta}{a_2}, \frac{\cos\phi}{a_3} \right). \tag{10.A.5}$$

For spheres ($a_1 = a_2 = a_3 = a$), we have $n = 1/a$. Thus,

$$P_{ijkl} = \frac{1}{a} I_{ijkl}, \qquad Q_{ijkl} = \frac{1}{5a}(2I_{ijkl} + \delta_{ij}\delta_{kl}). \tag{10.A.6}$$

For cylinders ($a_1 = a_2 = a$, $a_3 \to \infty$), the non-zero components are

$$P_{1111} = P_{2222} = 4P_{2323} = 4P_{1313} = 2P_{1212} = \frac{3\pi}{8a}, \tag{10.A.7}$$

$$Q_{1111} = Q_{2222} = 3Q_{2323} = 3Q_{1313} = 3Q_{1212} = \frac{9\pi}{32a}. \tag{10.A.8}$$

REFERENCES

Achenbach, J. D. and H. Zhu. (1989). Effects of Interfacial Zone on Mechanical Behavior and Failure of Fiber-Reinforced Composites, *J. Mech. Phys. Solids,* Vol. 37, pp. 381–393.

Budiansky, B. and Y. Cui. (1995). Toughening of Ceramics by Short Aligned Fibers, *Mech. Mater.,* Vol. 21, pp. 139–146.

Cui, Y. (1992). Interaction of Fiber and Transformation Toughening, *J. Mech. Phys. Solids,* Vol. 40, pp. 1837–1850.

Hutchinson, J. W. and Z. Suo. (1991). Mixed Mode Cracking in Layered Materials, *Adv. Appl. Mech.,* Vol. 29, pp. 62–191.

Mura, T. (1987). *Micromechanics of Defects in Solids,* Martinus Nijhoff, Boston.

Qu, J. (1993a). Eshelby Tensor for an Elastic Inclusion with Slightly Weakened Interface, *J. Appl. Mech.,* Vol. 60, pp. 1048–1050.

Qu, J. (1993b). Effects of Slightly Weakened Interfaces on the Overall Elastic Properties of Composite Materials, *Mech. Mater.,* Vol. 14, p. 269–281.

Qu, J. and J. L. Bassani. (1993). Interfacial Fracture Mechanics for Anisotropic Bimaterials, *J. Appl. Mech.,* Vol. 60, pp. 422–431.

Qu, J. and K. Hoiseth. (1998). Evolution of Transverse Matrix Cracking in Cross-Ply Laminates, *Int. J. Fatigue Fracture Eng. Mater. Structures,* Vol. 21, pp. 451–464.

Tada, H., P. Paris, and G. R. Irwin. (1985). *The Stress Analysis of Cracks Handbook,* Del Research, St. Louis.

Varna, J., and L. A. Berglund. (1991). Multiple Transverse Cracking and Stiffness Reduction in Cross-Play Laminates, *J. Composites Tech. Res., JCTRER,* Vol. 27, pp. 99–106.

Xia, Z. C., J. W. Hutchinson, A. G. Evans, and B. Budiansky. (1994). On Large Scale Sliding in Fiber-Reinforced Composite, *J. Mech. Phys. Solids,* Vol. 42, pp. 1139–1158.

SUGGESTED READINGS

Budiansky, B. and J. W. Hutchinson. (1986). Matrix Fracture in Fiber-Reinforced Ceramics, *J. Mech. Phys. Solids,* Vol. 34, pp. 167–189.

Christensen, R. M. (1979). *Mechanics of Composite Materials,* Wiley, New York.

Hashin, Z. (1985). Analysis of Cracked Laminates: A Variational Approach, *Mech. Mater.,* Vol. 4, pp. 121–136.

Hoiseth, K. and J. Qu. (2003). Cracking Paths at the Ply Interface in a Cross-Ply Laminate, *Composites Part B: Engr.,* Vol. 34, pp. 437–445.

Laws, N. and D. J. Dvorak. (1988). Progressive Transverse Cracking in Composite Laminates, *J. Composite Mater.,* Vol. 22, pp. 900–916.

Xia, Z. C., R. R. Carr, and J. W. Hutchinson. (1993). Transverse Cracking in Fiber-Reinforced Brittle Matrix, Cross-Ply Laminates, *Acta Metall.,* Vol. 41, pp. 2365–2376.

11

MEAN FIELD THEORY FOR NONLINEAR BEHAVIOR

In previous chapters, we have assumed that all the inhomogeneities in the heterogeneous material behave linear elastically. Therefore, the overall behavior of the heterogeneous material is also linear elastic. For linear elastic behavior, the effective stiffness tensor is usually sufficient to fully characterize the stress–strain relationship, or the material's response under load.

If one or more of the inhomogeneities in a heterogeneous medium behave nonlinearly (nonlinear elastic, viscoelastic, elastoplastic, elastoviscoplastic, etc.), the overall behavior, or the effective properties of the heterogeneous medium, will not be linear elastic. Unlike the effective stiffness and compliance for linear materials, the effective properties of nonlinear heterogeneous materials are much more complex to model and characterize. The main difficulty is the typical strong intraphase fluctuations of the stress and strain fields in nonlinear heterogeneous materials and in the hereditary nature of most inelastic behaviors. At a given length scale, the responses of the inhomogeneities can vary markedly at different load levels. For example, a two-phase elastoplastic composite material effectively behaves as a multiphase material, and phase averages have less predictive capabilities than in the linear elastic case. In this chapter, we will develop a general framework based on the mean field theories to describe the nonlinear behavior of heterogeneous media.

Historically, the mean field theory for inelastic behavior was developed mainly for crystalline materials initially. The early knowledge acquired through physical metallurgical studies of the plastic behavior of single crystals leads naturally to a certain interest in their extension to polycrystals, where the main challenge is to relate the mechanical response of an aggregate of crystals (known as a polycrystal) to the fundamental mechanisms of single-crystal deformation. Specifically, the problem of predicting the yield stress of a polycrystal as a function of the yield strength of the constituent single crystals was considered by Sachs (1928) and studied more extensively later by Taylor (1938), whereas Schmid and Boas (1935) originated their fundamental law regarding the crystallographic slip within single crystals. Note that these earlier theoretical frameworks are specific to crystal plasticity and are formulated within a context closer to physical metallurgy than to mechanics of materials. More rigorous mechanistic analysis was carries out by Bishop and Hill (1951), who generalized the Taylor model to describe the different loading paths and to show their extreme features for the yield stress of polycrystalline materials. Subsequently, the development of metal forming processing and related plastic anisotropy problems resulting from texture evolutions provided the driving force in extending the Taylor model to more accurate approaches.

In 1957, Eshelby introduced the concept of eigenstrains. The plastic deformation in each single crystal can be simulated by an eigenstrain distribution. Thus, the same homogenization techniques for linear elasticity can be followed to describe the nonlinear behavior of polycrystalline materials. This way of thinking was first adopted by Budiansky and Mangasarian (1960) and developed later by Kröner (1961). Their work constitutes the fundamental basis of the self-consistent mean field theories for nonlinear behavior of heterogeneous materials. Unlike the Sachs and Taylor models, the self-consistent framework ensures both the compatibility and equilibrium conditions across the grains interface in polycrystalline aggregate. However, as discussed later, Kröner's model is limited to spherical inhomogeneities and isotropic matrix with incompressible plasticity.

The next breakthrough was made by Hill (1965), who adopted the same concept as Kröner (in the sense of Eshelby) by introducing an original idea of describing the plastic flow by a sequence of linearizations leading to an incremental procedure of homogenization. In Hill's approach, the overall behavior of the polycrystal, as well as the behavior of each grain, is approximated by a "pseudo"-elastic behavior at each loading increment. Therefore, plastic unloading is accounted for. However, the nonlinear material behavior, compounded by the nonuniformity of stress and strain fields, renders the linearized properties nonuniform in each phase. This constitutes the major difficulty in using the Hill model without appropriate approximations.

The treatment of this difficulty was the center of different investigations leading to the emergence of mean field theories with different varieties of linearization sequences of the nonlinear behavior. Within these procedures, one can distinguish between the secant, the tangent, and affine approaches. The classical secant formulation was developed by Berveiller and Zaoui (1979) for crystalline materials and adapted later by Tandon and Weng (1988) to the case of two-phase elastoplastic composite materials. The secant formulation, which can be viewed as an intermediate method between the Kröner approach and the incremental method of Hill, reduces significantly the complexity of the Hill model by assuming isotropic homogeneous plastic flow in each phase. The first tangent mean field method was introduced by Hutchinson (1976) to describe steady-state creep behavior of crystalline materials. It turns out that Hutchinson's use of the power law creep also leads to another variant of the secant method (see below). More recently, Molinari et al. (1987) derived a tangent formulation for viscoplastic power law by adopting a sequence of linearization similar to linear thermoelasticity.

All of these self-consistent methods rely on the assumption of piecewise uniform properties so that the Eshelby inclusion solution could be combined with the linearization procedure to obtain the overall nonlinear properties. However, for most nonlinear behavior such as plastic flow, the fields fluctuate in each phase and their heterogeneity increases (e.g., with the plastic flow) at each loading increment. This makes it difficult to assess the predictive capabilities of these self-consistent methods and to select the accurate one. Such limitation was the main motivation behind the development of the variational principles for nonlinear behaviors in the mid-1990s.

Variational approaches for elastic and viscoelastic composites (Willis, 1983; Talbot and Willis, 1985; Castañeda, 1991; Suquet, 1993) combined with a linearization procedure enable us to compare some nonlinear mean field theories to rigorous bounds of the Hashin–Shtrikman type. It turned out (Gilormini, 1997) that both tangent and secant approaches lead to estimates that are too stiff and can even violate the bounds in some cases.

Different ways of thinking have emerged recently to overcome this deficiency of the mean field theories. Improvements can be made by modifying the secant approach to take into account a better description of stress and strain fluctuations at the microscale level. A first attempt in this direction was made for particulate-reinforced composite by Qiu and Weng (1992). It uses the average strain energy in the matrix to define its effective stress. Under this theory, the results are valid only for incompressible materials. A more general theory valid for arbitrary microstructure and compressible matrix has been developed by Suquet (1995). It was based on the second-order moments in each individual phase of the linear comparison solid. The use of second order moments has also been adopted independently by Hu (1996) and Buryachenko (1996). Another variety of the second-order estimates has been proposed by Castañeda (1996). It was based on Taylor's second-order developments of stress, strain, and resulting potentials.

An alternative in extending mean field theories to secant formulation plasticity has been introduced by Dvorak (1992) using the transformation field theory. The methodology models the plastic flow of multiphase materials by a uniform piecewise distribution of eigenstrains. Such approaches are in principle suitable for use as micromechanically based constitutive laws in finite element codes. They are also very attractive in terms of computational requirements but tend to strongly overestimate the nonlinear overall response of heterogeneous materials because they essentially use elastic accommodation of microstress and strains throughout the loading history.

In this chapter, several varieties of self-consistent mean field theories will be discussed in details. Illustrative results for face-center-cubic (FCC) elastoplastic polycrystalline materials using an incremental self-consistent scheme at small strains will be presented.

11.1 ESHELBY'S SOLUTION AND KRÖNER'S MODEL

As discussion above, the first attempts made in modeling the elastoplastic behavior of polycrystalline materials through a homogenization scheme had their principal inspiration from the Eshelby inclusion solution in linear elasticity. The idea of liking these two problems came from two considerations. First of all, it is possible to describe the plastic strain of each individual grain by a distribution of eigenstrains (or stress-free strain) as introduced by Eshelby. Second, the interactions between a typical grain and its surrounding grains can be simulated by an inclusion problem of a single inhomogeneity undergoing a stress-free plastic strain while being surrounded by an equivalent homogeneous elastoplastic medium with appropriate boundary conditions.

To introduce Kröner's model, let us consider a representative volume element of a polycrystalline material of total volume D comprised of N single crystals. It is assumed that D is much larger than the volume of each crystal in the representative volume element of the polycrystal aggregate so that $N >> 1$. The representative volume element may be subjected to either the traction boundary condition

$$\boldsymbol{\sigma} \bullet \mathbf{n}|_S = \overline{\boldsymbol{\sigma}} \bullet \mathbf{n} \tag{11.1.1}$$

or the rate form of displacement boundary condition

$$\dot{\mathbf{u}}|_S = \overline{\dot{\boldsymbol{\varepsilon}}} \bullet \mathbf{x}, \tag{11.1.2}$$

where S is the surface of D and $\mathbf{n}$ is the outward unit normal of S. The overdot indicates the rate, or increment. Note that the rate here is typically not the change with respect to time. Instead, it is the change (or increment) with respect to the load.

According to the average stress and average strain theorems introduced in Section 5.4, the volume averages of stress and strain tensors are

$$\overline{\sigma} = \frac{1}{D}\int_D \boldsymbol{\sigma}\, dV, \qquad \overline{\dot{\boldsymbol{\varepsilon}}} = \frac{1}{D}\int_D \dot{\boldsymbol{\varepsilon}}\, dV. \tag{11.1.3}$$

For the sake of simplicity, we will consider in this chapter small-strain deformation only. Therefore, it is possible to decompose the total strain and total strain rate into elastic and plastic parts, that is,

$$\boldsymbol{\varepsilon} = \boldsymbol{\varepsilon}^e + \boldsymbol{\varepsilon}^p, \qquad \dot{\boldsymbol{\varepsilon}} = \dot{\boldsymbol{\varepsilon}}^e + \dot{\boldsymbol{\varepsilon}}^p, \tag{11.1.4}$$

where $\boldsymbol{\varepsilon}^e$ and $\boldsymbol{\varepsilon}^p$ are the elastic and plastic parts of the strain tensor, respectively. The volume averages of strain and strain rate tensors can be decomposed similarly:

$$\overline{\boldsymbol{\varepsilon}} = \overline{\boldsymbol{\varepsilon}}^e + \overline{\boldsymbol{\varepsilon}}^p, \qquad \overline{\dot{\boldsymbol{\varepsilon}}} = \overline{\dot{\boldsymbol{\varepsilon}}}^e + \overline{\dot{\boldsymbol{\varepsilon}}}^p. \tag{11.1.5}$$

Note that, as discussed in Section 3.1, neither the elastic strain $\boldsymbol{\varepsilon}^e$ nor the plastic strain $\boldsymbol{\varepsilon}^p$ is compatible by itself, although the sum of these two fields is compatible. In other words, there is no, for example, such displacement field u_i^e such that $\varepsilon_{ij}^e = (u_{i,j}^e + u_{j,i}^e)/2$.

As discussed in Section 3.1, the elastic strain is related to the total stress through Hooke's law,

$$\boldsymbol{\sigma} = \mathbf{L}\boldsymbol{\varepsilon}^e = \mathbf{L}(\boldsymbol{\varepsilon} - \boldsymbol{\varepsilon}^p). \tag{11.1.6}$$

Single-phase polycrystals can be viewed as an assembly of essentially the same single crystals of different orientations and sizes. Therefore, an ensemble average of a field quantity can be defined as the average over all single crystals; see Section 5.3. For example, the ensemble averages of the stress and strain tensors are

$$\langle \boldsymbol{\sigma} \rangle = \sum_{r=1}^{N} c_r \boldsymbol{\sigma}_r, \qquad \langle \boldsymbol{\varepsilon} \rangle = \sum_{r=1}^{N} c_r \boldsymbol{\varepsilon}_r, \tag{11.1.7}$$

where c_r is the volume fraction of the rth crystal in the polycrystal assembly D. As discussed in Section 5.3, we consider only statistically homogeneous media in this book. Therefore, the volume average and ensemble average of a field quantity should be the same, for example,

$$\bar{\boldsymbol{\sigma}} \equiv \frac{1}{D} \int_D \boldsymbol{\sigma}\, dV = \langle \boldsymbol{\sigma} \rangle \equiv \sum_{r=1}^{N} c_r \boldsymbol{\sigma}_r, \tag{11.1.8}$$

$$\bar{\boldsymbol{\varepsilon}} \equiv \frac{1}{D} \int_D \boldsymbol{\varepsilon}\, dV = \langle \boldsymbol{\varepsilon} \rangle \equiv \sum_{r=1}^{N} c_r \boldsymbol{\varepsilon}_r. \tag{11.1.9}$$

Similarly,

$$\bar{\dot{\boldsymbol{\sigma}}} = \langle \dot{\boldsymbol{\sigma}} \rangle, \qquad \bar{\dot{\boldsymbol{\varepsilon}}} = \langle \dot{\boldsymbol{\varepsilon}} \rangle. \tag{11.1.10}$$

However, due to the interphase incompatibilities of the elastoplastic strain fields, the elastic and plastic strain tensors by themselves may not be ergodic, that is,

$$\bar{\boldsymbol{\varepsilon}}^e \neq \langle \boldsymbol{\varepsilon}^e \rangle, \qquad \bar{\boldsymbol{\varepsilon}}^p \neq \langle \boldsymbol{\varepsilon}^p \rangle, \qquad \bar{\dot{\boldsymbol{\varepsilon}}}^e \neq \langle \dot{\boldsymbol{\varepsilon}}^e \rangle, \qquad \bar{\dot{\boldsymbol{\varepsilon}}}^p \neq \langle \dot{\boldsymbol{\varepsilon}}^p \rangle. \tag{11.1.11}$$

To illustrate these, let us assume that the representative volume element D is subjected to the traction boundary condition (11.1.1). In the absence of body forces (so assumed for the rest of this chapter), the total stress field must satisfy

$$\nabla \cdot \boldsymbol{\sigma} = 0 \quad \text{in } D \tag{11.1.12}$$

and

$$\boldsymbol{\sigma} \cdot \mathbf{n}|_S = \overline{\boldsymbol{\sigma}} \cdot \mathbf{n}. \tag{11.1.13}$$

Next, let $\boldsymbol{\sigma}^e$ be the stress field in D induced by the traction boundary condition (11.1.1) when all the single crystals in the polycrystal assembly D are linear elastic, that is,

$$\nabla \cdot \boldsymbol{\sigma}^e = 0 \quad \text{in } D \tag{11.1.14}$$

and

$$\boldsymbol{\sigma}^e \cdot \mathbf{n}|_S = \overline{\boldsymbol{\sigma}} \cdot \mathbf{n}. \tag{11.1.15}$$

Clearly, the above equations imply that $\boldsymbol{\sigma}^e$ is a statically admissible stress field for the polycrystal assembly D subjected to the traction boundary condition (11.1.1), even when plastic deformation occurs. Further, since (11.1.14) and (11.1.15) describe a pure linear elastic problem, it then follows from (11.1.9) that

$$\overline{\boldsymbol{\sigma}^e} = \overline{\boldsymbol{\sigma}} = \langle \boldsymbol{\sigma} \rangle = \langle \boldsymbol{\sigma}^e \rangle. \tag{11.1.16}$$

The strain field corresponding to $\boldsymbol{\sigma}^e$ can be written as (11.1.17)

$$\beta_{ij} = \tfrac{1}{2}(u^e_{i,j} + u^e_{i,j}) = M_{ijkl}\sigma^e_{kl}, \tag{11.1.18}$$

where u^e_i is the displacement field when all the single crystals in the polycrystal assembly D behave linear elastically. Note that this β_{ij} is typically not the elastic part of the total strain field ε^e_{ij} introduced in (11.1.4), that is, $\beta_{ij} \neq \varepsilon^e_{ij}$.

Conceivably, $\boldsymbol{\sigma}^e$ and $\boldsymbol{\sigma}$ are different because of plastic deformation that occurred in D. So, we introduce the plastic residual stress field by

$$\boldsymbol{\sigma}^p = \boldsymbol{\sigma} - \boldsymbol{\sigma}^e \quad \text{or} \quad \boldsymbol{\sigma} = \boldsymbol{\sigma}^e + \boldsymbol{\sigma}^p. \tag{11.1.19}$$

It is so named because $\boldsymbol{\sigma}^p$ represents the stress field remained in the plastically deformed polycrystal assembly D after elastic unloading ($\boldsymbol{\sigma}^e = 0$). Further, it follows from (11.1.16) that

$$\overline{\boldsymbol{\sigma}^p} = \langle \boldsymbol{\sigma}^p \rangle = 0. \tag{11.1.20}$$

Equations (11.1.12)–(11.1.15) also imply that

$$\nabla \cdot \boldsymbol{\sigma}^p = 0 \quad \text{in } D \tag{11.1.21}$$

and

$$\boldsymbol{\sigma}^p \cdot \mathbf{n}|_S = 0. \tag{11.1.22}$$

We call such stress field self-equilibrium stress field. Note that generally

$$\boldsymbol{\sigma}^e \neq \mathbf{L}\boldsymbol{\varepsilon}^e, \tag{11.1.23}$$

although

$$\overline{\boldsymbol{\sigma}} = \overline{\boldsymbol{\sigma}^e} = \overline{\mathbf{L}\boldsymbol{\varepsilon}^e}. \tag{11.1.24}$$

These can be seen from (11.1.6) and (11.1.19).

Now, consider again the linear elastic problem. The stress field in the rth single-crystal Ω_r can be written as

$$\boldsymbol{\sigma}_r^e = \mathbf{B}_r\overline{\boldsymbol{\sigma}} \quad \text{in } \Omega_r \tag{11.1.25}$$

where $\mathbf{B}_r$ is the local elastic stress concentration tensor introduced in (5.7.9), which can be determined approximately by the homogenization schemes developed in Chapter 7. Note that

$$\langle \boldsymbol{\sigma}^e \rangle = \sum_{r=1}^{N} c_r \boldsymbol{\sigma}_r^e = \sum_{r=1}^{N} c_r \mathbf{B}_r \overline{\boldsymbol{\sigma}} = \sum_{r=1}^{N} c_r \mathbf{B}_r \langle \overline{\boldsymbol{\sigma}} \rangle = \sum_{r=1}^{N} c_r \mathbf{B}_r \langle \boldsymbol{\sigma}^e \rangle. \tag{11.1.26}$$

Thus,

$$\langle \mathbf{B}_r \rangle = \sum_{r=1}^{N} c_r \mathbf{B}_r = \mathbf{I}. \tag{11.1.27}$$

Substituting (11.1.25) into the second of (11.1.19) yields

$$\boldsymbol{\sigma}_r = \mathbf{B}_r\overline{\boldsymbol{\sigma}} + \boldsymbol{\sigma}_r^p \quad \text{in } \Omega_r. \tag{11.1.28}$$

It then follows from the decomposition (11.1.4) that

$$\langle \mathbf{B}^{\mathrm{T}}\boldsymbol{\varepsilon}\rangle = \langle \mathbf{B}^{\mathrm{T}}\boldsymbol{\varepsilon}^e\rangle + \langle \mathbf{B}^{\mathrm{T}}\boldsymbol{\varepsilon}^p\rangle, \tag{11.1.29}$$

or

$$\langle \mathbf{B}^{\mathrm{T}}\boldsymbol{\varepsilon}\rangle = \langle \boldsymbol{\sigma}\mathbf{M}\mathbf{B}\rangle + \langle \mathbf{B}^{\mathrm{T}}\boldsymbol{\varepsilon}^p\rangle. \tag{11.1.30}$$

In deriving (11.1.30), we have substituted the elastic strain by using Hooke's law in terms of elastic compliance $\mathbf{M}$. The diagonal symmetry of $\mathbf{M}$ allows us to replace $\mathbf{B}^{\mathrm{T}}$ with $\mathbf{B}$. In (11.1.30), we can show easily that the field $\mathbf{MB}$ is kinematically admissible. Thus, application of the Hill lemma to $\langle \boldsymbol{\sigma}\mathbf{MB}\rangle$ yields

$$\langle \boldsymbol{\sigma}\mathbf{M}\mathbf{B}\rangle = \langle \boldsymbol{\sigma}\rangle\langle \mathbf{M}\mathbf{B}\rangle = \overline{\boldsymbol{\sigma}}\overline{\mathbf{M}} = \overline{\boldsymbol{\varepsilon}}^e, \tag{11.1.31}$$

where $\overline{\mathbf{M}} = \langle \mathbf{MB}\rangle$ is the effective elastic compliance of the polycrystalline material.

On the other hand, since $\mathbf{B}^{\mathrm{T}}$ is statically admissible, one can write

$$\langle \mathbf{B}^{\mathrm{T}}\boldsymbol{\varepsilon}\rangle = \langle \mathbf{B}^{\mathrm{T}}\rangle\boldsymbol{\varepsilon} = \langle \boldsymbol{\varepsilon}\rangle = \overline{\boldsymbol{\varepsilon}}, \tag{11.1.32}$$

where (11.1.27) has been used. By combining (11.1.29), (11.1.31), and (11.1.32), one can conclude that

$$\overline{\boldsymbol{\varepsilon}}^e = \langle \mathbf{B}^{\mathrm{T}}\boldsymbol{\varepsilon}^e\rangle \neq \langle \boldsymbol{\varepsilon}^e\rangle, \qquad \overline{\boldsymbol{\varepsilon}}^p = \langle \mathbf{B}^{\mathrm{T}}\boldsymbol{\varepsilon}^p\rangle \neq \langle \boldsymbol{\varepsilon}^p\rangle. \tag{11.1.33}$$

The same statement can be made for the rates,

$$\overline{\dot{\boldsymbol{\varepsilon}}}^e = \langle \mathbf{B}^{\mathrm{T}}\boldsymbol{\varepsilon}^e\rangle \neq \langle \dot{\boldsymbol{\varepsilon}}^e\rangle, \qquad \overline{\dot{\boldsymbol{\varepsilon}}}^p = \langle \mathbf{B}^{\mathrm{T}}\dot{\boldsymbol{\varepsilon}}^p\rangle \neq \langle \dot{\boldsymbol{\varepsilon}}^p\rangle. \tag{11.1.34}$$

Based on the foregoing discussions, one can derive some interesting and well-known features of crystalline materials.

Stored Elastic Strain Energy

The total elastic strain energy stored in D is given by

$$U = \frac{1}{2}\int_D \boldsymbol{\sigma}\boldsymbol{\varepsilon}^e \, dV. \tag{11.1.35}$$

Making use of (11.1.6) and (11.1.19), we can rewrite this as

$$\begin{aligned} U &= \frac{1}{2}\int_D \boldsymbol{\sigma}\mathbf{M}\boldsymbol{\sigma} \, dV = \frac{1}{2}\int_D (\boldsymbol{\sigma}^e + \boldsymbol{\sigma}^p)\mathbf{M}(\boldsymbol{\sigma}^e + \boldsymbol{\sigma}^p) \, dV \\ &= \frac{1}{2}\int_D \boldsymbol{\sigma}^e\mathbf{M}\boldsymbol{\sigma}^e \, dV + \int_D \boldsymbol{\sigma}^p\mathbf{M}\boldsymbol{\sigma}^e \, dV + \frac{1}{2}\int_D \boldsymbol{\sigma}^p\mathbf{M}\boldsymbol{\sigma}^p \, dV. \end{aligned} \tag{11.1.36}$$

Note that the first term on the right-hand side is the strain energy in an elastic body. Thus, it follows from the discussions in Section 5.6 that

$$\int_D \boldsymbol{\sigma}^e\mathbf{M}\boldsymbol{\sigma}^e \, dV = D\overline{\boldsymbol{\sigma}}^e\overline{\mathbf{M}\boldsymbol{\sigma}}^e = D\overline{\boldsymbol{\sigma}}\overline{\mathbf{M}\boldsymbol{\sigma}}. \tag{11.1.37}$$

The second integral is

$$\int_D \boldsymbol{\sigma}^p\mathbf{M}\boldsymbol{\sigma}^e \, dV = \int_D \sigma_{ij}^p \, \beta_{ij} \, dV = \int_D \sigma_{ij}^p u_{i,j}^e \, dV = \int_S \sigma_{ij}^p u_i^e n_j \, dV = 0, \tag{11.1.38}$$

where we have used (11.1.18), (11.1.21), and (11.1.22). Thus, the strain energy stored in D when subjected to the traction boundary condition (11.1.13) is given by

$$U = \frac{D}{2}\overline{\boldsymbol{\sigma}}\overline{\mathbf{M}\boldsymbol{\sigma}} + \frac{1}{2}\int_D \boldsymbol{\sigma}^p\mathbf{M}\boldsymbol{\sigma}^p \, dV. \tag{11.1.39}$$

We see that the stored elastic strain energy contains, in addition to the macroscopic strain energy resulting from the applied load, a contribution from the plastic residual stresses.

Dissipated Energy

The dissipated energy due to plasticity, or the stress power from plasticity, is defined as [see (2.3.5)]

$$W = \frac{1}{2}\int_D \boldsymbol{\sigma}\dot{\boldsymbol{\varepsilon}}^p \, dV.$$

Making use of the strain decomposition, this can be rewritten as

$$W = \frac{1}{2}\int_D \boldsymbol{\sigma}\dot{\boldsymbol{\varepsilon}} \, dV - \frac{1}{2}\int_D \boldsymbol{\sigma}\dot{\boldsymbol{\varepsilon}}^e \, dV. \tag{11.1.40}$$

It follows from the Hill lemma (5.5.5) that the first term on the right-hand side of (11.1.40) can be written as

$$\int_D \boldsymbol{\sigma}\dot{\boldsymbol{\varepsilon}} \, dV = D\overline{\boldsymbol{\sigma}}\dot{\overline{\boldsymbol{\varepsilon}}} = D\overline{\boldsymbol{\sigma}}\dot{\overline{\boldsymbol{\varepsilon}}}^e + D\overline{\boldsymbol{\sigma}}\dot{\overline{\boldsymbol{\varepsilon}}}^p. \tag{11.1.41}$$

Following the same arguments used in deriving the strain energy, the second term on the right-hand side of (11.1.40) can be written as

$$\int_D \boldsymbol{\sigma}\dot{\boldsymbol{\varepsilon}}^e \, dV = \int_D \boldsymbol{\sigma}\mathbf{M}\dot{\boldsymbol{\sigma}} \, dV = D\overline{\boldsymbol{\sigma}}\dot{\overline{\boldsymbol{\varepsilon}}}^e + \int_D \boldsymbol{\sigma}^p\mathbf{M}\dot{\boldsymbol{\sigma}}^p \, dV. \tag{11.1.42}$$

Hence

$$\frac{D}{2}\overline{\boldsymbol{\sigma}}\dot{\overline{\boldsymbol{\varepsilon}}}^p = W + \frac{1}{2}\int_D \boldsymbol{\sigma}^p\mathbf{M}\dot{\boldsymbol{\sigma}}^p \, dV = \frac{1}{2}\int_D \boldsymbol{\sigma}\dot{\boldsymbol{\varepsilon}}^p \, dV + \frac{1}{2}\int_D \boldsymbol{\sigma}^p\mathbf{M}\dot{\boldsymbol{\sigma}}^p \, dV. \tag{11.1.43}$$

This clearly shows that the macroscopic plastic stress power $D\overline{\boldsymbol{\sigma}}\dot{\overline{\boldsymbol{\varepsilon}}}^p$ is not completely dissipative. The first term on the right-hand side of (11.1.43) is dissipative, while the second term resulting from the plastic residual stress is stored in the material. This contribution may explain the "interphase" work hardening observed in polycrystalline materials (see below).

Interphase Work Hardening

Following the same procedure used in calculating the dissipation energy, one can easily obtain the following:

$$\frac{D}{2}\overline{\dot{\boldsymbol{\sigma}}}\,\overline{\dot{\boldsymbol{\varepsilon}}^p} = \frac{1}{2}\int_D \dot{\boldsymbol{\sigma}}\dot{\boldsymbol{\varepsilon}}^p\,dV + \frac{1}{2}\int_D \dot{\boldsymbol{\sigma}}^p\mathbf{M}\dot{\boldsymbol{\sigma}}^p\,dV. \qquad (11.1.44)$$

If all the single crystals in the polycrystal assembly behave as elastic perfectly plastic materials (without work hardening), then the plastic flow of each crystal can be described using the normality rule as

$$\dot{\boldsymbol{\sigma}}\dot{\boldsymbol{\varepsilon}}^p = \dot{\lambda}\dot{f} = 0 \quad \text{in } D,$$

where $f(\boldsymbol{\sigma})$ is the yield surface and $\dot{\lambda}$ is the well-known plastic parameter. Therefore, we have

$$\frac{D}{2}\overline{\dot{\boldsymbol{\sigma}}}\,\overline{\dot{\boldsymbol{\varepsilon}}^p} = \frac{1}{2}\int_D \dot{\boldsymbol{\sigma}}^p\mathbf{M}\dot{\boldsymbol{\sigma}}^p\,dV \geq 0 \qquad (11.1.45)$$

because $\mathbf{M}$ is positive definite. Clearly, one can conclude from the inequality (11.1.45) that the macroscopic or homogenized behavior of polycrystal assembly display work hardening during its plastic flow even though all of its single crystals are perfectly plastic. In other words, the inhomogeneity of the material is one of the sources of its work hardening. Such prediction has been confirmed experimentally for polycrystalline materials. For example, for most single crystals, there is very little work hardening during the early stage (easy glide region) of deformation. However, during this stage, the polycrystal experiences a significant work hardening due to the intergranular interactions among the single crystals.

In general, a nonlinear heterogeneous material can be considered as a standard generalized material (in the sense of thermodynamics) by defining appropriate internal variables, state variables, and associated thermodynamic forces. Such formulation requires an infinite number of internal variables (e.g., the plastic strain field), which renders such a procedure not feasible for practical situations. This is why nonlinear heterogeneous materials require suitable approximations to model their overall behavior. The modeling of nonlinear behavior in inhomogeneous materials is still a challenging issue and has not been solved completely yet.

Kröner's Approach

The modeling of elastoplastic behavior of polycrystalline materials by the concept of Eshelby's elastic solutions was initiated by Budiansky and Mangasarian (1960). Their original idea was to model the first stage of the plastic deformation so that Eshelby's elastic inclusion solution is applicable without any major modifications. They argued that the favorably oriented grains that first experience a plastic deformation can be simulated by an ellipsoidal inclusion subjected to stress-free plastic strain (eigenstrain) in interaction within an elastic infinite medium representing the other grains that are still at the elastic regime. This approximation is also supported by the fact that the number of grains undergoing plastic deformation is low at the earlier stage of the plastic flow, and hence the homogenization procedure can be performed by the dilute approximation. In other words, the interactions between the plastically deformed grains can be neglected.

Subsequently, Kröner, who initiated the self-consistent scheme in elasticity, proposed a similar formulation that enables us to describe the elastoplastic behavior beyond the earlier stages of the plastic flow. Unlike Budiansky's approach, Kröner considered the infinite medium in the Eshelby scheme as the "unknown" homogeneous medium subjected to an average $\bar{\boldsymbol{\varepsilon}}^p$ at a certain stage of the plastic flow. To solve the interaction or localization problem for a given set of grains subjected to uniform plastic strain $\bar{\boldsymbol{\varepsilon}}^p$, whereas the polycrystals as a whole is plastically deformed by $\boldsymbol{\varepsilon}^p$, Kröner's approximation adopts the Eshelby solution by describing the set of grains by an ellipsoidal inclusion subjected to an eigenstrain $\boldsymbol{\varepsilon}^p$ and surrounded by an infinite medium, which in turn is subjected to a uniform deformation $\bar{\boldsymbol{\varepsilon}}^p$. One of the limitations of Kröner's model is that the framework was developed based on isotropic and spherical grains, and incompressible plasticity.

Let us first consider a homogeneous elastic matrix of stiffness $\mathbf{L}$ and volume D, subjected to a uniform strain $\bar{\boldsymbol{\varepsilon}}$. Let Ω_r be an ellipsoidal inclusion embedded within the matrix. It goes without saying that $\Omega_r << D$. To simulate the plasticity, let $\boldsymbol{\varepsilon}_r^p$ be an eigenstrain field distributed in Ω_r.

According to the Eshelby inclusion solution, the total strain on the inclusion is given by

$$\boldsymbol{\varepsilon}_r = \bar{\boldsymbol{\varepsilon}} + \mathbf{S}_r \boldsymbol{\varepsilon}_r^p \quad \text{in } \Omega_r, \tag{11.1.46}$$

where $\mathbf{S}_r$ is the Eshelby inclusion tensor for Ω_r. This gives

$$\boldsymbol{\varepsilon}_r^p = \mathbf{S}_r^{-1}(\boldsymbol{\varepsilon}_r - \bar{\boldsymbol{\varepsilon}}) \quad \text{in } \Omega_r. \tag{11.1.47}$$

The total stress field in the inclusion is thus given by [see (4.3.22)]

$$\boldsymbol{\sigma}_r = \mathbf{L}(\boldsymbol{\varepsilon}_r - \boldsymbol{\varepsilon}_r^p) \quad \text{in } \Omega_r. \tag{11.1.48}$$

Substituting (11.1.46) into (11.1.48) gives

$$\boldsymbol{\sigma}_r = \mathbf{L}\bar{\boldsymbol{\varepsilon}} + \mathbf{L}(\mathbf{S}_r - \mathbf{I})\boldsymbol{\varepsilon}_r^p = \bar{\boldsymbol{\sigma}} + \mathbf{L}(\mathbf{S}_r - \mathbf{I})\boldsymbol{\varepsilon}_r^p, \tag{11.1.49}$$

where $\bar{\boldsymbol{\sigma}} = \mathbf{L}\bar{\boldsymbol{\varepsilon}}$ is the stress tensor averaged over the entire polycrystal assembly D.

Recall that Eshelby's tensor $\mathbf{S}_r$ depends on the elastic stiffness tensor $\mathbf{L}$ and the shape and orientation of the inclusion. Therefore, in its general form, (11.1.49) is capable of capturing the plastic anisotropy resulting from morphological aspects related to the irregular shape of grains.

As mentioned above, Kröner initially considered the case of spherical inclusions in isotropic matrix. Under such conditions, one has [see (2.4.23)]

$$\mathbf{L} = 3K\mathbf{I}^h + 2\mu\mathbf{I}^d, \qquad \mathbf{S}_r = 3\gamma\mathbf{I}^h + 2\delta\mathbf{I}^d, \tag{11.1.50}$$

where

$$\gamma = \frac{K}{3K + 4\mu} = \frac{1 + \nu}{9(1 - \nu)}, \qquad \delta = \frac{3(K + 2\mu)}{5(3K + 4\mu)} = \frac{4 - 5\nu}{15(1 - \nu)}, \tag{11.1.51}$$

K and μ are, respectively, the bulk and shear moduli, and ν is the Poisson ratio of the isotropic matrix material. The symbolic notation introduced in Section 1.4 has been used for isotropic fourth-order tensors. Thus, we have

$$\mathbf{L}(\mathbf{S}_r - \mathbf{I}) = 3K(3\gamma - 1)\mathbf{I}^h + 2\mu(2\delta - 1)\mathbf{I}^d. \tag{11.1.52}$$

The incompressibility of the plastic deformation means

$$\mathbf{I}^h \boldsymbol{\varepsilon}_r^p = 0. \tag{11.1.53}$$

Thus, substituting (11.1.52) into (11.1.49) yields

$$\boldsymbol{\sigma}_r = \overline{\boldsymbol{\sigma}} + 2\mu(2\delta - 1)\mathbf{I}^d \boldsymbol{\varepsilon}_r^p. \tag{11.1.54}$$

To describe the plastic flow, (11.1.54) should be written in a rate or incremental form, that is,

$$\dot{\boldsymbol{\sigma}}_r = \dot{\overline{\boldsymbol{\sigma}}} + 2\mu(2\delta - 1)\mathbf{I}^d \dot{\boldsymbol{\varepsilon}}_r^p. \tag{11.1.55}$$

Substituting the second of (11.1.47) into (11.1.55), we arrive at

$$\dot{\boldsymbol{\sigma}}_r = \dot{\overline{\boldsymbol{\sigma}}} + \frac{\mu(2\delta - 1)}{2\delta} \mathbf{I}^d (\dot{\boldsymbol{\varepsilon}}_r - \dot{\overline{\boldsymbol{\varepsilon}}}). \tag{11.1.56}$$

Equation (11.1.56) relates local quantities $\dot{\boldsymbol{\sigma}}_r$ and $\dot{\boldsymbol{\varepsilon}}_r$ to the macroscopic quantities $\dot{\overline{\boldsymbol{\sigma}}}$ and $\dot{\overline{\boldsymbol{\varepsilon}}}$. It constitutes the first step for a homogenization scheme crucial for an accurate prediction of the macroscopic behavior. It is clearly seen from (11.1.56) that the interaction between the different quantities is purely elastic. This results from the description of the plastic strain as an eigenstrain leading to a purely heterogeneous elastic problem. In reality, however, during the plastic flow, constraints exerted by the aggregate on a single grain become softer than in the elastic regime and change with the plastic deformation. Since the Kröner model is based on elastic constraints that remain elastic during the plastic flow, it will result in stiffer predictions of the overall behavior.

Let us now introduce the instantaneous or tangent modulus of the polycrystal $\overline{\mathbf{L}}^t$ and the tangent modulus of the single-crystal $\mathbf{L}^t$ through

$$\dot{\overline{\boldsymbol{\sigma}}} = \overline{\mathbf{L}}^t \dot{\overline{\boldsymbol{\varepsilon}}}, \qquad \dot{\boldsymbol{\sigma}}_r = \mathbf{L}_r^t \dot{\boldsymbol{\varepsilon}}_r. \tag{11.1.57}$$

Substituting (11.1.57) into (11.1.56) leads to

$$\left[\mathbf{L}_r^t - \frac{\mu(2\delta - 1)}{2\delta} \mathbf{I}^d\right] \dot{\boldsymbol{\varepsilon}}_r = \left[\overline{\mathbf{L}}^t - \frac{\mu(2\delta - 1)}{2\delta} \mathbf{I}^d\right] \dot{\overline{\boldsymbol{\varepsilon}}} \tag{11.1.58}$$

or

$$\dot{\boldsymbol{\varepsilon}}_r = \mathbf{A}_r\dot{\overline{\boldsymbol{\varepsilon}}}, \tag{11.1.59}$$

where

$$\mathbf{A}_r = \left[\mathbf{L}_r^t - \frac{\mu(2\delta - 1)}{2\delta}\mathbf{I}^d\right]^{-1}\left[\overline{\mathbf{L}}^t - \frac{\mu(2\delta - 1)}{2\delta}\mathbf{I}^d\right] \tag{11.1.60}$$

defines the strain rate concentration tensor. Substitution of (11.1.60) into the second of (11.1.57) leads to

$$\dot{\overline{\boldsymbol{\sigma}}} = \sum_{r=1}^{N} c_r\dot{\boldsymbol{\sigma}}_r = \sum_{r=1}^{N} \mathbf{L}_r^t\mathbf{A}_r\dot{\overline{\boldsymbol{\varepsilon}}}. \tag{11.1.61}$$

One can then conclude that

$$\overline{\mathbf{L}}^t = \sum_{r=1}^{N} \mathbf{L}_r^t\mathbf{A}_r \tag{11.1.62}$$

or

$$\overline{\mathbf{L}}^t = \sum_{r=1}^{N} \mathbf{L}_r^t\left[\mathbf{L}_r^t - \frac{\mu(2\delta - 1)}{2\delta}\mathbf{I}^d\right]^{-1}\left[\overline{\mathbf{L}}^t - \frac{\mu(2\delta - 1)}{2\delta}\mathbf{I}^d\right]. \tag{11.1.63}$$

Equation (11.1.63) inherits the implicit character of the self-consistent scheme as is the case in linear elasticity. Clearly the nonlinearity is captured in (11.1.63) since the local tangent modulus depends on the plastic strain $\boldsymbol{\varepsilon}^p$.

If we further assume that the instantaneous tangent modulus tensor is also isotropic, that is,

$$\mathbf{L}_r^t = 3K_r^t\mathbf{I}^h + 2\mu_r^t\mathbf{I}^d, \qquad \overline{\mathbf{L}}^t = 3\overline{K}^t\mathbf{I}^h + 2\overline{\mu}^t\mathbf{I}^d, \tag{11.1.64}$$

then

$$\mathbf{L}_r^t\left[\mathbf{L}_r^t - \frac{\mu(2\delta - 1)}{2\delta}\mathbf{I}^d\right]^{-1}\left[\overline{\mathbf{L}}^t - \frac{\mu(2\delta - 1)}{2\delta}\mathbf{I}^d\right]$$
$$= 3\overline{K}_r^t\mathbf{I}^h + \frac{2\mu_r^t[4\delta\overline{\mu}^t - \mu(2\delta - 1)]}{4\delta\mu_r^t - \mu(2\delta - 1)}\mathbf{I}^d. \tag{11.1.65}$$

Making use of this in (11.1.63) gives us

$$\overline{\mu}^t = \sum_{r=1}^{N} \frac{c_r \mu_r^t [4\delta \overline{\mu}^t - \mu(2\delta - 1)]}{4\delta \mu_r^t - \mu(2\delta - 1)}. \tag{11.1.66}$$

To see some insight of the Kröner model prediction, we consider the case

$$\mu_r^t << \mu, \qquad \overline{\mu}^t << \mu. \tag{11.1.67}$$

This is approximately true for later stages of plastic deformation. Under this assumption, (11.1.66) reduces to

$$\overline{\mu}^t \approx \sum_{r=1}^{N} c_r \mu_r^t. \tag{11.1.68}$$

This last equation is the Taylor–Lin bound, which is equivalent to the Voigt model in linear elasticity. In fact, the Taylor–Lin model (Taylor, 1938; Payne et al. 1958) assumes homogeneous strain in the polycrystalline aggregate so that $\boldsymbol{\varepsilon} = \overline{\boldsymbol{\varepsilon}}$. Therefore, the effective tangent modulus predicted by the Taylor–Lin model simply leads to (11.1.68).

Treating the interactions between grains as elastic instead of elastoplastic is one of the major limitations of the Kröner model. Additionally, the consistency attributed to the Kröner approach is not really true since in his procedure the equivalent homogeneous medium is taken to be elastic even though with assigned plastic deformation. These shortcomings are taken into consideration in the Hill self-consistent model to be discussed next.

In closing this section, we mention that although the Kröner model presented here is in terms of a summation over all the single crystals, the results can be easily recast into integral forms (ensemble average), as discussed in Section 7.5.

Hill's Self-Consistent Model

To solve the nonlinear elastoplastic deformation, Hill proposed solving successive linear problems at each loading increment. To accomplish this, Hill introduced the following local and global constitutive laws:

$$\overline{\dot{\boldsymbol{\sigma}}}_r = \mathbf{L}_r^t \overline{\dot{\boldsymbol{\varepsilon}}}_r, \qquad \overline{\dot{\boldsymbol{\sigma}}} = \overline{\mathbf{L}}^t \overline{\dot{\boldsymbol{\varepsilon}}}, \tag{11.1.69}$$

where $\mathbf{L}_r^t$ is the tangent modulus of the rth grain, $\overline{\mathbf{L}}^t$ is the effective tangent modulus of the polycrystal assembly, and $\overline{\dot{\boldsymbol{\sigma}}}_r$ and $\overline{\dot{\boldsymbol{\varepsilon}}}_r$ are, respectively, the stress and strain rates averaged over the rth grain, while $\bar{\dot{\boldsymbol{\sigma}}}$ and $\bar{\dot{\boldsymbol{\varepsilon}}}$ are, respectively, the stress and strain rates averaged over the entire polycrystal assembly. The inverse of these equations are

$$\bar{\dot{\boldsymbol{\varepsilon}}}_r = \mathbf{M}_r^t \bar{\dot{\boldsymbol{\sigma}}}_r, \qquad \bar{\dot{\boldsymbol{\varepsilon}}} = \overline{\mathbf{M}}{}^t \bar{\dot{\boldsymbol{\sigma}}}, \tag{11.1.70}$$

where $\mathbf{M}_r^t$ is the tangent compliance of the rth grain and $\overline{\mathbf{M}}{}^t$ is the effective tangent compliance tensor of the polycrystal assembly.

To solve for $\overline{\mathbf{L}}{}^t$, we assume that the rth grain in the polycrystal assembly is simulated by an ellipsoidal inclusion Ω_r embedded in a homogeneous matrix with effective modulus tensor $\overline{\mathbf{L}}{}^t$, which is yet to be determined. As in the linear elastic case, such self-consistent approach accounts for, at least partially, the interactions between the grains.

Making use of the rate form of (4.5.19), we have

$$\bar{\dot{\boldsymbol{\sigma}}}_r = \bar{\dot{\boldsymbol{\sigma}}} + \overline{\mathbf{H}}_r(\bar{\dot{\boldsymbol{\varepsilon}}} - \bar{\dot{\boldsymbol{\varepsilon}}}_r), \tag{11.1.71}$$

where the concentration tensor $\overline{\mathbf{H}}_r$ is given by [see (4.5.13)]

$$\overline{\mathbf{H}}_r = \overline{\mathbf{L}}{}^t(\overline{\mathbf{S}}_r^{-1} - \mathbf{I}). \tag{11.1.72}$$

In the above, $\overline{\mathbf{S}}_r$ is the Eshelby inclusion tensor computed for the inclusion Ω_r embedded in a matrix with elastic stiffness tensor $\overline{\mathbf{L}}{}^t$. It depends on the shape of Ω_r and on the overall property $\overline{\mathbf{L}}{}^t$. Note that, although (11.1.71) and (11.1.72) are similar to the linear elastic case, the effective property to be determined here is the effective tangent modulus of the heterogeneous material, not the effective elastic modulus.

Substituting (11.1.69) into (11.1.71), one obtains

$$(\mathbf{L}_r^t + \overline{\mathbf{H}}_r)\bar{\dot{\boldsymbol{\varepsilon}}}_r = (\overline{\mathbf{L}}{}^t + \overline{\mathbf{H}}_r)\bar{\dot{\boldsymbol{\varepsilon}}}. \tag{11.1.73}$$

From this equation, the average strain rate on the rth grain is expressed in terms of the macroscopic strain rate as

$$\bar{\dot{\boldsymbol{\varepsilon}}}_r = \mathbf{A}_r \bar{\dot{\boldsymbol{\varepsilon}}}, \tag{11.1.74}$$

where

$$\mathbf{A}_r = (\mathbf{L}_r^t + \overline{\mathbf{H}}_r)^{-1}\,(\overline{\mathbf{L}}{}^t + \overline{\mathbf{H}}_r), \tag{11.1.75}$$

is the strain rate concentration tensor in elastoplastic deformation, which depends on the effective tangent modulus and geometrical aspects of the grain in consideration Ω_r.

Similarly, making use of (11.1.70) in (11.1.71) gives the average stress rate on the rth grain in terms of the global stress rate,

$$\overline{\dot{\boldsymbol{\sigma}}}_r = \mathbf{B}_r \overline{\dot{\boldsymbol{\sigma}}}, \tag{11.1.76}$$

where

$$\mathbf{B}_r = (\overline{\mathbf{H}}_r^{-1} + \mathbf{M}_r^t)^{-1} (\overline{\mathbf{H}}_r^{-1} + \overline{\mathbf{M}}^t). \tag{11.1.77}$$

It then follows from (11.1.74)–(11.1.76) and (11.1.69) and (11.1.70) that

$$\overline{\dot{\boldsymbol{\sigma}}} = \sum_{r=1}^{N} c_r \overline{\dot{\boldsymbol{\sigma}}}_r = \sum_{r=1}^{N} c_r \mathbf{L}_r^t \overline{\dot{\boldsymbol{\varepsilon}}}_r = \sum_{r=1}^{N} c_r \mathbf{L}_r^t \mathbf{A}_r \overline{\dot{\boldsymbol{\varepsilon}}}, \tag{11.1.78}$$

$$\overline{\dot{\boldsymbol{\varepsilon}}} = \sum_{r=1}^{N} c_r \overline{\dot{\boldsymbol{\varepsilon}}}_r = \sum_{r=1}^{N} c_r \mathbf{M}_r^t \overline{\dot{\boldsymbol{\sigma}}}_r = \sum_{r=1}^{N} c_r \mathbf{M}_r^t \mathbf{B}_r \overline{\dot{\boldsymbol{\sigma}}}. \tag{11.1.79}$$

These lead to

$$\overline{\mathbf{L}}^t = \sum_{r=1}^{N} c_r \mathbf{L}_r^t \mathbf{A}_r, \qquad \overline{\mathbf{M}}^t = \sum_{r=1}^{N} c_r \mathbf{M}_r^t \mathbf{B}_r. \tag{11.1.80}$$

The last two equations are the Hill self-consistent scheme to evaluate the effective tangent modulus and effective tangent compliance tenors of a polycrystal assembly. Note that the second of (11.1.80) requires that the tangent compliance tensor $\mathbf{M}^t$ for each grain to exist. When this is the case, one can show that (see Problem 11.6)

$$\mathbf{L}^t \mathbf{A}_r = \mathbf{B}_r \overline{\mathbf{L}}^t \quad \text{and} \quad \mathbf{M}^t \mathbf{B}_r = \mathbf{A}_r \overline{\mathbf{M}}^t. \tag{11.1.81}$$

This leads to $\overline{\mathbf{M}}\overline{\mathbf{L}} = \overline{\mathbf{L}}\overline{\mathbf{M}} = \mathbf{I}$; see Problem 11.6. Therefore, it is sufficient to solve either the first or the second of (11.1.80). In practice, the first of (11.1.80) is more widely used because $\mathbf{M}^t$ is not always well defined for all single crystals for a given polycrystal assembly.

As discussed in Section 7.5, (11.1.80) can be cast into an integral form. We will not repeat the discussion here. Readers are encouraged to formulate (11.1.80) into an integral form on their own.

Note that (11.1.80) reduces to the self-consistent estimates (7.5.13) and (7.5.14) if the tangent modulus and tangent compliance tensors are replaced by their corresponding elastic stiffness and compliance tensors. This implies that the linear elastic behavior is a special case of the Hill model for elastoplastic behavior. Generally, (11.1.80) gives more accurate results in comparison with Kröner's model because Hill's model captures at least partially the fluctuations of strains between each grain. However, the Hill model is still an approximation because of the use of piecewise uniform mechanical properties on each grain. Therefore, any intraphase fluctuation arising naturally from the elastoplastic behavior is disregarded in this approach.

It should also be noted that obtaining $\overline{\mathbf{L}}^t$ from (11.1.80) is not an easy task. This is due to the implicit nature of the equation and the anisotropy of the linearized constitutive laws (11.1.69), which makes the calculation of the Hill constraint tensors a complicated task. In general, the determination of $\overline{\mathbf{L}}^t$ through (11.1.80) requires an iterative procedure. First, for a given state of deformation, an initial guess of $\overline{\mathbf{L}}^t$ is used on the right hand of (11.1.80) to calculate an improved value for $\overline{\mathbf{L}}^t$. Then, this improved $\overline{\mathbf{L}}^t$ is substituted back into the right-hand side of the equation again to obtain the next improvement. This procedure is repeated sufficiently until a convergent value is obtained for $\overline{\mathbf{L}}^t$.

Another major difficulty behind the Hill self-consistent model is related to how the effective tangent modulus $\overline{\mathbf{L}}^t$ depends on the prescribed values of strain rate $\dot{\boldsymbol{\varepsilon}}$. Unlike in the single-crystal case where only a finite number of branches exist, for polycrystals $\overline{\mathbf{L}}^t$ varies continuously as the direction of prescribed strain rate varies in the strain rate space. In other words, $\overline{\mathbf{L}}^t$ is a homogeneous function of degree zero of $\dot{\boldsymbol{\varepsilon}}$ (Hutchinson, 1970). In practice, additional assumptions are usually made regarding the anisotropy of the tangent modulus in order to simplify the calculations. Examples of this will be discussed in the next section.

11.2 APPLICATIONS

The main purpose of this section is to explore the feasibility using the Hill self-consistent method to predict stress–strain behavior of polycrystalline materials from the elastoplastic properties of single-crystal constituents. We will focus on FCC metallic materials by presenting briefly the main features of plastic deformation at the continuum level of single crystals under conventional loading conditions of strain rates and temperature. Our attention is not to describe exhaustively the various mechanisms of plastic deformation from a physical metallurgy point of view, for such a purpose the readers are referred to more specialized textbooks and technical papers that are recommended at the end of this chapter. Instead, we will follow a mechanistic procedure to bridge the scales between the single-crystal level and the polycrystalline level.

Plastic Deformation of Single Crystals

The physical understanding of single-crystal plasticity was established during the earlier years of the last century, in 1900 to 1938, with the contribution of Ewing and Rosenhain (1900), Bragg and Bragg (1933), Taylor and co-workers (1923, 1925, 1934, 1938), Polanyi (1922), Schmid and Boas (1935), among others. Their experimental measurements established that at room temperature the major source of plastic deformation is the dislocation movements through the crystal lattices. These motions occur on certain crystal planes in certain crystallographic directions, and the crystal structure of metals is not altered by the plastic flow. Mathematical descriptions of these physical phenomena of plastic deformation in single crystals were presented by Taylor (1938), when he investigated the plastic deformation of polycrystalline materials in terms of single-crystal deformation. More rigorous and rational formulations were given by Hill (1965), Hill and Rice (1972), Asaro and Rice (1977), and by Hill and Havner (1982). A comprehensive review of this subject can be found in Azaro (1983).

The kinematics of single-crystal deformation and the resulting elastoplastic constitutive laws are based on an idealization of dislocation motion as a collective movement of the atoms leading to slips in certain directions on specific crystallographic planes. This process occurs when the resolved shear stress on one or more of these slip systems reaches a critical value. As plastic deformation proceeds, the critical yield stresses associated with the slip systems increases. This contributes to the strain hardening of the polycrystalline aggregate.

Consider a single crystal with N possible slip systems. Each system g is characterized by the unit normal $\mathbf{n}^g$ to the slip plane on which the collective movement of dislocations occurs, and by the direction $\mathbf{m}^g$ of dislocation gliding, which is colinear with the Burger vector $\mathbf{b}^g$ of gliding dislocations on the system g, so that $\mathbf{b}^g = b\mathbf{m}^g$, where b is the magnitude of the Burger vector. The mathematical tool treating the collective movement of dislocations considers each dislocation line as the boundary of a cutting surface S^g with a unit normal $\mathbf{n}^g$, across which the discontinuities of the displacement vector are uniform and characterized by the Burger vector $\mathbf{b}^g$ so that $b_i^g n_i^g = 0$. This transformation can be described at each material point by a second-order tensor $\beta^{\mathrm{p}}(\mathbf{r})$ as

$$\beta_{ij}^{\mathrm{p}}(\mathbf{r}) = b_i^g n_j^g \delta(S^g), \tag{11.2.1}$$

where $\delta(S^g)$ is the surface Dirac function given by

$$\delta(S^g) = \int_{S^g} \delta(\mathbf{r} - \mathbf{r})\, dS'. \tag{11.2.2}$$

If many dislocations with the same Burger vector $\mathbf{b}^g$ and the same cutting surfaces are present in the single-crystal volume V, one can define an average transformation β^p by

$$\beta_{ij}^p = bm_i^g n_j^g \frac{1}{V} \int_V \delta(S^g)\, dV. \tag{11.2.3}$$

Introducing the average plastic shear γ^g by

$$\gamma^g = b \frac{1}{V} \int_V \delta(S^g)\, dV \tag{11.2.4}$$

leads to

$$\beta_{ij}^p = \gamma^g m_i^g n_j^g. \tag{11.2.5}$$

If we account for all dislocations present at the slip systems, (11.2.5) can be extended to

$$\beta_{ij}^p = \sum_g \gamma^g m_i^g n_j^g. \tag{11.2.6}$$

The shear rate $\dot{\gamma}^g$ is calculated from (11.2.4) as

$$\dot{\gamma}^g = b \frac{\partial}{\partial t} \left\{ \frac{1}{V} \int_V \delta(S^g) \, dV \right\}. \tag{11.2.7}$$

Equation (11.2.7) describing the creation and movement of dislocations at the continuum level corresponds to the Orowan relation.

The rate of plastic distortion $\dot{\boldsymbol{\beta}}^p$ is the sum of the contributions of the shear rates $\dot{\gamma}^g$ from all the active slip systems. That is

$$\dot{\beta}^p_{ij} = \sum_g \dot{\gamma}^g m^g_i n^g_j. \tag{11.2.8}$$

The symmetric part of the plastic distortion gives the plastic strain rate

$$\dot{\varepsilon}^p_{ij} = \frac{1}{2} (\dot{\beta}^p_{ij} + \dot{\beta}^p_{ji}) = \sum_g \dot{\gamma}^g R^g_{ij}, \tag{11.2.9}$$

where

$$R^g_{ij} = \tfrac{1}{2}(m^g_i n^g_j + m^g_j n^g_i). \tag{11.2.10}$$

The antisymmetric part of the plastic distortion defines the plastic spin,

$$\dot{w}^p_{ij} = \frac{1}{2} (\dot{\beta}^p_{ij} - \dot{\beta}^p_{ji}) = \sum_g \dot{\gamma}^g \tilde{R}^g_{ij}, \tag{11.2.11}$$

where

$$\tilde{R}^g_{ij} = \tfrac{1}{2}(m^g_i n^g_j - m^g_j n^g_i). \tag{11.2.12}$$

The second-order tensors $\mathbf{R}^g$ and $\tilde{\mathbf{R}}^g$ are also called the orientation tensors because they depend only on the orientation of the single crystal in consideration.

Let $\boldsymbol{\sigma}$ denote the stress in the single crystal. The so-called resolved shear stress on a slip system g is given by

$$\tau^g = \sigma_{ij} R^g_{ij}. \tag{11.2.13}$$

Within the framework of time-independent plasticity (any viscous effect is neglected), a slip system g is considered to be active if the resolved shear stress τ^g reaches a critical value τ^g_c, which depends on the previous deformation history of the single crystal, leading to a strain hardening state. It is generally assumed that the deformation history of a given slip system g depends on only the amplitude of shear strain associated with N active systems, so that one can write

$$\tau_c^g = \tilde{F}^g(\gamma^1, \gamma^2, \ldots, \gamma^N). \tag{11.2.14}$$

When the amount of shear is small enough, we can use a linear approximation of (11.2.14),

$$\tau_c^g \approx \tilde{F}^g(0, 0, \ldots, 0) + \sum_h \frac{\partial \tilde{F}^g}{\partial \gamma^h}(0, 0, \ldots, 0)\gamma^h, \tag{11.2.15}$$

where $\tilde{F}^g(0, 0, \ldots, 0)$ can be viewed as the initial critical shear stress of the slip system g. It is generally assumed the same for all slip systems. Therefore, (11.2.15) can be expressed as

$$\tau_c^g = \tau^0 + \sum_h H^{gh}\gamma^h \tag{11.2.16}$$

where $\tau^0 = \tilde{F}^g(0, 0, \ldots, 0)$, and $H^{gh} = \partial\tilde{F}^g/\partial\gamma^h(0, 0, \ldots, 0)$ is the strain hardening matrix, which describes the hardening interactions between the different slip systems. Note that the diagonal components of the matrix H^{gh} define the self-hardening due to the plastic shear in the same system, while the nondiagonal components correspond to the latent hardening due to shear slip on the other systems. The matrix coefficients can be evaluated by experimental characterization performed on single crystals.

At any stage of the deformation process, the rate of changes of critical shear stress is deduced from (11.2.16) as

$$\dot{\tau}_c^g = \sum_h H^{gh}\dot{\gamma}^h. \tag{11.2.17}$$

It follows from the above definitions that a slip system is potentially active if $\tau^g = \tau_c^g$ and loading or unloading, respectively, depending on whether

$$\dot{\tau}^g = \dot{\tau}_c^g \quad \text{with} \quad \dot{\gamma}^g \geq 0 \tag{11.2.18}$$

or

$$\dot{\tau}^g < \dot{\tau}_c^g \quad \text{with} \quad \dot{\gamma}^g = 0. \tag{11.2.19}$$

A system is inactive if $\tau^g < \tau_c^g$ and $\dot{\gamma}^g = 0$.

Relation (11.2.17) is known as the consistency condition whose resolution for each active system g determines the shear rate $\dot{\gamma}^g$ on this system. Taking into account (11.2.13) and (11.2.18), the consistency condition writes

$$\dot{\sigma}_{ij} R_{ij}^g = \sum_h H^{gh} \dot{\gamma}^h. \tag{11.2.20}$$

From the definition (11.2.9) of the plastic strain rate, the total strain rate is the sum of the elastic and plastic parts, that is,

$$\dot{\boldsymbol{\varepsilon}} = \dot{\boldsymbol{\varepsilon}}^e + \dot{\boldsymbol{\varepsilon}}^p = \mathbf{L}^{-1}\dot{\boldsymbol{\sigma}} + \sum_g \dot{\gamma}^g \mathbf{R}^g \tag{11.2.21}$$

or

$$\dot{\boldsymbol{\sigma}} = \mathbf{L}\left(\dot{\boldsymbol{\varepsilon}} - \sum_g \dot{\gamma}^g \mathbf{R}^g\right). \tag{11.2.22}$$

Note that for a given state of stress $\boldsymbol{\sigma}$, $\dot{\boldsymbol{\sigma}}$ is uniquely related to $\dot{\boldsymbol{\varepsilon}}$ if the hardening matrix H^{gh}, governing the shear rates in different slip systems, is positive semidefinite (Hill, 1966), while only for certain hardening laws are the shear rates $\dot{\gamma}^g$ always unique. At least one set of shear rates exists that satisfies the constitutive relations (11.2.9) and (11.2.17) through (11.2.21) for a prescribed strain rate $\dot{\boldsymbol{\varepsilon}}$ (or prescribed stress $\dot{\boldsymbol{\sigma}}$). If there are N nonzero $\dot{\gamma}^g$, they would satisfy N equations resulting from the combination of the consistency condition $\dot{\tau}^g = \dot{\tau}_c^g$ and the constitutive relations (11.2.22). In fact, substituting (11.2.22) into (11.2.20) yields the following set of equations:

$$\sum_h Q^{gh} \dot{\gamma}^h = \mathbf{R}^g \mathbf{L} \dot{\boldsymbol{\varepsilon}}, \tag{11.2.23}$$

where

$$Q^{gh} = H^{gh} + \mathbf{R}^g \mathbf{L} \mathbf{R}^h. \tag{11.2.24}$$

These equations are associated with the loading system together with the constraints $\dot{\gamma}^h \geq 0$.

Only for certain hardening laws will the $N \times N$ matrix Q^{gh} be necessarily nonsingular. For perfect plasticity ($H^{gh} = 0$), for example, it is always possible to choose at least one set of linearly independent slip systems among the potentially active ones such that this matrix is nonsingular and the auxiliary equations (11.2.20) are satisfied. Thus, for perfect plasticity the dimension of Q^{gh} is never greater than 5×5. If its inverse is denoted by $\tilde{Q}^{gh}$, the N nonzero shear rates for this choice of active slip systems are expressed by

$$\dot{\gamma}^g = \widehat{\dot{\boldsymbol{\varepsilon}}^g}\dot{\boldsymbol{\varepsilon}} \quad \text{with} \quad \widehat{\dot{\boldsymbol{\varepsilon}}^g} = \sum_h \tilde{Q}^{gh}\mathbf{L}\mathbf{R}^h. \tag{11.2.25}$$

Recall that the tangent modulus and compliance tensors of a single crystal are given, respectively, by (11.1.69), that is,

$$\dot{\boldsymbol{\sigma}} = \mathbf{L}^t\dot{\boldsymbol{\varepsilon}}, \qquad \dot{\boldsymbol{\varepsilon}} = \mathbf{M}^t\dot{\boldsymbol{\sigma}}. \tag{11.2.26}$$

From the foregoing kinematics of single-crystal plastic deformation, the main feature of the tangent modulus and compliance is that they depend on the set of active slip systems, which in turn depends on the prescribed strain rate $\dot{\boldsymbol{\varepsilon}}$ (or stress $\dot{\boldsymbol{\sigma}}$). It is well known in crystal plasticity that the definitions of tangent modulus and tangent compliance lead to a multibranch description. It also should be noted that, regarding (11.2.26), the inverse of tangent modulus does not exist in all situations. In other words $\mathbf{L}^t$ may possess some singularities. Perfect plastic behavior is one of such cases, where the problem of homogenization is typically carried out by using directly the tangent modulus rather than its inverse, since there is no restriction on the strain rate, whereas the stress rate is subjected to certain conditions regarding the regions in stress rate space.

Substituting (11.2.25) into (11.2.22) and in comparison to (11.2.26), one obtains the following expression for single-crystal tangent modulus:

$$L^t_{ijmn} = L_{ijkl}\left(I_{klmn} - \sum_g R^g_{kl}\widehat{\dot{\varepsilon}^g_{mn}}\right). \tag{11.2.27}$$

Using (11.2.25) and (11.2.24), (11.2.27) can be explicitly rewritten as

$$L^t_{ijkl} = L_{ijmn}\left(I_{mnkl} - \sum_{g,h} (H^{gh} + \mathbf{R}^g\mathbf{L}\mathbf{R}^h)^{-1} R^g_{mn}L_{klpq}R^h_{pg}\right). \tag{11.2.28}$$

It can be readily shown that the tangent modulus $\mathbf{L}^t$ as given by (11.2.28) satisfies the symmetries

$$L^t_{ijkl} = L^t_{klij} \quad \text{if} \quad H^{gh} = H^{hg}. \tag{11.2.29}$$

If the single crystal is assumed to be elastically isotropic, then we have [see (1.4.5)]

$$\mathbf{L} = 3K\mathbf{I}^h + 2\mu\mathbf{I}^d, \tag{11.2.30}$$

where

$$I^h_{ijkl} = \tfrac{1}{3}\delta_{ij}\delta_{kl}, \qquad I^d_{ijkl} = \tfrac{1}{2}(\delta_{ik}\delta_{jl} + \delta_{il}\delta_{jk} - \tfrac{2}{3}\delta_{ij}\delta_{kl}).$$

The tangent modulus thus becomes

$$L^t_{ijkl} = 3KI^h_{ijkl} + 2\mu I^d_{ijkl} - 4\mu^2 \sum_{g,h} R^g_{ij}\,(H^{gh} + 2\mu R^g_{pq}R^h_{pq})^{-1}\,R^h_{kl}, \tag{11.2.31}$$

where the plastic incompressibility is used.

In summary, the single-crystal tangent modulus described by (11.2.28) and (11.2.31) is unique for a given strain rate $\dot{\boldsymbol{\varepsilon}}$ even if the strain rates $\dot{\gamma}^g$ are not. Referring to (11.2.31), one can remark that even though the elasticity is approximated as isotropic, the tangent moduli are anisotropic in nature. This results from the typical process of the plastic flow activated on discrete slip systems and their interactions governing the strain hardening behavior.

Plastic Deformation of Polycrystalline Materials

In the previous section, we presented a possible way to determine the tangent modulus of a single crystal from the theory of homogeneous and continuum slip on well-defined slip systems. Let us now consider a polycrystalline material where the current configuration has undergone certain plastic deformation, and the potentially active slip systems as well as the stress distribution in each grain are known. In this stage the polycrystal is subjected to a loading increment characterized, for example, by a strain rate $\dot{\overline{\boldsymbol{\varepsilon}}}$. To determine the effective tangent modulus $\overline{\mathbf{L}}^t$ using the Hill self-consistent model, we need to evaluate the instantaneous stress and strain rates, $\dot{\boldsymbol{\sigma}}$ and $\dot{\boldsymbol{\varepsilon}}$, as well as the tangent moduli $\mathbf{L}^t$, for a given prescribed macroscopic stress rate $\dot{\overline{\boldsymbol{\sigma}}}$ or strain rate $\dot{\overline{\boldsymbol{\varepsilon}}}$.

To this end, an iterative procedure is needed. First, a tentative guess is made for $\overline{\mathbf{L}}^t$ so that the Hill constraint tensor given by (11.1.72) can be determined. Then, for each active set of slip systems, necessarily a subset of potentially active systems, the tangent modulus for the grain considered is calculated from (11.2.28) or (11.2.31) in the case of isotropic elasticity. Next, the strain concentration tensor $\mathbf{A}$ is computed for this grain using (11.1.75). To ensure that the assumed set of active slip systems do constitute the correct branch of $\mathbf{L}^t$ for the prescribed $\dot{\bar{\boldsymbol{\varepsilon}}}$ (or $\dot{\bar{\boldsymbol{\sigma}}}$), the auxiliary conditions (11.2.20) should be checked by taking into account Eqs. (11.1.75) and (11.2.26). If condition (11.2.20) is fulfilled, then $\mathbf{L}^t$ is correct for this iteration; if not, a new set of potentially active systems should be chosen until the correct $\mathbf{L}^t$ and $\mathbf{A}$ are found. This procedure is carried out for each grain orientation. The final step in the first iteration is to calculate an improved estimate (in comparison to the initial guess) of $\overline{\mathbf{L}}^t$ from (11.1.80). The entire procedure is then repeated until a convergent $\overline{\mathbf{L}}^t$ is reached.

In the next two sections, the method is illustrated by simulating a monotonically increasing uniaxial load. In this case, the tangent modulus tensor is transversely isotropic. The corresponding Eshelby inclusion tensor and the Hill constraint tensors are obtained for computing the strain or stress concentration tensors.

Hutchinson's Calculations

Hutchinson (1970) implemented the Hill self-consistent model to study the behavior of a polycrystal consisting of randomly oriented FCC single crystals. In his computational procedure, the grains are considered spherical in shape and their orientations are assumed equally represented so that the average procedure required for the homogenization steps can be performed over all orientations.

The elasticity of each single crystal is described by three independent Voigt elastic constants C_{11}, C_{12}, and C_{44} in the classical way of cubic symmetries. The random distribution of single crystals leads to isotropic overall elastic constant of the polycrystalline aggregate characterized by its bulk modulus K and shear modulus μ. Using the linear elastic self-consistent method for randomly distributed cubic crystal, the elastic moduli of the polycrystal can be obtained (Problem 11.4):

$$K = \tfrac{1}{3}(C_{11} + 2C_{12}), \tag{11.2.32}$$

$$\begin{aligned} 8\mu^3 + (5C_{11} + 4C_{12})\,\mu^2 - C_{44}\,(7C_{11} - 4C_{12})\,\mu \\ - C_{44}\,(C_{11} - C_{12})(C_{11} + 2C_{12}) = 0. \end{aligned} \tag{11.2.33}$$

An FCC single crystal has crystallographically similar systems whose unit normals are the four [1,1,1]-type directions, relative to the crystal axes, and whose slip directions are the [1,1,0] types. As for elastic deformation, the self-consistent method can be used to express the initial critical macroscopic stress $\overline{\tau}_c^0$ (the yield stress) in terms of the initial shear stress τ_c^0 of a slip system introduced by (11.2.16). That is (see Problem 11.5)

$$\overline{\tau}_c^0 = \tau_c^0 \sqrt{\left(\frac{3}{2\rho_2^2 + \rho_3^2}\right)}, \tag{11.2.34}$$

where

$$\rho_2 = \frac{C_{11} - C_{12}}{2\mu(1 - 2\delta) + 2\delta(C_{11} - C_{12})}, \qquad \rho_3 = \frac{C_{44}}{\mu(1 - 2\delta) + 2\delta C_{44}}, \tag{11.2.35}$$

and δ is given by (4.3.38).

For the case of isotropic elasticity where $2C_{44} = C_{11} - C_{12}$, it can be readily shown that $\rho_2 = \rho_3 = 1$. Hence $\overline{\tau}_c^0 = \tau_c^0$.

Note that expression (11.2.33) is equivalent to assuming a Tresca criterion for the overall yield stress. This is only valid if no residual stresses are present in the polycrystalline materials.

During a monotonic uniaxial tensile test in the x_3 direction, the tangent moduli $\overline{L}$ display transverse isotropy with respect to the macroscopic coordinate system, so that Hooke's law can be written in terms of the Voigt elastic constants:

$$\begin{pmatrix} \dot{\overline{\sigma}}_{11} \\ \dot{\overline{\sigma}}_{22} \\ \dot{\overline{\sigma}}_{33} \\ \dot{\overline{\sigma}}_{13} \\ \dot{\overline{\sigma}}_{23} \\ \dot{\overline{\sigma}}_{12} \end{pmatrix} = \begin{pmatrix} \overline{C}_{11}^t & \overline{C}_{12}^t & \overline{C}_{13}^t & 0 & 0 & 0 \\ \overline{C}_{11}^t & \overline{C}_{11}^t & \overline{C}_{13}^t & 0 & 0 & 0 \\ \overline{C}_{11}^t & \overline{C}_{11}^t & \overline{C}_{33}^t & 0 & 0 & 0 \\ 0 & 0 & 0 & \overline{C}_{44}^t & 0 & 0 \\ 0 & 0 & 0 & 0 & \overline{C}_{44}^t & 0 \\ 0 & 0 & 0 & 0 & 0 & \overline{C}_{66}^t \end{pmatrix} \begin{pmatrix} \dot{\overline{\varepsilon}}_{11} \\ \dot{\overline{\varepsilon}}_{22} \\ \dot{\overline{\varepsilon}}_{33} \\ 2\dot{\overline{\varepsilon}}_{13} \\ 2\dot{\overline{\varepsilon}}_{23} \\ 2\dot{\overline{\varepsilon}}_{12} \end{pmatrix}, \tag{11.2.36}$$

where

$$\overline{C}_{66}^t = \tfrac{1}{2}(\overline{C}_{11}^t - \overline{C}_{12}^t). \tag{11.2.37}$$

Copper is selected for this simulation because its single crystal shows strong elastic anisotropy. The elastic moduli for single-crystal copper used in our calculations are $C_{12}/C_{11} = 0.722$ and $C_{44}/C_{11} = 0.447$. The self-consistent predictions using Eqs. (11.2.32)–(11.2.34) give, respectively, the polycrystalline elastic moduli and initial critical shear stress:

$$K = 0.815C_{11}, \qquad \mu = 0.285C_{11}, \qquad \bar{\tau}_c^0 = 1.129\tau_c^0 \qquad (11.2.38)$$

or

$$E = 0.722C_{11}, \qquad \nu = 0.343, \qquad (11.2.39)$$

where E is Young's modulus and ν Poisson's ratio.

Furthermore, we pay particular attention to the single-crystal strain hardening and its effect on the polycrystalline hardening, by considering both perfect plasticity for each grain, as well as different forms of the hardening matrix H^{gh}.

As shown by Kneer (1965), the transverse isotropy of the overall tangent modulus leads to analytical expressions of the Eshelby tensor. It follows from the Kneer calculations that the Eshelby tensor can be expressed as

$$S_{ijkl} = T_{ijmn}\bar{L}_{mnkl},$$

where

$$T_{ijmn} = \sum_{m=0}^{3} K^{(m)}\tilde{T}^{(m)}_{ijmn},$$

$$K^{(m)} = \frac{1}{8}\int_0^1 \frac{z^{2m}}{a_0 + a_1z^2 + a_2z^4 + a_3z^6}\,dz, \qquad (m = 0, 3).$$

The coefficients a_m and $\tilde{T}^{(m)}_{ijmn}$ are a polynomial functions of the components $\bar{C}_{ij}$ of the transverse isotropic tangent modulus. Details of this calculation can be found in (Kneer, 1965).

We first consider the case without strain hardening in single crystals. The results are presented in Figure 11.1 for isotropic elasticity and anisotropic elasticity with the moduli of copper listed in (11.2.38). Clearly, the results show that the elastic anisotropy increases the initial

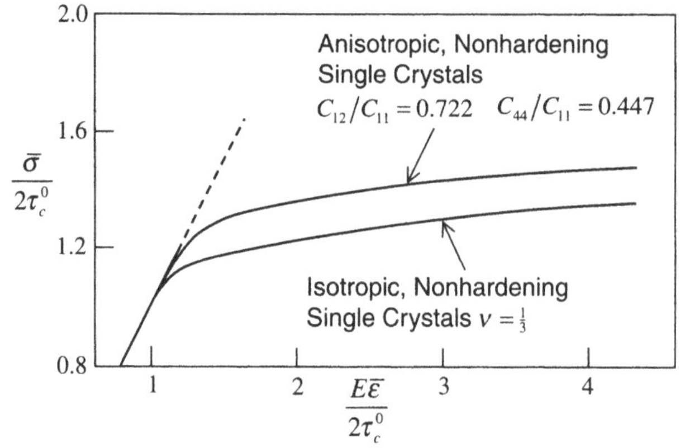

Figure 11.1 Tensile stress–strain curves for FCC polycrystals with randomly oriented and perfectly plastic single crystals.

yield stress of the polycrystal. This is not expected because the inhomogeneous elasticity arising from single-crystal anisotropy should decrease, rather than increase, the yield strength of the polycrystal yields. This might be explained by the fact that stresses in each grain are estimated by treating them as spherical inclusions. This is one of the limitations of the self-consistent model.

As discussed previously, the polycrystalline aggregate displays a strain hardening behavior even if every single crystal in the aggregate behaves as a perfect plastic material. This arises from the intergranular interactions.

To show the effect of single-crystal strain hardening, the interaction matrix H^{gh} should be estimated. In general, this procedure requires an experimental analysis performed on single crystals. Different forms of strain hardening matrix have been proposed in the literature. Taylor's (1938) isotropic hardening law is the simplest and the most widely used one. This law states that the yield stresses of all slip systems remain equal and increase in proportion to the total shear. So that $H^{gh} = h$ for all g and h, we have $\dot{\tau}_c^g = h\Sigma_h \dot{\gamma}^h$.

Taylor's simple hardening law is not able to reproduce the principal features of a single-crystal hardening, nor can it capture the polycrystalline hardening behavior, especially, at later stages of plastic deformation where multiple activations of slip systems render the interactions highly anisotropic.

Nevertheless, the Taylor's hardening law is used in the calculations presented in Figure 11.2 to show at least at the earlier stage of the

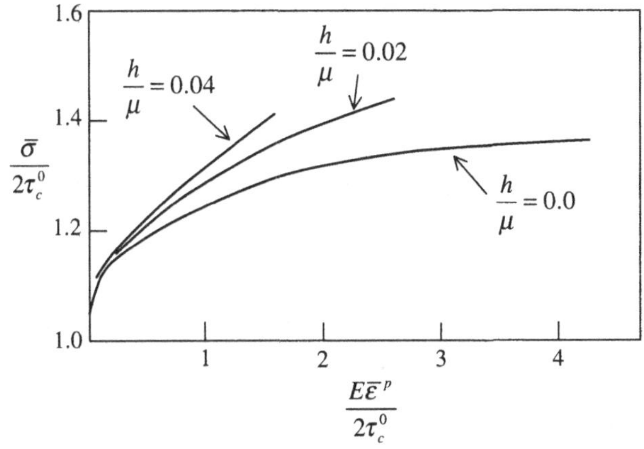

Figure 11.2 Effect of single-crystal hardening.

plastic flow the effect of single crystals hardening. The computations are performed for different hardening parameters h.

Finally, under the same conditions without single-crystal hardening, the stress–strain relationship for polycrystal copper is estimated using the upper bound of the Taylor–Lin model and the Kröner's self-consistent model. The comparison is shown in Figure 11.3. As discussed throughout this chapter, the stress–strain curve obtained from Kröner's model lies above the one obtained by the Hill method. This is expected because in the Kröner inclusion problem, the matrix exerts

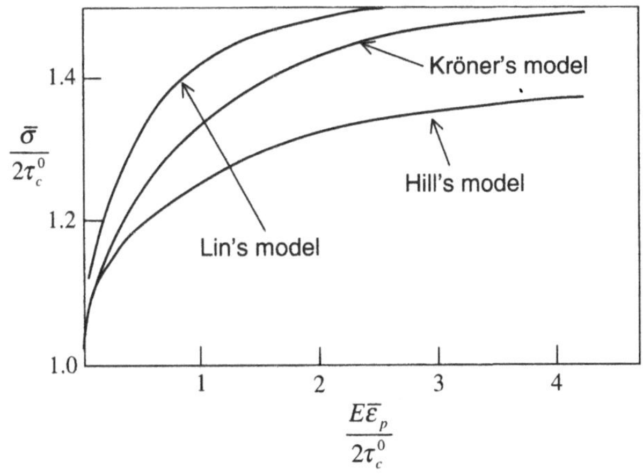

Figure 11.3 Upper and lower bounds on stress–strain curves.

an elastic constraint, whereas in the Hill model the matrix provides a much softer constraint because of plastic deformation.

11.3 TIME-DEPENDENT BEHAVIOR OF POLYCRYSTALLINE MATERIALS: SECANT APPROACH

Time-dependent behavior of polycrystalline materials was first modeled by Hutchinson (1976) by assuming a power law describing the shear rate on slip systems for a given single crystal. The analysis was performed under small-strain condition leading to an elastoviscoplastic type behavior.

Hutchinson (1976) assumed a steady-state creep behavior for single crystals for which the shear rate induced in a slip system g by a given resolved shear stress τ^g is described by a rate-sensitive criterion, which is equivalent to assuming a nonlinear viscous behavior:

$$\dot{\gamma}^g = \dot{\gamma}^0 \left(\frac{\tau^g}{\tau_c^g}\right)^n, \tag{11.3.1}$$

where $\dot{\gamma}^0$ is a reference rate and n is the inverse of rate sensitivity. When $n >> 1$, the shear rate increase is negligible unless τ^g is very close to τ_c^g. This statement is equivalent to conditions (11.2.19) and (11.2.20) for slip systems activation in time-independent plasticity. The critical shear stress τ_c^g, often called reference stress in time-dependent plasticity, is strongly dependent on temperature. The exponent n depends also on temperature, although somewhat less strongly, and usually falls between 3 and 8 for metals. A survey on the temperature ranges where the steady-state creep of polycrystal and single crystal can be approximated by a power law was given by Ashby and Frost (1975).

If N is the total number of slip systems, the total strain rate is the sum of the contributions from all these systems:

$$\dot{\varepsilon}_{ij} = \sum_{g=1}^{N} \dot{\gamma}^g R_{ij}^g = M_{ijpq}^s \sigma_{pq}, \tag{11.3.2}$$

where

$$M^s_{ijpq} = \sum_{g=1}^{N} \left(\frac{\dot{\gamma}^0}{\tau^g_c}\right)\left(\frac{\tau^g}{\tau^g_c}\right)^{n-1} R^g_{ij}R^g_{pq} \tag{11.3.3}$$

with $\mathbf{M}^s$ being the so-called secant viscoplastic compliance of the single crystal. As reported by Hutchinson (1976), the compliance tensor is homogeneous of degree $n - 1$ in stress so that

$$\mathbf{M}^s\,(\lambda\boldsymbol{\sigma}) = \lambda^{n-1}\mathbf{M}^s\,(\boldsymbol{\sigma}). \tag{11.3.4}$$

Let us now define the stress potential $\psi(\boldsymbol{\sigma})$ and strain rate potential $\phi(\dot{\boldsymbol{\varepsilon}})$ such that

$$\dot{\boldsymbol{\varepsilon}} = \frac{\partial\psi}{\partial\boldsymbol{\sigma}} \quad \text{and} \quad \boldsymbol{\sigma} = \frac{\partial\phi}{\partial\dot{\boldsymbol{\varepsilon}}}. \tag{11.3.5}$$

The viscoplastic constitutive law (11.3.2) leads to the following typical relationships between the dissipation $\boldsymbol{\sigma}\dot{\boldsymbol{\varepsilon}}$, $\psi(\boldsymbol{\sigma})$, and $\phi(\dot{\boldsymbol{\varepsilon}})$:

$$\boldsymbol{\sigma}\dot{\boldsymbol{\varepsilon}} = (n+1)\psi(\boldsymbol{\sigma}) = \frac{n+1}{n}\,\phi(\dot{\boldsymbol{\varepsilon}}) = \sum_{g=1}^{n} \tau^g\dot{\gamma}^g. \tag{11.3.6}$$

Substituting (11.3.5) into (11.3.6) and by differentiating (11.3.6) with respect to the stresses, one obtains

$$\frac{\partial}{\partial\sigma_{ij}}\,(\sigma_{kl}\dot{\varepsilon}_{kl}) = (n+1)\,\frac{\partial\psi}{\partial\sigma_{ij}} = \frac{\partial\psi}{\partial\sigma_{ij}} + \sigma_{kl}\,\frac{\partial^2\psi}{\partial\sigma_{ij}\,\partial\sigma_{kl}}. \tag{11.3.7}$$

This last equation leads to

$$n\,\frac{\partial\psi}{\partial\sigma_{ij}} = \sigma_{kl}\,\frac{\partial^2\psi}{\partial\sigma_{ij}\,\partial\sigma_{kl}}. \tag{11.3.8}$$

Combining (11.3.8) with (11.3.2), we can readily show that

$$M^s_{ijkl} = \frac{1}{n}\,\frac{\partial^2\psi}{\partial\sigma_{ij}\,\partial\sigma_{kl}}, \tag{11.3.9}$$

where

$$\frac{\partial^2 \psi}{\partial \sigma_{ij}\, \partial \sigma_{kl}} = \frac{\partial \dot{\varepsilon}_{ij}}{\partial \sigma_{ij}} = M^t_{ijkl}. \tag{11.3.10}$$

In Eq. (11.3.10), $\mathbf{M}^t$ is the tangent compliance tensor.

On the other hand, a Taylor expansion of (11.3.2) at the vicinity of a point $\tilde{\boldsymbol{\sigma}} - \dot{\boldsymbol{\varepsilon}}$ can be written as

$$\dot{\varepsilon}_{ij} = \left.\frac{\partial \dot{\varepsilon}_{ij}}{\partial \sigma_{kl}}\right|_{\sigma=\tilde{\sigma}} \sigma_{kl} + \tilde{\dot{\varepsilon}}_{ij} = M^t_{ijkl}\,(\dot{\boldsymbol{\sigma}})\, \sigma_{kl} + \tilde{\dot{\varepsilon}}_{ij}, \tag{11.3.11}$$

where $\tilde{\dot{\varepsilon}}_{ij}$ is called the back-extrapolated strain rate.

From (11.3.9) and (11.3.10), the relation between the grain's secant and tangent compliances are

$$\mathbf{M}^t = n\mathbf{M}^s. \tag{11.3.12}$$

At the polycrystal level, the macroscopic constitutive law is assumed to be similar to that of single crystals, so that one can write

$$\bar{\dot{\varepsilon}}_{ij} = \overline{M}^s_{ijpq}\bar{\sigma}_{pq}, \tag{11.3.13}$$

where $\overline{\mathbf{M}}^s$ is the macroscopic secant compliance tensor. In the same way as in the single-crystal case, Taylor expansion of (11.3.13) at the vicinity of the macroscopic stress leads to the definition of a macroscopic tangent modulus $\overline{\mathbf{M}}^t$ as

$$\bar{\dot{\varepsilon}}_{ij} = \overline{M}^t_{ijpq}\,(\bar{\boldsymbol{\sigma}})\, \bar{\sigma}_{pq} + \dot{\varepsilon}^0_{ij}, \tag{11.3.14}$$

where $\overline{M}^t_{ijkl}\,(\bar{\boldsymbol{\sigma}}) = \partial\bar{\dot{\varepsilon}}_{ij}/\partial\bar{\sigma}_{kl}|_{\bar{\sigma}=\bar{\sigma}}$ and $\dot{\varepsilon}^0_{ij}$ is a macroscopic extrapolated strain rate.

Hutchinson (1976) has shown that the macroscopic tangent and secant compliance tensors are linked by a relation similar to the one for single crystals, that is, $\overline{\mathbf{M}}^t = n\overline{\mathbf{M}}^s$. This is derived by defining the macroscopic dissipation $\bar{\boldsymbol{\sigma}}\bar{\dot{\boldsymbol{\varepsilon}}}$, macroscopic strain rate potential $\Phi(\bar{\dot{\boldsymbol{\varepsilon}}})$, and stress potential $\Psi(\bar{\boldsymbol{\sigma}})$ such that

$$\bar{\dot{\boldsymbol{\varepsilon}}} = \frac{\partial \Psi}{\partial \bar{\boldsymbol{\sigma}}} \quad \text{and} \quad \bar{\boldsymbol{\sigma}} = \frac{\partial \Phi}{\partial \bar{\dot{\boldsymbol{\varepsilon}}}}.$$

Note that Eqs. (11.3.11) and (11.3.14) are exact only when they describe the strain rate associated with the stress used as a reference for the expansion. Otherwise they are only approximate. This will not present a limitation for the treatment of the grains since the stress and the strain rates are assumed to be uniform within the framework of the self-consistent scheme. As a result, the actual value of stress in the grain considered can be selected to perform the expansion.

Starting from the linearized equation (11.3.14), the macroscopic tangent compliance tensor $\overline{\mathbf{M}}^t$ can be estimated by adopting a Hill-type self-consistent method. In other words, one can consider each grain with tangent compliance $\mathbf{M}^t$ and prescribed reference strain rate $\tilde{\dot{\boldsymbol{\varepsilon}}}$ embedded in a homogenized effective medium having the properties of $\overline{\mathbf{M}}^t$ and prescribed reference strain rate $\dot{\boldsymbol{\varepsilon}}^0$.

Following the same procedure as for Eqs. (11.1.71) and (11.1.72), the Eshelby solution extended to a tangent formulation gives the interaction relation linking the local to the macroscopic quantities [see (4.5.19)],

$$\dot{\boldsymbol{\varepsilon}} = \overline{\dot{\boldsymbol{\varepsilon}}} + \tilde{\mathbf{H}}(\overline{\boldsymbol{\sigma}} - \boldsymbol{\sigma}), \tag{11.3.15}$$

where $\tilde{\mathbf{H}}$ is the inverse of the Hill constant tensor,

$$\tilde{\mathbf{H}} = (\mathbf{S}^{-1} - \mathbf{I})^{-1}\, \overline{\mathbf{M}}^t. \tag{11.3.16}$$

Note that Eshelby's tensor in (11.3.16) depends on the tangent compliance tensor together with the shape of the grain in consideration. As reported by Lebensohn and Tomé (1993), the relation $\overline{\mathbf{M}}^t = n\overline{\mathbf{M}}^s$ enables us to express the equations in terms of the secant compliance tensor as

$$\tilde{\mathbf{H}} = n(\mathbf{S}^{-1} - \mathbf{I})^{-1}\, \overline{\mathbf{M}}^s. \tag{11.3.17}$$

Substituting (11.3.2) and (11.3.12) into (11.3.15) yields

$$\boldsymbol{\sigma} = \mathbf{B}\overline{\boldsymbol{\sigma}}, \tag{11.3.18}$$

where

$$\mathbf{B} = (\mathbf{M}^s + \tilde{\mathbf{H}})^{-1}\, (\overline{\mathbf{M}}^s + \tilde{\mathbf{H}}).$$

Finally, the homogenization procedure using $\overline{\boldsymbol{\sigma}} = \langle \boldsymbol{\sigma} \rangle$ and $\overline{\mathbf{M}}^s = \langle \mathbf{M}^s\mathbf{B} \rangle$ leads to

$$\overline{\mathbf{M}}^s = \langle \mathbf{M}^s : (\mathbf{M}^s + \tilde{\mathbf{H}})^{-1}\, (\overline{\mathbf{M}}^s + \tilde{\mathbf{H}}) \rangle. \tag{11.3.19}$$

In summary, we have introduced in this section the viscoplastic self-consistent model initially developed by Hutchinson (1976) for modeling steady-state creep of polycrystalline materials, and have shown how to relate a secant formulation to a tangent formulation. We have also shown how the Eshelby solution required for the self-consistent scheme can be used to combine both the secant and tangent moduli for solving interaction problems.

Since Hutchinson's work (1976), the viscoplastic self-consistent model has been adopted by many authors as an alternative strategy to tackle the problem of large plastic deformations by simply neglecting the elastic deformation. This way of thinking was successively adopted by Molinari et al. (1987) to describe the texture development in cubic polycrystals. For more information regarding the numerical implementation and limitations of the method, the readers are referred to the work of Lebensohn and Tomé (1993).

PROBLEMS

11.1 Prove Eq. (11.1.30).

11.2 Check the consistency $\overline{L} = (\overline{M})^{-1}$ of the prediction from (11.1.80).

11.3 Show that, for perfect plasticity, the dimension of Q^{gh} is never greater than 5×5.

11.4 For cubic symmetries, if we assume that all single crystals are randomly oriented, show that a self-consistent estimation of the elastic constant of the polycrystal gives isotropic moduli expressed by Eqs. (11.2.32) and (11.2.33).

11.5 For cubic symmetries, if we assume that all single crystals are randomly oriented, show that the overall critical shear stress is given by (11.2.38).

11.6 Show the following relationships:

$$\mathbf{L}'\mathbf{A}_r = \mathbf{B}_r\overline{\mathbf{L}} \text{ and } \mathbf{M}'\mathbf{B}_r = \mathbf{A}_r\overline{\mathbf{M}}.$$

REFERENCES

Asaro, R. J. (1983). Crystal Plasticity, *J. Appl. Mech.,* Vol. 50, pp. 921–934.

Asaro, R. J. and J. R. Rice (1977). Strain Localization in Ductile Single Crystals, *J. Mech. Physics Solids,* Vol. 25, pp. 309–338.

Ashby, M. F. and Frost, H. J. (1975). The Kinematics of Inelastic Deformation Above 0 K, in *Constitutive Equations in Plasticity,* A. S. Argon, ed., MIT Press, Cambridge. pp. 117.

Berveiller, M. and A. Zaoui (1979). An Extension of the Self-Consistent Scheme to Plastically-Flowing Polycrystals, *J. Mech. Phys. Solids,* Vol. 26, pp. 325–344.

Bishop, J. F. W. and R. Hill (1951). A Theoretical Derivation of the Plastic Properties of a Polycrystalline Facecentered Metal, *Phil. Mag*. Vol. 42, pp. 1298–1307.

Bragg, W. H. and W. L. Bragg (1933). *The Crystalline State,* Bell, London.

Budiansky, B. and O. L. Mangasarian (1960). Plastic Stress Concentration at Circular Hole in Infinite Sheet Subjected to Equal Biaxial Tension, *J. Appl. Mech.*, Vol. 27, pp. 59–64.

Buryachenko, V. (1996). The Overall Elastoplastic Behavior of Multiphase Materials with Isotropic Components, *Acta Mech.,* Vol. 119, pp. 93–117.

Castañeda, P. (1991). The Effective Mechanical Properties of Nonlinear Isotropic Composites, *J. Mech. Phys. Solids,* Vol. 39, pp. 45–71.

Castañeda, P. (1996). Exact Second-Order Estimates for the Effective Mechanical Properties of Nonlinear Composite Materials, *J. Mech. Phys. Solids,* Vol. 44, pp. 827–862.

Dvorak, G. J. (1992). Transformation-Field Analysis of Inelastic Composite Materials, *Proc. R. Soc. London,* Ser. A, Vol. 431, pp. 89–110.

Ewing, J. A. and W. Rosenhain (1900). Experiments In Micro-Metallurgy: Effects of Strain. Preliminary Notice, *Philos. Trans. R. Soc. Lond.,* Vol. A199, pp. 85–90.

Eshelby, J. D. (1957). The Determination of the Elastic Field of an Ellipsoidal Inclusion and Related Problems, *Proc. Roy. Soc. Lond.* Vol. A241, pp. 376–396.

Gilormini, P. (1997). Shortcomings of Several Extensions of the Self Consistent Model for Heterogeneous, Nonlinear Solids, *Revue de Metallurgie,* Vol. 94, pp. 1081–1087.

Hill, R. (1965). Continuum Micro-mechanics of Elastoplastic Polycrystals, *J. Mech. Phys. Solids,* Vol. 13, pp. 89–101.

Hill, R. (1966). Generalized Constitutive Relations for Incremental Deformation of Metals by Multislip, *J. Mech. Phys. Solid,* Vol. 14, p. 99.

Hill, R. and K. S. Havner (1982). Perspectives in the Mechanics of Elastoplastic Crystals, *J. Mech. Phys. Solids,* Vol. 30, pp. 5–22.

Hill, R. and J. R. Rice (1972). Constitutive Analysis of Elasto-Plastic Crystals at Arbitrary Strains, *J. Mech. Phys. Solids,* Vol. 20, pp. 401–413.

Hu, G. (1996). A Method of Plasticity for General Aligned Spheroidal Void of Fiber-Reinforced Composites, *Int. J. Plasticity,* Vol. 12, pp. 439–449.

Hutchinson, J. W. (1970). Elastic-Plastic Behaviour of Polycrystalline Metals and Composites, *Proc. R. Soc. Lond.,* Vol. A319, pp. 247–272.

Hutchinson, J. W. (1976). Bounds of Self-Consistent Estimates for Creep of Polycrystalline Materials, *Proc. R. Soc. Lond.*, Vol. A348, pp. 101–127.

Kröner, E. (1961). Zur plastischen Verformung des Vielkristalls, *Acta Metall.*, Vol. 9, pp. 155–161.

Lebensohn, R. and C. N. Tomé (1993). A Self-Consistent Anisotropic Approach for the Simulation of Plastic Deformation and Texture Development of Polycrystals Applications to Zirconium Alloys, *Acta Metall. Mater.*, Vol. 41, pp. 2611–2624.

Masson, R. and A. Zaoui (1999). Self-Consistent Estimates for the Rate-Dependent Elastoplastic Behaviour of Polycrystalline Materials, *J. Mech. Phys. Solids*, Vol. 47, pp. 1543–1568.

Molinari, A., G. R. Canova, and S. Ahzi (1987). A Self-Consistent Approach of the Large Deformation Polycrystal Viscoplasticity, *Acta Metall.*, Vol. 35, pp. 2983–2994.

Kneer, G. (1965). Uber die Berechnung der Elastizitatsmoduln vielkristalliner Aggregate mit Textur, *Phys. Stat. Sol.*, Vol. 9, pp. 825–838.

Polanyi, von M. (1922). Rontgenographische Bestimmung von Ksistallanordnunge, *Naturwissenschaften*, Vol. 10, p. 411.

Qui, Y. P. and G. J. Weng (1992). A Theory of Plasticity for Porous Materials and Particle-Reinforced Composites, *J. Appl. Mech.*, Vol. 59, pp. 261–268.

Schmid, E. and W. Boas (1935). *Kristallplastizität*, Springer, Berlin, p. 64.

Sachs, G. (1928). Zur ableilung einer fleissbedingung (stresses causing flow), *VDI Z.*, Vol. 72, pp. 734–736.

Suquet, P. (1993). Overall Potentials and Extremal Surfaces of Power Law or Ideally Plastic Materials, *J. Mech. Phys. Solids*, Vol. 41, pp. 981–1002.

Suquet, P. (1995). Overall Properties of Nonlinear Composites: A Modified Secant Moduli Theory and Its Link with Ponte Castañeda's Nonlinear Variational Procedure, *C.R. Acad. Sci. Paris*, Vol. 320 (Série IIb), pp. 563–571.

Talbot, D. and J. Willis (1985). Variational Principles for Inhomogeneous Nonlinear Media," *IMA J. Appl. Math.*, Vol. 35, pp. 39–54.

Tandon, G. P. and G. J. Weng (1988). A Theory of Particle-Reinforced Plasticity, *J. Appl. Mech.*, Vol. 55, pp. 126–135.

Taylor, G. I. (1934). Plastic Deformation of Crystals, *Proc. Roy. Soc.*, Vol. A145, pp. 362–404.

Taylor, G. I. (1938). Plastic Strain in Metals, *J. Inst. Metals*, Vol. 62, p. 307.

Taylor, G. I. and C. F. Elam (1923). The Distortion of an Aluminum Crystal During a Tensile Test, *Proc. Roy. Soc.*, Vol. A102, p. 647.

Taylor, G. I. and C. F. Elam (1925). The Plastic Extension and Fracture of Aluminum Crystals, *Proc. Roy. Soc.*, Vol. A108, pp. 28–51.

Willis, J. R. (1983). The Overall Response of Composite Materials, *J. Appl. Mech.*, Vol. 50, pp. 1202–1209.

12

NONLINEAR PROPERTIES OF COMPOSITES MATERIALS: THERMODYNAMIC APPROACHES

Thermodynamic approaches in micromechanics consist of solving the field equations by minimizing a free energy, where the minimum found corresponds to the desired macroscopic potential $W(\overline{\boldsymbol{\varepsilon}})$, which represents the mechanical energy stored in the domain V of the representative volume element (RVE) subjected to a macroscopic strain $\overline{\boldsymbol{\varepsilon}}$. As mentioned in the linear case through the introduction of variational principles, which leads to different bounds, an exact solution to these field equations is rather difficult to obtain and becomes impossible to derive in the nonlinear case. This is mainly due to highly fluctuating fields resulting from the nonlinear behavior of the constituents.

It has been shown that approximated solutions for the local problem can be obtained in the linear case, leading to pertinent estimations of the macroscopic behavior, which are capable of accounting for the influence of morphological parameters and phase spatial distribution on the global behavior. Unfortunately, due to the nonapplicability of the superposition principle, on which most developments are based in linear cases (e.g., use of elementary solutions such as Eshelby's one), the philosophy behind these approaches cannot be transported directly to nonlinear behaviors.

To take advantage of the knowledge acquired in linear cases, linearization of the local constitutive laws could be one of the interesting approaches. The procedure consists of replacing the nonlinear field equations in the RVE by linear equations in the same RVE, whose solution can be evaluated exactly or approximately via the use of the tools developed for linear materials. The linearization is completed with a set of complimentary equations, which characterizes the parameters defining the linear problem (e.g., elasticity moduli). Typically, these equations are nonlinear so that the character of the initial problem is conserved. However, in this case, we will not deal with field equations but with equations involving a finite set of variables which can be solved with the appropriate numerical tools. In most cases, simple algorithms lead to a solution.

The methodology presented in the above is referred to as the "nonlinear extension" of a linear model. The local linear problem resulting from the linearization procedure is identical to the homogenization problem for linear composites, referred to as linear comparison composite (LCC). This virtual LCC results solely from the linearization step and has no physical existence. Further, its moduli are distinct from the initial elasticity moduli of the real nonlinear composite. Although it is often the case, the LCC does not necessarily have the same spatial distribution of phases, or the same number of constituents, as the real nonlinear composite.

One of the difficulties of this method lies in the use of the linear model to obtain the nonlinear macroscopic behavior. Typically, linear models do not provide detailed description of the local fields in the LCC but only averaged strains or stresses in the phases. This information is sufficient in the case of linear problems since the macroscopic stress can be obtained from averaged strains in the phases via the following equation:

$$\bar{\boldsymbol{\sigma}} = \sum_r c_r \bar{\boldsymbol{\sigma}}_r = \sum_r c_r \mathbf{L}_r \bar{\boldsymbol{\varepsilon}}_r.$$

Due to the nonlinear behavior of the constituents, this property does not hold for nonlinear composites where the macroscopic stress is given by

$$\bar{\boldsymbol{\sigma}} = \sum_r c_r \bar{\boldsymbol{\sigma}}_r = \sum_r c_r \left\langle \frac{\partial \omega(\boldsymbol{\varepsilon})}{\partial \boldsymbol{\varepsilon}} \right\rangle_{V_r} \neq \sum_r c_r \frac{\partial \omega(\bar{\boldsymbol{\varepsilon}}_r)}{\partial \bar{\boldsymbol{\varepsilon}}_r}$$

where $\omega(\boldsymbol{\varepsilon})$ is the local free energy, c_r the volume fraction of each phase, and $\mathbf{L}_r$ their stiffness.

The above general principles will be addressed in this chapter, in the particular case where the LCC is obtained from the secant moduli of the nonlinear constituents. Two approaches, the "classical" approach and the "modified" approach, will be presented and compared. These two approaches are both relatively simple to use and differ only from the set of complimentary equations characterizing the LCC.

In the first part, the nature of the constitutive laws will be presented and illustrated with examples. Both tangent and secant linearization methods for the constitutive laws will be defined. The variational formulation that defines the homogenized behavior of heterogeneous materials will be presented in the second part. The general method of "nonlinear extension" of linear models is presented at the end and applied to the case of the classical secant linearization, which shows inconsistencies related to the lack of information regarding the intraphase fluctuations of local fields. An attempt to estimate these heterogeneities is made through the modified-secant formulation based on the second-order deformation moments within the phases, correcting the deficiencies of the classical extension. The two approaches will be illustrated and compared through the different problems given at the end of this chapter. Recent methodologies to describe the overall behavior of nonlinear composites adopt variational procedures. These approaches are not addressed here. The reader may refer to the Suggested Readings at the end of the chapter.

12.1 NONLINEAR BEHAVIOR OF CONSTITUENTS

Thermodynamic Potentials

The behavior of the constituents treated in this chapter can be described by two thermodynamic potentials: the free energy ω and the dissipation potential φ, which are both expressed as a function of strain $\boldsymbol{\varepsilon}$ and of the appropriate internal variables $\boldsymbol{\alpha}$ (scalar or tensorial):

Constitutive laws: $$\boldsymbol{\sigma}^{\text{rev}} = \frac{\partial \omega(\boldsymbol{\varepsilon}, \boldsymbol{\alpha})}{\partial \boldsymbol{\varepsilon}}, \qquad \mathbf{a} = -\frac{\partial \omega(\boldsymbol{\varepsilon}, \boldsymbol{\alpha})}{\partial \boldsymbol{\alpha}}, \tag{12.1.1}$$

Complimentary laws: $$\boldsymbol{\sigma}^{\text{irr}} = \frac{\partial \varphi(\dot{\boldsymbol{\varepsilon}}, \dot{\boldsymbol{\alpha}})}{\partial \dot{\boldsymbol{\varepsilon}}}, \qquad \mathbf{a} = -\frac{\partial \varphi(\dot{\boldsymbol{\varepsilon}}, \dot{\boldsymbol{\alpha}})}{\partial \dot{\boldsymbol{\alpha}}}, \tag{12.1.2}$$

with $$\boldsymbol{\sigma} = \boldsymbol{\sigma}^{\text{rev}} + \boldsymbol{\sigma}^{\text{irr}}. \tag{12.1.3}$$

The convexity of φ and its nullity at $\dot{\boldsymbol{\varepsilon}} = 0$ impose a positive dissipation and, consequently, the second law is satisfied. The convexity of ω is associated with the stability of the material. Although many approximate formulations have been proposed, there is no homogenization theory giving rigorous solutions to the homogenized behavior of phase mixtures described by (12.1.1)–(12.1.3). However, theories based on rigorous bounds have recently been developed in the simple case where the behavior of the phases can be described with only one potential, the free energy or the dissipation potential, the remaining potential being equal to zero.

Behaviors Described by a Single Potential

We assume the behavior of each phase is described with a single convex potential, $\omega(\boldsymbol{\varepsilon})$ or $\varphi(\dot{\boldsymbol{\varepsilon}})$, the remaining potential is equal to zero, such that the stress–strain or stress–strain rate relation can be written as follows:

$$\boldsymbol{\sigma} = \boldsymbol{\sigma}^{\text{rev}} = \frac{\partial \omega(\boldsymbol{\varepsilon})}{\partial \boldsymbol{\varepsilon}}, \qquad \varphi = 0. \tag{12.1.4}$$

Or

$$\boldsymbol{\sigma} = \boldsymbol{\sigma}^{\text{irr}} = \frac{\partial \varphi(\dot{\boldsymbol{\varepsilon}})}{\partial \dot{\boldsymbol{\varepsilon}}}, \qquad \omega = 0. \tag{12.1.5}$$

Equation (12.1.4) describes a nonlinear elastic behavior under small-strain deformation, while (12.1.5) corresponds to a purely viscous behavior where deformation can be finite (the problem is purely Eulerian). For the sake of simplicity of the notations, only the first expression will be used (nonlinear elasticity). The results will be directly transferable to the second case by considering $\boldsymbol{\varepsilon}$ as a strain rate.

Equation (12.1.4) can be inverted to obtain

$$\boldsymbol{\varepsilon} = \frac{\partial \tilde{\omega}(\boldsymbol{\sigma})}{\partial \boldsymbol{\sigma}}, \tag{12.1.6}$$

where $\tilde{\omega}(\boldsymbol{\sigma})$ is the complimentary energy of the material. Mathematically, the relation between $\tilde{\omega}(\boldsymbol{\sigma})$ and $\omega(\boldsymbol{\varepsilon})$ is obtained via the Legendre transformation:

$$\tilde{\omega}(\boldsymbol{\sigma}) = \sup_{\boldsymbol{\varepsilon}}\{\boldsymbol{\sigma}:\boldsymbol{\varepsilon} - \omega(\boldsymbol{\varepsilon})\}. \tag{12.1.7}$$

Let us now consider a few examples.

Isotropic Materials

Most isotropic materials are linear under hydrostatic load and nonlinear under shear; their behavior can be written with the following expressions:

$$\sigma_m = 3k\varepsilon_m, \qquad \mathbf{s} = 2\mu^{\text{sec}}(\varepsilon_{\text{eq}})\mathbf{e}, \qquad \mu^{\text{sec}}(\varepsilon_{\text{eq}}) = \frac{\sigma_{\text{eq}}}{3\varepsilon_{\text{eq}}}, \tag{12.1.8}$$

where

$$\sigma_m = \frac{\sigma_{kk}}{3}, \qquad \varepsilon_m = \frac{\varepsilon_{kk}}{3}, \qquad \sigma_{\text{eq}} = (\tfrac{3}{2}\mathbf{s}:\mathbf{s})^{1/2}, \qquad s_{ij} = \sigma_{ij} - \sigma_m\delta_{ij}, \tag{12.1.9}$$

$$\varepsilon_{\text{eq}} = (\tfrac{2}{3}\boldsymbol{\varepsilon}:\boldsymbol{\varepsilon})^{1/2}, \qquad e_{ij} = \varepsilon_{ij} - \varepsilon_m\delta_{ij}. \tag{12.1.10}$$

Therefore, the bulk modulus K is constant while the shear modulus μ^{sec} depends on strain (hence the nonlinearity of the deformation). This behavior is derived from the following free energy:

$$\omega(\boldsymbol{\varepsilon}) = \tfrac{9}{2}k\varepsilon_m^2 + \omega^{\text{eq}}(\varepsilon_{\text{eq}}), \qquad \omega^{\text{eq}}(\varepsilon^{\text{eq}}) = \int_0^{\varepsilon_{\text{eq}}} \sigma_{\text{eq}}\, d\varepsilon_{\text{eq}}, \tag{12.1.11}$$

where ω^{eq} is the area under the curve in the $\sigma_{\text{eq}}(\varepsilon_{\text{eq}})$ diagram. The associated complimentary energy $\tilde{\omega}(\boldsymbol{\sigma})$ is written as follows:

$$\tilde{\omega}(\boldsymbol{\sigma}) = \frac{1}{2K}\sigma_m^2 + \tilde{\omega}^{\text{eq}}(\sigma_{\text{eq}}). \tag{12.1.12}$$

Here $\tilde{\omega}^{\text{eq}}$ is the dual potential of ω^{eq}.

Hencky's Model in Plasticity

The elastoplastic behavior is rigorously described with incremental relations that enable us to distinguish plastic and elastic loads. In the case of isotropic hardening (where the hardening variable is λ) with the Von Mises criterion, these relations are written as follows:

$$\boldsymbol{\varepsilon} = \boldsymbol{\varepsilon}^e + \boldsymbol{\varepsilon}^p, \qquad \boldsymbol{\varepsilon}^e = \frac{\sigma_m}{3K}\mathbf{I} + \frac{1}{2\mu}\mathbf{s}, \qquad \dot{\boldsymbol{\varepsilon}}^p = \frac{3}{2}\dot{\lambda}\frac{\mathbf{s}}{\sigma_{\text{eq}}},$$
$$\dot{\lambda} = \frac{3}{2}\frac{\mathbf{s}:\dot{\mathbf{s}}}{\sigma_{\text{eq}}}\left(\frac{d\sigma^y}{d\lambda}\right)^{-1} \quad \text{if} \quad \sigma_{\text{eq}} = \sigma^y(\lambda), \qquad \dot{\lambda} = 0 \quad \text{if} \quad \sigma_{\text{eq}} < \sigma^y(\lambda). \tag{12.1.13}$$

Hencky's model, also referred to as the deformation theory of plasticity, neglects the incremental nature of plasticity and replaces the equation describing the evolution of the plastic deformation (12.1.13) by

$$\boldsymbol{\varepsilon} = \boldsymbol{\varepsilon}^e + \boldsymbol{\varepsilon}^p, \qquad \boldsymbol{\varepsilon}^e = \frac{\sigma_m}{3K}\mathbf{I} + \frac{1}{2\mu}\mathbf{s}, \qquad \boldsymbol{\varepsilon}^p = \frac{3}{2}\lambda\frac{\mathbf{s}}{\sigma_{\text{eq}}},$$
$$\lambda = (\sigma^y)^{-1}(\sigma_{\text{eq}}). \tag{12.1.14}$$

For example, the model proposed by Ramberg and Osgood describes a power law hardening $\sigma^y(\lambda) = \sigma_0(\lambda/\varepsilon_0)^{1/n}$. This model can be written in the three-dimensional case:

$$\boldsymbol{\varepsilon} = \boldsymbol{\varepsilon}^e + \boldsymbol{\varepsilon}^p, \qquad \boldsymbol{\varepsilon}^e = \frac{\sigma_m}{3K}\mathbf{I} + \frac{1}{2\mu}\mathbf{s}, \qquad \boldsymbol{\varepsilon}^p = \frac{3\varepsilon_0}{2}\left(\frac{\sigma_{\text{eq}}}{\sigma_0}\right)\frac{\mathbf{s}}{\sigma_{\text{eq}}}, \tag{12.1.15}$$

which is derived from the following potential:

$$\tilde{\omega}(\boldsymbol{\sigma}) = \frac{1}{2K}\sigma_m^2 + \frac{1}{6\mu}(\sigma_{\text{eq}})^2 + \frac{\sigma_0\varepsilon_0}{n+1}\left(\frac{\sigma_{\text{eq}}}{\sigma_0}\right)^{n+1}, \tag{12.1.16}$$

where μ, $n \geq 1$, σ_0, and ε_0 are the elastic shear modulus, the hardening exponent, and reference material constants, respectively.

In the case of monotonic loading, Hencky's model is a useful approximation. Unfortunately, it is inapplicable in the case of complex loadings. It can be shown that the incremental method and the deformation theory give the same response under monotonic and radial (the principle stress directions are time independent) loading. The adoption of the deformation theory in the case of heterogeneous materials supposes the stress state to be monotonic and radial at any point of the RVE. This hypothesis is more than likely not satisfied. However, in simple cases (e.g., elastoplastic matrix reinforced with simply shaped elastic inclusions), it can be observed that under a monotonic and radial macroscopic load, deviations from this hypothesis are small. Therefore, the approximation of the incremental theory by a deformation theory is fairly good. For a more detailed discussion on the relationship between the incremental plasticity and the deformation theory, one can refer to Hutchinson (1976).

Creep Behavior

The high-temperature creep of metals can often be treated by neglecting elastic deformation. It is typically expressed by a power law relation between strain rate and Cauchy's stress:

$$\boldsymbol{\sigma} = \sigma_m \mathbf{I} + \mathbf{s}, \qquad \frac{\mathbf{s}}{\sigma_{\text{eq}}} = \frac{2}{3}\frac{\dot{\mathbf{e}}}{\dot{\varepsilon}_{\text{eq}}}, \qquad \frac{\sigma_{\text{eq}}}{\sigma_0} = \frac{\dot{\varepsilon}_{\text{eq}}}{\dot{\varepsilon}_0}, \qquad \dot{\varepsilon}_m = 0. \tag{12.1.17}$$

The potentials φ and $\tilde{\varphi}$ describing the dissipation are written as follows:

$$\varphi(\dot{\boldsymbol{\varepsilon}}) = \frac{\sigma_0 \dot{\varepsilon}_0}{m+1}\left(\frac{\dot{\varepsilon}_{\text{eq}}}{\dot{\varepsilon}_0}\right)^{m+1}, \qquad \tilde{\varphi}(\boldsymbol{\sigma}) = \frac{\sigma_0 \dot{\varepsilon}_0}{n+1}\left(\frac{\sigma_{\text{eq}}}{\sigma_0}\right)^{n+1}. \tag{12.1.18}$$

Here $m = 1/n$, and $\dot{\varepsilon}_0$ and σ_0 are the strain rate sensitivity exponent and the reference strain rate and stress, respectively.

Continuum Description of Single Crystals

The general framework of thermodynamic potentials can also be applied to crystalline materials. The total strain rate, $\dot{\boldsymbol{\varepsilon}}$, in a single crystal is given by a superposition of the elementary shear rates in each slip system

$$\dot{\boldsymbol{\varepsilon}} = \sum_{g=1}^{N} \dot{\gamma}^g \mathbf{R}^g, \tag{12.1.19}$$

where the shear rate $\dot{\gamma}^g$ in the slip system g depends only on the critical shear stress τ^g in this system, such that

$$\dot{\gamma}^g = \frac{\partial \tilde{\varphi}^g}{\partial \tau^g}(\tau^g), \tag{12.1.20}$$

where the potential $\tilde{\varphi}^g$ is convex and typically given by a power law as follows:

$$\tilde{\varphi}^g(\tau^g) = \frac{\tau_0^g \dot{\gamma}_0}{n^g + 1} \left(\frac{|\tau^g|}{\tau_0^g} \right)^{n^g+1}. \tag{12.1.21}$$

In the above, $n^g \geq 1$ is the creep exponent, τ_0^g is the critical shear stress for the considered system, and $\dot{\gamma}_0$ is a reference shear rate.

The constitutive law described by Eqs. (12.1.19) and (12.1.20) are derived from a convex potential for the single crystal:

$$\dot{\boldsymbol{\varepsilon}} = \frac{\partial \tilde{\varphi}(\boldsymbol{\sigma})}{\partial \boldsymbol{\sigma}}, \qquad \tilde{\varphi}(\boldsymbol{\sigma}) = \sum_{g=1}^{N} \tilde{\varphi}^g(\tau^g) = \sum_{g=1}^{N} \tilde{\varphi}^g(\mathbf{R}^g{:}\boldsymbol{\sigma}). \tag{12.1.22}$$

Secant Moduli Description of the Behavior

In what follows, only one notation of the behavior will be used, based on the elastic potentials, ω and $\tilde{\omega}$. The constitutive law (12.1.8) can be written in a more compact form as follows:

$$\boldsymbol{\sigma} = \mathbf{L}^{\text{sec}}(\boldsymbol{\varepsilon}){:}\boldsymbol{\varepsilon}, \tag{12.1.23}$$

where $\mathbf{L}^{\text{sec}}$ is the secant modulus tensor. In the case of isotropic materials, it is a function of the projection tensors $\mathbf{I}^h$ and $\mathbf{I}^d$:

$$\mathbf{L}^{\text{sec}}(\boldsymbol{\varepsilon}) = 3K\mathbf{I}^h + 2\mu^{\text{sec}}(\varepsilon_{\text{eq}})\mathbf{I}^d. \tag{12.1.24}$$

Note that $\mathbf{L}^{\text{sec}}$ is not uniquely defined. For example, one can add an anisotropic component to the secant tensor without changing the constitutive law:

$$\tilde{L}_{ijkl}^{\text{sec}} = L_{ijkl}^{\text{sec}} + \alpha(I_{ijkl}^d - \tfrac{2}{3}\tilde{e}_{ij}\tilde{e}_{kl}), \qquad \tilde{e}_{ij} = \frac{e_{ij}}{\varepsilon_{\text{eq}}} \tag{12.1.25}$$

such that $\mathbf{L}^{\text{sec}}(\boldsymbol{\varepsilon}){:}\boldsymbol{\varepsilon} = \tilde{\mathbf{L}}^{\text{sec}}(\boldsymbol{\varepsilon}){:}\ \boldsymbol{\varepsilon}$ for all α. And $\tilde{\mathbf{L}}^{\text{sec}}$ is a secant tensor in the sense described by (12.1.23), but $\mathbf{L}^{\text{sec}}$ given by (12.1.24) is the only isotropic tensor verifying this relation.

Incremental Formulation and Tangent Moduli

We now discuss homogenization methods based on tangent moduli. It results from an incremental form of the constitutive law given by (12.1.4), obtained with a differentiation with respect to time:

$$\dot{\boldsymbol{\sigma}} = \mathbf{L}^{\text{tg}}(\boldsymbol{\varepsilon}):\dot{\boldsymbol{\varepsilon}}, \qquad L^{\text{tg}}_{ijkl}(\boldsymbol{\varepsilon}) = \frac{\partial^2 \omega(\boldsymbol{\varepsilon})}{\partial \varepsilon_{ij} \partial \varepsilon_{kl}} = \frac{\partial \sigma_{ij}(\boldsymbol{\varepsilon})}{\partial \varepsilon_{kl}}, \qquad (12.1.26)$$

where $\mathbf{L}^{\text{tg}}(\boldsymbol{\varepsilon})$ is the tangent modulus tensor, which is typically anisotropic, even when the material is isotropic. Accordingly, in the case of an isotropic material described by Eqs. (12.1.23) and (12.1.24), the tangent tensor is given by

$$L^{\text{tg}}_{ijkl}(\boldsymbol{\varepsilon}) = 3KI^h_{ijkl} + 2\mu^{\text{sec}}(\varepsilon_{\text{eq}})I^d_{ijkl} + \frac{4}{3}\frac{d\mu^{\text{sec}}}{d\varepsilon_{\text{eq}}}(\varepsilon_{\text{eq}})\varepsilon_{\text{eq}}\tilde{e}_{ij}\tilde{e}_{kl}. \qquad (12.1.27)$$

For convenience, $\mathbf{L}^{\text{tg}}$ can be expressed as

$$L^{\text{tg}}_{ijkl}(\boldsymbol{\varepsilon}) = 3KI^h_{ijkl} + 2\mu^{\text{sec}}(\varepsilon_{\text{eq}})E^1_{ijkl} + 2\mu^{\text{tg}}(\varepsilon_{\text{eq}})E^2_{ijkl}, \qquad (12.1.28)$$

where

$$E^2_{ijkl} = \tfrac{2}{3}\tilde{e}_{ij}\tilde{e}_{kl}, \qquad \mathbf{E}^1 = \mathbf{I}^d - \mathbf{E}^2,$$

$$\mu^{\text{tg}} = \mu^{\text{sec}} + \frac{d\mu^{\text{sec}}(\varepsilon_{\text{eq}})}{d\varepsilon_{\text{eq}}}\varepsilon_{\text{eq}} = \frac{1}{3}\frac{d\sigma_{\text{eq}}(\varepsilon_{\text{eq}})}{d\varepsilon_{\text{eq}}}\varepsilon_{\text{eq}}. \qquad (12.1.29)$$

The tensors $\mathbf{E}^1$ and $\mathbf{E}^2$ have the following properties:

$$\mathbf{E}^2:\mathbf{E}^2 = \mathbf{E}^2, \qquad \mathbf{E}^1:\mathbf{E}^1 = \mathbf{E}^1, \qquad \mathbf{E}^2:\mathbf{E}^1 = \mathbf{0},$$
$$\mathbf{E}^2:\mathbf{I}^h = \mathbf{0}, \qquad \mathbf{E}^1:\mathbf{I}^h = \mathbf{0}. \qquad (12.1.30)$$

The shear modulus μ^{tg} gives the slope of the curve $\sigma_{\text{eq}}(\varepsilon_{\text{eq}})$.

In the case of isotropic incompressible materials described by a power law, the secant tensor $\mathbf{L}^{\text{sec}}$ (12.1.24) and the tangent tensor $\mathbf{L}^{\text{tg}}$ (12.1.28) are given by

$$\mathbf{L}^{\text{sec}}(\boldsymbol{\varepsilon}) = +\infty\mathbf{I}^h + 2\mu^{\text{sec}}(\varepsilon_{\text{eq}})\mathbf{I}^d, \qquad \mu^{\text{sec}}(\varepsilon_{\text{eq}}) = \frac{1}{3}\frac{\sigma_0}{\varepsilon_0}\left(\frac{\varepsilon_{\text{eq}}}{\varepsilon_0}\right)^{m-1}, \qquad (12.1.31)$$

and

$$\mathbf{L}^{\text{tg}}(\boldsymbol{\varepsilon}) = +\infty\mathbf{I}^h + 2\mu^{\text{sec}}(\varepsilon_{\text{eq}})(\mathbf{E}^1 + m\mathbf{E}^2), \qquad \mu^{\text{tg}}(\varepsilon_{\text{eq}}) = m\mu^{\text{sec}}(\varepsilon_{\text{eq}}). \qquad (12.1.32)$$

To illustrate the nonuniqueness of the secant moduli, it can be easily verified that the anisotropic tensor $(1/m)\mathbf{L}^{\mathrm{tg}}$ is also the secant tensor for isotropic materials described by a power law see Problem 12.1.

The tangent modulus tensor is uniquely defined with equation (12.1.26). However, in some cases, it is approximated, to simplify the algebra, by an isotropic tensor,

$$^{\mathrm{iso}}\mathbf{L}^{\mathrm{tg}}(\boldsymbol{\varepsilon}) = +\infty\mathbf{I}^h + 2\mu^{\mathrm{tg}}(\varepsilon_{\mathrm{eq}})\mathbf{I}^d. \tag{12.1.33}$$

Another isotropic tangent tensor is expressed by

$$^{\mathrm{iso}}\mathbf{L}^{\mathrm{tg}}(\boldsymbol{\varepsilon}) = +\infty\mathbf{I}^h + \frac{1}{5}(\mathbf{L}^{\mathrm{tg}}::\mathbf{I}^d)\mathbf{I}^d = +\infty\mathbf{I}^h + 2\,\frac{4+m}{5}\,\mu^{\mathrm{sec}}(\varepsilon_{\mathrm{eq}})\mathbf{I}^d. \tag{12.1.34}$$

It should be noted that the above approximation may lead to severe discrepancies in the homogenization schemes adopting a tangent linearization of local constitutive laws.

12.2 EFFECTIVE POTENTIALS

Let us now consider a RVE with volume V and boundary S representing a nonlinear heterogeneous material. The boundary conditions on S are given in terms of combined loading: traction boundary condition $\mathbf{p}^0$ on S_σ and displacement boundary condition $\mathbf{u}^0$ on S_u with $S = S_\sigma + S_u$.

We are to find the fields $\mathbf{u}$, $\boldsymbol{\varepsilon}$, and $\boldsymbol{\sigma}$ so that $\mathbf{u}$ and $\boldsymbol{\varepsilon}$ fulfill the conditions of kinematics:

$$\begin{cases} \boldsymbol{\varepsilon} = \frac{1}{2}(\nabla\mathbf{u} + (\nabla\mathbf{u})^T) & \text{within } V \\ \mathbf{u} & \text{is continuous in } V. \\ \mathbf{u} = \mathbf{u}^0 & \text{on } S_u \end{cases} \tag{12.2.1}$$

Under the conditions (12.2.1), $\mathbf{u}$ and $\boldsymbol{\varepsilon}$ are called kinematically admissible:

$$\begin{cases} \operatorname{div}\boldsymbol{\sigma} + \mathbf{f} = 0 & \text{inside } V \\ \boldsymbol{\sigma}\cdot\mathbf{n} = \mathbf{p}^0 & \text{on } S_u \end{cases}. \tag{12.2.2}$$

Under the conditions (12.2.2), $\boldsymbol{\sigma}$ is called statically admissible.

By introducing the thermodynamics potentials, $\boldsymbol{\varepsilon}$ and $\boldsymbol{\sigma}$ are related at each material point $\mathbf{r}$ of V by the constitutive laws:

$$\boldsymbol{\sigma}(\mathbf{r}) = \frac{\partial \omega(\boldsymbol{\varepsilon}(\mathbf{r}), \mathbf{r})}{\partial \boldsymbol{\varepsilon}}, \qquad \boldsymbol{\varepsilon}(\mathbf{r}) = \frac{\partial \tilde{\omega}(\boldsymbol{\sigma}(\mathbf{r}), \mathbf{r})}{\partial \boldsymbol{\varepsilon}}. \tag{12.2.3}$$

The solution to the nonlinear problem requires solving one of the following problems:

Under displacement boundary conditions related to an applied macroscopic strain $\bar{\boldsymbol{\varepsilon}}$, the problem consists of solving the following problem for statistically admissible stress:

$$\begin{aligned} \forall\, \mathbf{r} \in V \qquad \sigma(\mathbf{r}) &= \frac{\partial \omega(\boldsymbol{\varepsilon}(\mathbf{r}), \mathbf{r})}{\partial \boldsymbol{\varepsilon}}, \\ \langle \boldsymbol{\varepsilon}(\mathbf{r}) \rangle_V &= \bar{\boldsymbol{\varepsilon}}. \end{aligned} \tag{12.2.4}$$

Under traction boundary conditions related to an applied macroscopic stress $\bar{\boldsymbol{\sigma}}$, the problem consists of solving the following problem for kinematically admissible strain:

$$\begin{aligned} \forall\, \mathbf{r} \in V \qquad \boldsymbol{\sigma}(\mathbf{r}) &= \frac{\partial \omega(\boldsymbol{\varepsilon}(\mathbf{r}), \mathbf{r})}{\partial \boldsymbol{\varepsilon}}, \\ \langle \boldsymbol{\sigma}(\mathbf{r}) \rangle_V &= \bar{\boldsymbol{\sigma}}. \end{aligned} \tag{12.2.5}$$

Solutions to (12.1.1) or (12.1.2) lead to the definition of local stress and strain fields fulfilling the following variational properties, which result from the convexity of the local potentials ω and $\tilde{\omega}$. This can be stated as:

The solution of strain field for the first problem minimizes the free energy ω, so that

$$\langle \omega(\boldsymbol{\varepsilon}) \rangle = \inf \langle \omega(\tilde{\boldsymbol{\varepsilon}}) \rangle, \tag{12.2.6}$$

where $\tilde{\boldsymbol{\varepsilon}}$ is a kinematically admissible strain with $\langle \tilde{\boldsymbol{\varepsilon}} \rangle = \bar{\boldsymbol{\varepsilon}}$.

The solution of stress field for the second problem minimizes the complementary free energy $\tilde{\omega}$, so that

$$\langle \tilde{\omega}(\boldsymbol{\sigma}) \rangle = \inf \langle \tilde{\omega}(\tilde{\boldsymbol{\sigma}}) \rangle, \tag{12.2.7}$$

where $\tilde{\boldsymbol{\sigma}}$ is a trial statically admissible stress with $\langle \tilde{\boldsymbol{\sigma}} \rangle = \bar{\boldsymbol{\sigma}}$.

Definition (12.2.6) corresponds to the macroscopic or effective free energy of the representative volume element, denoted by $W(\overline{\boldsymbol{\varepsilon}})$, so that the macroscopic constitutive relation writes

$$\overline{\boldsymbol{\sigma}} = \frac{\partial W(\overline{\boldsymbol{\varepsilon}})}{\partial \overline{\boldsymbol{\varepsilon}}}, \tag{12.2.8}$$

with

$$W(\overline{\boldsymbol{\varepsilon}}) = \inf\langle \omega(\tilde{\boldsymbol{\varepsilon}})\rangle. \tag{12.2.9}$$

Equation (12.2.8) can be shown from (12.2.4) and (12.2.6) by using the Hill lemma.

Similarly, one can define a macroscopic complementary energy as

$$\overline{\boldsymbol{\varepsilon}} = \frac{\partial \tilde{W}(\overline{\boldsymbol{\sigma}})}{\partial \overline{\boldsymbol{\sigma}}} \quad \text{with } \tilde{W}(\overline{\boldsymbol{\sigma}}) = \inf\langle \tilde{\omega}(\tilde{\boldsymbol{\sigma}})\rangle. \tag{12.2.10}$$

Note that the potentials W and $\tilde{W}$ are dual convex functions, such that with the help of the Hill lemma, one has

$$W(\overline{\boldsymbol{\varepsilon}}) + \tilde{W}(\overline{\boldsymbol{\sigma}}) = \langle \omega(\boldsymbol{\varepsilon})\rangle + \langle \tilde{\omega}(\boldsymbol{\sigma})\rangle = \langle \boldsymbol{\sigma}:\boldsymbol{\varepsilon}\rangle = \overline{\boldsymbol{\sigma}}:\overline{\boldsymbol{\varepsilon}}. \tag{12.2.11}$$

The principles of minimum energy (12.2.6) and (12.2.7) may be used to set bounds for the effective potentials W and $\tilde{W}$. For example, by choosing the solution of problem (12.1.1) or (12.1.2) corresponding to a uniform stress $\boldsymbol{\sigma} = \overline{\boldsymbol{\sigma}}$ and uniform strain field $\boldsymbol{\varepsilon} = \overline{\boldsymbol{\varepsilon}}$, respectively, one obtains the equivalent Voigt and Reuss bounds for the effective potentials. By setting $\boldsymbol{\varepsilon} = \overline{\boldsymbol{\varepsilon}}$ in (12.2.9), one has

$$W(\overline{\boldsymbol{\varepsilon}}) \le \langle \omega(\overline{\boldsymbol{\varepsilon}})\rangle = \sum_{r=1}^{n} c_r \omega_r(\overline{\boldsymbol{\varepsilon}}). \tag{12.2.12}$$

Similarly

$$\tilde{W}(\overline{\boldsymbol{\sigma}}) \le \langle \tilde{\omega}(\overline{\boldsymbol{\sigma}})\rangle = \sum_{r=1}^{n} c_r \tilde{\omega}_r(\overline{\boldsymbol{\sigma}}). \tag{12.2.13}$$

As an application for incompressible power law materials defined by (12.1.18), the Voigt and Reuss bounds read (see Problem 12.2)

$$\frac{\sigma_0^{\text{Reuss}}\varepsilon_0}{m+1}\left(\frac{\varepsilon_{\text{eq}}}{\varepsilon_0}\right)^{m+1} \leq W(\overline{\boldsymbol{\varepsilon}}) \leq \frac{\sigma_0^{\text{Voigt}}\varepsilon_0}{m+1}\left(\frac{\varepsilon_{\text{eq}}}{\varepsilon_0}\right)^{m+1}, \qquad (12.2.14)$$

where

$$\sigma_0^{\text{Reuss}} = \langle \sigma_0^{-n} \rangle^{-m}, \qquad \sigma_0^{\text{Voigt}} = \langle \sigma_0 \rangle. \qquad (12.2.15)$$

12.3 THE SECANT APPROACH

For general purposes, the secant method solves the following field equations:

$$\begin{aligned} \boldsymbol{\varepsilon} &= \tfrac{1}{2}(\nabla \mathbf{u} + (\nabla \mathbf{u})^{\mathrm{T}}), \\ \operatorname{div} \boldsymbol{\sigma} &= \mathbf{0}, \\ \langle \boldsymbol{\varepsilon} \rangle &= \overline{\boldsymbol{\varepsilon}}, \\ \boldsymbol{\sigma}(\mathbf{r}) &= \mathbf{L}^{\text{sec}}(\mathbf{r}, \boldsymbol{\varepsilon}(\mathbf{r})):\boldsymbol{\varepsilon}(\mathbf{r}), \end{aligned} \qquad (12.3.1)$$

where $\mathbf{L}^{\text{sec}}$ is the local secant modulus tensor, which depends on the phase considered. It typically fluctuates within a phase due to the fluctuation of the local strain $\boldsymbol{\varepsilon}(\mathbf{r})$. Therefore, the secant modulus tensor is highly heterogeneous. Its fluctuation results from the nonlinearity of the problem associated with its dependency on the local strain.

Indeed, the heterogeneity of secant modulus tensor depends on which type of nonlinear behavior is displayed by the constituents and also on the amount of applied strain. To set up a direct homogenization procedure of such highly heterogeneous materials is very difficult unless systematic approximations are used, which, in general, relies on a linearization procedure with appropriate complementary laws.

As a first attempt and within a general procedure, the problem could be seen at a given strain state as a linear problem with the following local constitutive law:

$$\boldsymbol{\sigma}(\mathbf{r}) = \mathbf{L}^{\text{lin}}(\mathbf{r}):\boldsymbol{\varepsilon}(\mathbf{r}), \qquad (12.3.2)$$

with

$$\mathbf{L}^{\text{lin}}(\mathbf{r}) = \mathbf{L}^{\text{sec}}(\mathbf{r}, \boldsymbol{\varepsilon}(\mathbf{r})). \qquad (12.3.3)$$

Definition (12.3.2) is the linear model required for the linearization procedure of the local behavior, whereas definition (12.3.3) corresponds to additional or complementary relationships, so that the nonlinear behavior can be captured. When these two steps are accomplished, the problem becomes a classical one. Then an appropriate "classical" linear homogenization scheme can be chosen to obtain the nonlinear macroscopic behavior. However, (12.3.2) and (12.3.3) are still not suitable for analytical calculations due to the infinite number of complementary equations required for definition (12.3.3).

Therefore, approximations are needed, which, clearly, need to render a finite number of complementary equations with a certain accuracy in describing the heterogeneous nature of the nonlinear behavior. For such a purpose, approximations are introduced both in the step of linearization and complementary equations. The linear model in Eq. (12.3.2) may be assumed piecewise uniform for the stiffness tensor $\mathbf{L}^{\text{lin}}(\mathbf{r})$, so that, for a given phase $r(r = 1, \ldots, n)$ one has $\mathbf{L}^{\text{lin}}(\mathbf{r}) = \mathbf{L}_r$. In addition, the complementary equations are reduced to a finite number corresponding to the identified number of constituents or phases, which lead to a definition of stiffness tensors $\mathbf{L}_r$ at some effective piecewise uniform strains $\tilde{\boldsymbol{\varepsilon}}_r$, representing the strain distribution in each phase, and therefore requiring an accurate model to be determined. The n complementary equations read

$$\mathbf{L}_r = \mathbf{L}_r^{\text{sec}}(\tilde{\boldsymbol{\varepsilon}}_r), \tag{12.3.4}$$

where the nonlinearity of the problem lies in the dependency of each individual effective strain $\tilde{\boldsymbol{\varepsilon}}_r$ on the stiffness $\mathbf{L}_r$ of the different phases, so that n nonlinear problems have to be solved, requiring in general simple iterative procedures. Once the tensors $\mathbf{L}_r$ are determined, the problem becomes a classical one by taking advantage of the homogenization approaches developed in linear elasticity.

The choice of the appropriate linear homogenization scheme to describe the microstructure of the real nonlinear composite material defines the so-called linear comparison composite, for which the overall effective stiffness $\mathbf{L}$ is expressed formally as

$$\mathbf{L} = \mathbf{L}(c_r, \mathbf{L}_r, \ldots) \equiv \mathbf{L}(\bar{\boldsymbol{\varepsilon}}), \tag{12.3.5}$$

which depends on the stiffness of each constituent and some morphological aspects related to the microstructure. The overall constitutive law is then nonlinear and formally given by

$$\overline{\boldsymbol{\sigma}} = \overline{\mathbf{L}}(\overline{\boldsymbol{\varepsilon}}):\overline{\boldsymbol{\varepsilon}}. \tag{12.3.6}$$

In the following, two methods to define the effective strains $\tilde{\boldsymbol{\varepsilon}}_r$ are discussed and compared. The first approach is known as the classical secant method. It simply defines the effective strains as the average strain in each phase. The second method, called the "modified" secant method, could be seen as a refinement of the first method by introducing the second-order moment of the strain field.

Classical Method

The classical method has been extensively used to deal with the nonlinear behavior of composite materials. It consists of defining the effective strain $\tilde{\boldsymbol{\varepsilon}}_r$ as the mean value of the local strain field over the considered phase. That is

$$\tilde{\boldsymbol{\varepsilon}}_r = \langle \boldsymbol{\varepsilon}(\mathbf{r}) \rangle_{V_r}. \tag{12.3.7}$$

The main advantage behind assumption (12.3.7) lies in the expression of effective strains $\tilde{\boldsymbol{\varepsilon}}_r$ as a function of the applied strain $\overline{\boldsymbol{\varepsilon}}$ by means of the global strain concentration $\mathbf{A}_r$:

$$\tilde{\boldsymbol{\varepsilon}}_r = \mathbf{A}_r:\overline{\boldsymbol{\varepsilon}}, \; r(r = 1, \ldots, n), \tag{12.3.8}$$

which are determined by appropriate explicit or implicit linear mean field theories (see Chapter 7), which give the n concentration tensors in terms of the stiffness tensors $\mathbf{L}_r$ of each phase for explicit schemes and the overall stiffness $\overline{\mathbf{L}}$ for implicit schemes. That is

$$\mathbf{A}_r = \mathbf{A}_r(\overline{\mathbf{L}}, \mathbf{L}_r, s = 1, \ldots, n). \tag{12.3.9}$$

Finally, since $\mathbf{L}_r$ depends on the corresponding effective strain $\tilde{\boldsymbol{\varepsilon}}_r$ through Eq. (12.3.4), Eq. (12.3.8) together with expression (12.3.9) provide n nonlinear equations, whose solutions determine the overall nonlinear property $\overline{\mathbf{L}}$ by means of the constitutive equation (12.3.6). As noticed before, such a scheme requires in general iterative procedure and suitable convergence criteria.

Note that through the definition of a linear comparison composite, upper and lower bounds of the effective properties $\overline{\mathbf{L}}$ can be found using the linear variational principles presented in Chapter 6.

The classical secant method can be illustrated in case of two-phase materials. In fact, combination of

$$\langle \boldsymbol{\varepsilon}(\mathbf{r}) \rangle_V = \overline{\boldsymbol{\varepsilon}}, \qquad \tilde{\boldsymbol{\varepsilon}}_1 = \mathbf{A}_1 : \overline{\boldsymbol{\varepsilon}}, \qquad \tilde{\boldsymbol{\varepsilon}}_2 = \mathbf{A}_2 : \overline{\boldsymbol{\varepsilon}} \tag{12.3.10}$$

leads to

$$c_1 \mathbf{A}_1 + c_2 \mathbf{A}_2 = \mathbf{I}. \tag{12.3.11}$$

In addition, from the constitutive law of each constituent, we have

$$\overline{\boldsymbol{\sigma}}_1 = \mathbf{L}_1 : \tilde{\boldsymbol{\varepsilon}}_1, \qquad \overline{\boldsymbol{\sigma}}_2 = \mathbf{L}_2 : \tilde{\boldsymbol{\varepsilon}}_2, \tag{12.3.12}$$

and

$$\overline{\boldsymbol{\sigma}} = c_1 \overline{\boldsymbol{\sigma}}_1 + c_2 \overline{\boldsymbol{\sigma}}_2. \tag{12.3.13}$$

Equation (12.3.10) gives

$$\overline{\mathbf{L}} = c_1 \mathbf{L}_1 : \mathbf{A}_1 + c_2 \mathbf{L}_2 : \mathbf{A}_2. \tag{12.3.14}$$

Then substituting (12.3.11) in (12.3.14) yields

$$\mathbf{A}_1 = \frac{1}{c_1} (\mathbf{L}_1 - \mathbf{L}_2)^{-1} (\overline{\mathbf{L}} - \mathbf{L}_2), \qquad \mathbf{A}_2 = \frac{1}{c_2} (\mathbf{L}_2 - \mathbf{L}_1)^{-1} (\overline{\mathbf{L}} - \mathbf{L}_1). \tag{12.3.15}$$

Therefore, when the linear homogenization model is identified to obtain the effective stiffness as

$$\overline{\mathbf{L}} = \overline{\mathbf{L}}(c_1, \mathbf{L}_1, \mathbf{L}_2), \tag{12.3.16}$$

the solution to the nonlinear problem is given by the following set of equations:

$$\tilde{\boldsymbol{\varepsilon}}_1 = \frac{1}{c_1}(\mathbf{L}_1 - \mathbf{L}_2)^{-1}(\overline{\mathbf{L}} - \mathbf{L}_2):\overline{\boldsymbol{\varepsilon}},$$

$$\tilde{\boldsymbol{\varepsilon}}_2 = \frac{1}{c_2}(\mathbf{L}_2 - \mathbf{L}_1)^{-1}(\overline{\mathbf{L}} - \mathbf{L}_1):\overline{\boldsymbol{\varepsilon}}, \qquad (12.3.17)$$

$$\mathbf{L}_1 = \mathbf{L}_1^{\text{sec}}(\tilde{\boldsymbol{\varepsilon}}_1), \qquad \mathbf{L}_2 = \mathbf{L}_2^{\text{sec}}(\tilde{\boldsymbol{\varepsilon}}_2)$$

When the two phases are isotropic,

$$\mathbf{L}_1(\tilde{\boldsymbol{\varepsilon}}_1) = 3K_1\mathbf{I}^h + 2\mu_1^{\text{sec}}(\tilde{\varepsilon}_{\text{eq}}^{(1)})\mathbf{I}^d \quad \text{and} \quad \mathbf{L}_2(\boldsymbol{\varepsilon}_2) = 3K_2\mathbf{I}^h + 2\mu_2^{\text{sec}}(\tilde{\varepsilon}_{\text{eq}}^{(2)})\mathbf{I}^d, \qquad (12.3.18)$$

and the linear comparison composite displays an overall isotropy such that

$$\overline{\mathbf{L}}(\overline{\boldsymbol{\varepsilon}}) = 3\overline{K}\mathbf{I}^h + 2\overline{\mu}(\overline{\varepsilon}_{\text{eq}})\mathbf{I}^d, \qquad (12.3.19)$$

where the linear homogenization scheme gives

$$\overline{K} = \overline{K}(K_1, K_2, \mu_1, \mu_2, c_1), \qquad \overline{\mu} = \overline{\mu}(K_1, K_2, \mu_1, \mu_2, c_1). \qquad (12.3.20)$$

The set of nonlinear equations (12.3.17) is reduced to

$$\tilde{\varepsilon}_m^{(1)} = A_m^{(1)}\overline{\varepsilon}_m, \qquad \tilde{\varepsilon}_m^{(2)} = A_m^{(2)}\overline{\varepsilon}_m, \qquad \tilde{\varepsilon}_{\text{eq}}^{(1)} = A_{\text{eq}}^{(1)}\overline{\varepsilon}_{\text{eq}}, \qquad \tilde{\varepsilon}_{\text{eq}}^{(2)} = A_{\text{eq}}^{(2)}\overline{\varepsilon}_{\text{eq}},$$

$$A_m^{(1)} = \frac{1}{c_1}\frac{\overline{K} - K_2}{K_1 - K_2}, \qquad A_{\text{eq}}^{(1)} = \frac{1}{c_1}\frac{\overline{\mu} - \mu_2}{\mu_1 - \mu_2},$$

$$A_m^{(2)} = \frac{1}{c_2}\frac{\overline{K} - K_1}{K_2 - K_1}, \qquad A_{\text{eq}}^{(2)} = \frac{1}{c_2}\frac{\overline{\mu} - \mu_1}{\mu_2 - \mu_1}, \qquad (12.3.21)$$

$$\mu_1 = \mu_1^{\text{sec}}(\tilde{\varepsilon}_{\text{eq}}^{(1)}), \qquad \mu_2 = \mu_2^{\text{sec}}(\tilde{\varepsilon}_{\text{eq}}^{(2)}).$$

Further, if the materials are incompressible, the linear homogenization model gives

$$\overline{\mu} = \overline{\mu}(\mu_1, \mu_2, c_1), \qquad (12.3.22)$$

and the nonlinear set of equations become

$$
\begin{aligned}
\tilde{\varepsilon}_{\mathrm{eq}}^{(1)} &= A_{\mathrm{eq}}^{(1)}\bar{\varepsilon}_{\mathrm{eq}}, & \tilde{\varepsilon}_{\mathrm{eq}}^{(2)} &= A_{\mathrm{eq}}^{(2)}\bar{\varepsilon}_{\mathrm{eq}}, \\
A_{\mathrm{eq}}^{(1)} &= \frac{1}{c_1}\frac{\bar{\mu}-\mu_2}{\mu_1-\mu_2}, & A_{\mathrm{eq}}^{(2)} &= \frac{1}{c_2}\frac{\bar{\mu}-\mu_1}{\mu_2-\mu_1}, \\
\mu_1 &= \mu_1^{\mathrm{sec}}(\tilde{\varepsilon}_{\mathrm{eq}}^{(1)}), & \mu_2 &= \mu_2^{\mathrm{sec}}(\tilde{\varepsilon}_{\mathrm{eq}}^{(2)}).
\end{aligned}
\tag{12.3.23}
$$

As discussed above, when the appropriate linear homogenization scheme is chosen, the classical method becomes relatively easy to implement through an iterative algorithm. A few examples are treated as problems at the end of this chapter.

Modified Secant Method

The classical secant method for describing the nonlinear behavior of composite materials assumes basically homogeneous strain field within each phase and therefore neglects any intraphase fluctuations of local fields. This results in few discrepancies and limitations, which was behind the principal motivations in developing the modified secant approach.

Let us first recall the basis of the classical secant method, which leads to a certain number of inconsistencies. As shown above, the classical method derives the average stress $\bar{\boldsymbol{\sigma}}_r$ over a phase r ($r = 1, \ldots, n$) as

$$
\bar{\boldsymbol{\sigma}}_r = \mathbf{L}_r{:}\bar{\boldsymbol{\varepsilon}}_r = \mathbf{L}_r^{\mathrm{sec}}(\bar{\boldsymbol{\varepsilon}}_r){:}\bar{\boldsymbol{\varepsilon}}_r, \tag{12.3.24}
$$

which implies the existence of a phase strain energy potential $\omega^{(r)}(\bar{\boldsymbol{\varepsilon}}^{(r)})$ determined with respect to the average strain $\bar{\boldsymbol{\varepsilon}}_r$, such that

$$
\bar{\boldsymbol{\sigma}}_r = \frac{\partial \omega_r(\bar{\boldsymbol{\varepsilon}}_r)}{\partial \bar{\boldsymbol{\varepsilon}}_r}. \tag{12.3.25}
$$

In the case of incompressible materials, (12.1.24) reads

$$
\bar{\boldsymbol{\sigma}}_{\mathrm{eq}}^{(r)} = \frac{\partial \omega_{\mathrm{eq}}^{(r)}(\bar{\varepsilon}_{\mathrm{eq}}^{(r)})}{\partial \bar{\varepsilon}_{\mathrm{eq}}^{(r)}}, \tag{12.3.26}
$$

where

$$\overline{\sigma}_{\text{eq}}^{(r)} = \left[\frac{1}{V_r}\int_{V_r} \boldsymbol{\sigma}(\mathbf{r})\, dV\right]_{\text{eq}}, \qquad \overline{\varepsilon}_{\text{eq}}^{(r)} = \left[\frac{1}{V_r}\int_{V} \boldsymbol{\varepsilon}(\mathbf{r})\, dV\right]_{\text{eq}}. \tag{12.3.27}$$

One can also define the following equivalent average strain as

$$\overline{\overline{\varepsilon}}_{\text{eq}}^{(r)} = \frac{1}{V_r}\int_{V_r} \varepsilon_{\text{eq}}(\mathbf{r})\, dV. \tag{12.3.28}$$

The first discrepancy of the classical method results from the equalities (12.3.25) and (12.3.26), which are satisfied only if the strain field is homogeneous in each phase. Or, in general, the nonlinear behavior leads to highly intraphase fluctuations, and as a result one can show that

$$\overline{\boldsymbol{\sigma}}_r = \frac{1}{V_r}\int_{V_r} \sigma(\mathbf{r})\, dV = \frac{1}{V_r}\int_{V_r} \frac{\partial \omega_r(\boldsymbol{\varepsilon})}{\partial \boldsymbol{\varepsilon}}\, dV \neq \frac{\partial \omega_r(\overline{\boldsymbol{\varepsilon}}_r)}{\partial \overline{\boldsymbol{\varepsilon}}_r}, \tag{12.3.29}$$

or in the case of incompressible materials

$$\overline{\boldsymbol{\sigma}}_{\text{eq}}^{(r)} \neq \frac{\partial \omega_{\text{eq}}^{(r)}(\overline{\boldsymbol{\varepsilon}}_{\text{eq}}^{(r)})}{\partial \overline{\boldsymbol{\varepsilon}}_{\text{eq}}^{(r)}}. \tag{12.3.30}$$

The result (12.3.30) is shown in the following.

In fact, one has

$$\begin{aligned}\overline{\boldsymbol{\sigma}}_{\text{eq}}^{(r)} &= \left[\frac{1}{V_r}\int_{V_r} \boldsymbol{\sigma}(\mathbf{r})\, dV\right]_{\text{eq}} = \left[\frac{1}{V_r}\int_{V_r} \frac{\partial \omega_r(\boldsymbol{\varepsilon})}{\partial \boldsymbol{\varepsilon}}\, dV\right]_{\text{eq}} \\ &= \frac{1}{V_r}\int_{V_r} \frac{\partial \omega_{\text{eq}}^{(r)}(\varepsilon_{\text{eq}})}{\partial \varepsilon_{\text{eq}}} \frac{\partial \varepsilon_{\text{eq}}(\boldsymbol{\varepsilon})}{\partial \boldsymbol{\varepsilon}}\, dV\Bigg]_{\text{eq}}\end{aligned} \tag{12.3.31}$$

which leads to

$$\overline{\boldsymbol{\sigma}}_{\text{eq}}^{(r)} = \left[\frac{1}{V_r}\int_{V_r} \frac{\partial \omega_{\text{eq}}^{(r)}(\varepsilon_{\text{eq}})}{\partial \boldsymbol{\varepsilon}_{\text{eq}}} \frac{2\mathbf{e}(\boldsymbol{\varepsilon})}{3\boldsymbol{\varepsilon}_{\text{eq}}}\, dV\right]_{\text{eq}}. \tag{12.3.32}$$

On the other hand, the convexity of the function $\mathbf{e}{:}\mathbf{e}$ leads to the following inequality:

$$\overline{\sigma}_{\mathrm{eq}}^{(r)} = \left[\frac{1}{V_r}\int_{V_r} \frac{\partial \omega_{\mathrm{eq}}^{(r)}(\varepsilon_{\mathrm{eq}})}{\partial \varepsilon_{\mathrm{eq}}} \frac{2e(\varepsilon)}{3\varepsilon_{\mathrm{eq}}}\, dV\right]_{\mathrm{eq}} \leq \frac{1}{V_r}\int_{V_r} \frac{\partial \omega_{\mathrm{eq}}^{(r)}(\varepsilon_{\mathrm{eq}})}{\partial \varepsilon_{\mathrm{eq}}}\, dV. \qquad (12.3.33)$$

Since for most nonlinear composite the function $[\partial \omega_{\mathrm{eq}}^{(r)}(\varepsilon_{\mathrm{eq}})/\partial \varepsilon_{\mathrm{eq}}]$ is concave, one can easily show that

$$\frac{1}{V_r}\int_{V_r} \frac{\partial \omega_{\mathrm{eq}}^{(r)}(\varepsilon_{\mathrm{eq}})}{\partial \varepsilon_{\mathrm{eq}}}\, dV < \frac{\partial \omega_{\mathrm{eq}}^{(r)}(\overline{\overline{\varepsilon}}_{\mathrm{eq}}^{(r)})}{\partial \varepsilon_{\mathrm{eq}}} = \frac{\partial \omega_{\mathrm{eq}}^{(r)}(\overline{\varepsilon}_{\mathrm{eq}}^{(r)})}{\partial \varepsilon_{\mathrm{eq}}}, \qquad (12.3.34)$$

where we further assume that $\overline{\overline{\varepsilon}}_{\mathrm{eq}}^{(r)} = \overline{\varepsilon}_{\mathrm{eq}}^{(r)}$. With (12.3.34), statement (12.3.30) is proved.

Another limitation of the classical method results from the fact that the definition of the macroscopic properties does not necessarily rely on the definition of a macroscopic $W(\overline{\boldsymbol{\varepsilon}})$, so that

$$\overline{\boldsymbol{\sigma}} = \frac{\partial W(\overline{\boldsymbol{\varepsilon}})}{\partial \overline{\boldsymbol{\varepsilon}}}. \qquad (12.3.35)$$

In fact, it turned out that in some cases of nonlinear composite materials, the following property of the macroscopic potential of isotropic materials

$$\frac{\partial^2 W(\overline{\varepsilon}_{\mathrm{eq}}, \overline{\varepsilon}_m)}{\partial \overline{\varepsilon}_{\mathrm{eq}}\, \partial \overline{\varepsilon}_m} = \frac{\partial \overline{\sigma}_m}{\partial \overline{\varepsilon}_{\mathrm{eq}}} = \frac{\partial^2 W(\overline{\varepsilon}_{\mathrm{eq}}, \overline{\varepsilon}_m)}{\partial \overline{\varepsilon}_m\, \partial \overline{\varepsilon}_{\mathrm{eq}}} = \frac{\partial \overline{\sigma}_{\mathrm{eq}}}{\partial \overline{\varepsilon}_m}, \qquad (12.3.36)$$

is not fulfilled by the classical secant approach, see Problem 12.4.

The modified secant method took its inspiration from the above statement. It was developed in accordance to the following. The first step is the definition of the macroscopic potential from the Hill lemma:

$$W(\overline{\boldsymbol{\varepsilon}}) = \frac{1}{V}\int_V \boldsymbol{\varepsilon}{:}\mathbf{L}(\mathbf{r}){:}\boldsymbol{\varepsilon}\, dV = \overline{\boldsymbol{\varepsilon}}{:}\overline{\mathbf{L}}{:}\overline{\boldsymbol{\varepsilon}}, \qquad (12.3.37)$$

and its derivative with respect to the stiffness $\mathbf{L}_r$ of the r phase in the linear comparison composite:

$$\frac{\partial W(\overline{\boldsymbol{\varepsilon}})}{\partial L_{ijkl}^{(r)}} = \overline{\boldsymbol{\varepsilon}}{:}\frac{\partial \overline{\mathbf{L}}}{\partial L_{ijkl}^{(r)}}{:}\overline{\boldsymbol{\varepsilon}} = \frac{1}{V}\int_V \boldsymbol{\varepsilon}{:}\frac{\partial \mathbf{L}(\mathbf{r})}{\partial L_{ijkl}^{(r)}}{:}\boldsymbol{\varepsilon}\, dV + \frac{2}{V}\int_V \boldsymbol{\varepsilon}{:}\mathbf{L}(\mathbf{r}){:}\frac{\partial \boldsymbol{\varepsilon}}{\partial L_{ijkl}^{(r)}}\, dV, \qquad (12.3.38)$$

where the local stiffness $\mathbf{L}_r(\mathbf{r})$ is assumed to be piecewise uniform:

$$\mathbf{L}_r(\mathbf{r}) = \sum_{r=1}^{n} \mathbf{L}_r f_r(\mathbf{r}), \tag{12.3.39}$$

where f_r is the characteristic function defined in (5.3.8). With (12.3.39), the first term on the right-hand side of (12.3.38) yields

$$\frac{1}{V}\int_V \boldsymbol{\varepsilon}:\frac{\partial \mathbf{L}(\mathbf{r})}{\partial L_{ijkl}^{(r)}}:\boldsymbol{\varepsilon}\, dV = c_r \frac{1}{V_r}\int_{V_r} \varepsilon_{ij}\varepsilon_{kl}\, dV = c_r\langle \varepsilon_{ij}\varepsilon_{kl}\rangle_{V_r}, \tag{12.3.40}$$

while the second term writes

$$\frac{1}{V}\int_V \boldsymbol{\varepsilon}:\mathbf{L}(\mathbf{r}):\frac{\partial \boldsymbol{\varepsilon}}{\partial L_{ijkl}^{(r)}}\, dV = \left\{\frac{1}{V}\int_V \sigma(\mathbf{r})\, dV\right\}:\left\{\frac{1}{V}\int_V \frac{\partial \boldsymbol{\varepsilon}}{\partial L_{ijkl}^{(r)}}\, dV\right\} = 0. \tag{12.3.41}$$

To establish (12.3.41) we used the Hill lemma in accordance to the fact that the strain field $\partial \boldsymbol{\varepsilon}/\partial L_{ijkl}^{(r)}$ is kinematically admissible, so that

$$\int_V \frac{\partial \boldsymbol{\varepsilon}}{\partial L_{ijkl}^{(r)}}\, dV = 0.$$

According to (12.3.40) and (12.3.41), (12.3.38) is reduced to

$$\langle \varepsilon_{ij}\varepsilon_{kl}\rangle_{V_r} = \frac{1}{c_r}\bar{\boldsymbol{\varepsilon}}:\frac{\partial \overline{\mathbf{L}}}{\partial L_{ijkl}^{(r)}}:\bar{\boldsymbol{\varepsilon}}. \tag{12.3.42}$$

The fourth-order tensor $\langle \varepsilon_{ij}\varepsilon_{kl}\rangle_{V_r}$ corresponds to the second-order moment of the strain field over the r phase in the linear comparison composite material. It is calculated by (12.3.42) and therefore requires the definition of the linear homogenization scheme to express the macroscopic properties in terms of the local ones.

The diagonal term $\langle \boldsymbol{\varepsilon}:\boldsymbol{\varepsilon}\rangle_{V_r}$ of the second-order moment can be adopted in the modified second method as an alternative way to measure the intraphase fluctuation of the strain field better than the classical method. This comes from the convexity of the function $\boldsymbol{\varepsilon}:\boldsymbol{\varepsilon}$:

$$\langle \boldsymbol{\varepsilon}:\boldsymbol{\varepsilon}\rangle_{V_r} \geq \langle \boldsymbol{\varepsilon}\rangle_{V_r}:\langle \boldsymbol{\varepsilon}\rangle_{V_r}, \tag{12.3.43}$$

where the equal sign holds only if the strain field is homogeneous.

In the case of isotropic materials, the second-order moment uses the equivalent strain such that

$$\frac{1}{c_r}\,\bar{\boldsymbol{\varepsilon}}:\frac{\partial \overline{\mathbf{L}}}{\partial \mu_r}:\bar{\boldsymbol{\varepsilon}} = \frac{1}{c_r}\,\bar{\boldsymbol{\varepsilon}}:\frac{\partial \overline{\mathbf{L}}}{\partial L^{(r)}_{ijkl}}:\bar{\boldsymbol{\varepsilon}}\,\frac{\partial L^{(r)}_{ijkl}}{\partial \mu_r} = 2\langle \varepsilon_{ij}\varepsilon_{kl}\rangle_{V_r}\, I^{d}_{ijkl} = 3(\overline{\overline{\varepsilon}}{}^{(r)}_{\text{eq}})^2, \quad (12.3.44)$$

where $\overline{\overline{\varepsilon}}{}^{(r)}_{\text{eq}}$ is given by (12.3.28).

Finally, by adopting the second-order moment of strain, the modified secant method involves the following steps:

- The identification of the appropriate linear homogenization scheme, which give the overall stiffness $\overline{\mathbf{L}}$ as function of the phase stiffness $L^{(r)}_{ijkl}$ in the linear comparison composite. Then the derivatives in (12.3.44) can be accomplished.
- The resolution of the following n nonlinear set of equations

$$L^{(r)}_{ijkl} = L^{(r)}_{ijkl}(\overline{\overline{\varepsilon}}{}^{(r)}_{\text{eq}}), \qquad \overline{\overline{\varepsilon}}{}^{(r)}_{\text{eq}} = \left(\frac{1}{3c_r}\,\bar{\boldsymbol{\varepsilon}}:\frac{\partial \overline{\mathbf{L}}}{\partial \mu_r}:\bar{\boldsymbol{\varepsilon}}\right)^{1/2}, \quad (12.3.45)$$

which gives the n unknown secant tensors $L^{(r)}_{ijkl}(\overline{\overline{\varepsilon}}{}^{(r)}_{\text{eq}})$.

As in the case of the classical method, the modified method requires simple iterative algorithm to derive the overall properties of the nonlinear composite.

If the linear comparison composite has overall isotropy, one can easily show from (12.3.45) that

$$\overline{\overline{\varepsilon}}{}^{(r)}_{\text{eq}} = \left[\frac{1}{c_r}\left(\frac{1}{3}\frac{\partial \overline{K}}{\partial \mu_r}\,\bar{\varepsilon}^2_m + \frac{\partial \bar{\mu}}{\partial \mu_r}\,\bar{\varepsilon}^2_{\text{eq}}\right)\right]^{1/2}, \quad (12.3.46)$$

where $\overline{K}$ and $\bar{\mu}$ are computed by a linear homogenization approach.

Let us illustrate the method in the case of a two-phase isotropic composite material, where phase (1) is softer and dispersed in phase (2). Suppose that Hashin–Shtrikman lower bounds are appropriate to derive the overall properties of the linear composite. One has (see Example 6.2)

$$\overline{K} = K_2 + \frac{c_1}{\dfrac{1}{K_1 - K_2} + \dfrac{c_2}{K_2 + \frac{4}{3}\mu_2}}, \qquad (12.3.47)$$

$$\overline{\mu} = \mu_2 + \frac{c_1}{\dfrac{1}{\mu_1 - \mu_2} + \dfrac{6c_2(K_2 + 2\mu_2)\mu_2}{5(3K_2 + 4\mu_2)}},$$

from which one can derive explicit expressions for the second-order moment of strain required to compute the tensor of secant moduli in each phase. The results are

$$\overline{\overline{\varepsilon}}_{\text{eq}}^{(1)} = \frac{1}{c_1}\frac{\overline{\mu} - \mu_2}{\mu_1 - \mu_2}\,\overline{\varepsilon}_{\text{eq}}, \qquad \overline{\overline{\varepsilon}}_{\text{eq}}^{(2)} = (N\overline{\varepsilon}_m^2 + M\overline{\varepsilon}_{\text{eq}}^2)^{1/2}, \quad (12.3.48)$$

with

$$N = \frac{1}{3c_2\mu_2}\left[\overline{K} - c_1K_1\left(\frac{1}{c_1}\frac{\overline{K} - K_2}{K_1 - K_2}\right)^2 - c_2K_2\left(\frac{1}{c_2}\frac{\overline{K} - K_1}{K_2 - K_1}\right)^2\right], \qquad (12.3.49)$$

$$M = \frac{1}{c_2\mu_2}\left[\overline{\mu} - c_1\mu^{(1)}\left(\frac{1}{c_1}\frac{\overline{\mu} - \mu_2}{\mu_1 - \mu_2}\right)^2 - \frac{12}{5}c_1c_2k_2\left(\frac{1}{c_1}\frac{\overline{\mu} - \mu_2}{\mu_1 - \mu_2}\right)^2\left(\frac{\mu_1 - \mu_2}{3K_2 + 4\mu_2}\right)^2\right]. \qquad (12.3.50)$$

Concluding Remarks

In the classical and modified secant nonlinear extensions presented in the above, the phase distribution is the same in the LCC and in the nonlinear composite. As explained in previous sections, this results from the choice of a particular linearization scheme. This option is pertinent and does not lead to any ambiguity in the choice of the linear homogenization model used to describe the morphology of the LCC.

However, another richer strategy can be used in which the homogeneous domain for the secant moduli tensors does not correspond to the domain occupied by the constitutive phases. For example, one could define LCCs with more phases than the nonlinear composite. One can easily anticipate that this richer description of the local heterogeneity of the secant moduli will be closer to the real distribution of the moduli in the nonlinear composite. Hence, the prediction will be more suited. However, the evaluation of a large number of internal variables, the critical choice of a linear model, and the difficulty related to the larger number of considered phases complicate the use of this approach. There is a configuration where this description can be naturally called upon, at least theoretically; when the phase distribution of the nonlinear composite can be appropriately described with morphological patterns. Let us consider the simple case of Hashin's composite spheres assembly. In this case, the linear isotropic behavior of the microstructure can be well described with a three-phase self-consistent scheme based on the analytical solution of the problem of a composite inclusion embedded in an infinite medium.

PROBLEMS

12.1 Show that $(1/m)\mathbf{L}^{\text{tg}}$ is a secant modulus for power law materials defined by constitutive equations (12.1.18), where $\mathbf{L}^{\text{tg}}$ is the tensor of tangent moduli given by (12.1.26).

12.2 Show equation (12.2.15).

12.3 Consider a two-phase composite material made of two power law isotropic materials. The two phases obey constitutive relations (12.1.8). They have the same components n and m, the same reference strain ε_0, but differ by their flow stress denoted by $\sigma_0^{(1)}$ for phase (1) and $\sigma_0^{(2)}$ for phase (2).

By combining the nonlinear problem (12.3.23) with a lower bound Hashin–Shtrikman type of estimation of the effective shear modulus $\overline{\mu}$, the average strain distribution in each phase does not depend on the volume fractions of the constituents, so that

$$\frac{\overline{\varepsilon}_{\text{eq}}^{(1)}}{\overline{\varepsilon}_{\text{eq}}^{(2)}} = \frac{1}{1 + \dfrac{2}{5}\left(\dfrac{\mu_1}{\mu_2} - 1\right)}, \quad \text{with} \quad \frac{\mu_1}{\mu_2} = \frac{\sigma_0^{(1)}}{\sigma_0^{(2)}}\left(\frac{\overline{\varepsilon}_{\text{eq}}^{(1)}}{\overline{\varepsilon}_{\text{eq}}^{(2)}}\right)^{m-1}.$$

When phase (1) is rigid show from the average condition

$$c_1\overline{\varepsilon}_{\text{eq}}^{(1)} + c_2\overline{\varepsilon}_{\text{eq}}^{(2)} = \overline{\varepsilon}_{\text{eq}}$$

the following results:

$$\lim_{\sigma_0^{(1)} \to +\infty} \overline{\varepsilon}_{\rm eq}^{(1)} = 0, \qquad \lim_{\sigma_0^{(1)} \to +\infty} \overline{\varepsilon}_{\rm eq}^{(2)} = \frac{\overline{\varepsilon}_{\rm eq}}{c_2},$$

$$\lim_{\sigma_0^{(1)} \to +\infty} \sigma_0^{(1)} (\overline{\varepsilon}_{\rm eq}^{(1)})^m = \tfrac{5}{2} \sigma_0^{(2)} (\overline{\varepsilon}_{\rm eq}^{(2)})^m.$$

From the definition of the overall stress deviator $\overline{\mathbf{s}}$

$$\overline{\mathbf{s}} = c_1 \overline{\mathbf{s}}_1 + c_2 \overline{\mathbf{s}}_2$$

and the constitutive equations of each phase

$$\overline{\mathbf{s}}_1 = \lim_{\sigma_0^{(1)} \to +\infty} \frac{2}{3} \sigma_0^{(1)} \left(\frac{\overline{\varepsilon}_{\rm eq}^{(1)}}{\varepsilon_0} \right)^m \frac{\mathbf{e}_1}{\overline{\varepsilon}_{\rm eq}^{(1)}}, \qquad \overline{\mathbf{s}}_2 = \lim_{\sigma_0^{(1)} \to +\infty} \frac{2}{3} \sigma_0^{(2)} \left(\frac{\overline{\varepsilon}_{\rm eq}^{(2)}}{\varepsilon_0} \right)^m \frac{\mathbf{e}_2}{\overline{\varepsilon}_{\rm eq}^{(2)}}$$

show that the overall constitutive equation reads

$$\overline{\mathbf{s}} = \frac{2}{3} \frac{\overline{\sigma}_0}{\varepsilon_0} \left(\frac{\overline{\varepsilon}_{\rm eq}}{\varepsilon_0} \right)^{m-1} \mathbf{e}, \quad \text{with} \quad \overline{\sigma}_0 = \overline{\sigma}_0^{(2)} \frac{(1 + \frac{3}{2} c_1)}{(c_2)^m}.$$

12.4 Consider the above problem when the inclusions [phase (1)] are voids. Show that the effective properties of the linear comparison composite are given by

$$\overline{K} = \frac{4}{3} \frac{c_2}{c_1} \mu_2, \qquad \overline{\mu} = \mu_2 \left(\frac{c_2}{1 + \frac{2}{3} c_1} \right)$$

and the solution of the nonlinear problem is reduced to

$$\overline{\varepsilon}_{\rm eq}^{(2)} = \frac{\overline{\varepsilon}_{\rm eq}}{1 + \frac{2}{3} c_1}.$$

Express the overall constitutive laws for the spherical and deviatoric parts and show that the following condition

$$\frac{\partial^2 W(\overline{\varepsilon}_{\rm eq}, \overline{\varepsilon}_m)}{\partial \overline{\varepsilon}_{\rm eq} \, \partial \overline{\varepsilon}_m} = \frac{\partial \overline{\sigma}_m}{\partial \overline{\varepsilon}_{\rm eq}} = \frac{\partial^2 W(\overline{\varepsilon}_{\rm eq}, \overline{\varepsilon}_m)}{\partial \overline{\varepsilon}_m \, \partial \overline{\varepsilon}_{\rm eq}} = \frac{\partial \overline{\sigma}_{\rm eq}}{\partial \overline{\varepsilon}_m}$$

is not satisfied.

12.5 Consider Problem 12.4 when both phases are rigid plastic ($m = 0$) and show that the nonlinear problem is reduced to the following relations:

$$\frac{\overline{\varepsilon}^{(1)}_{eq}}{\overline{\varepsilon}^{(2)}_{eq}} = \frac{5}{3}\left(1 - \frac{2}{5}\frac{\sigma_0^{(1)}}{\sigma_0^{(2)}}\right) \quad \text{if} \quad \frac{\sigma_0^{(1)}}{\sigma_0^{(2)}} \leq \frac{5}{2},$$

$$\frac{\overline{\varepsilon}^{(1)}_{eq}}{\overline{\varepsilon}^{(2)}_{eq}} = 0 \quad \text{if} \quad \frac{\sigma_0^{(1)}}{\sigma_0^{(2)}} \leq \frac{5}{2}.$$

If we denote by $\overline{\sigma}_0$ the flow stress of the rigid plastic composite, show that

$$\overline{\sigma}_0 = c_1\overline{\sigma}_0^{(1)} + c_2\overline{\sigma}_0^{(2)} \quad \text{if} \quad \frac{\sigma_0^{(1)}}{\sigma_0^{(2)}} \leq \frac{5}{2},$$

$$\overline{\sigma}_0 = \overline{\sigma}_0^{(2)}(1 + \tfrac{3}{2}c_1) \quad \text{if} \quad \frac{\sigma_0^{(1)}}{\sigma_0^{(2)}} > \frac{5}{2}.$$

12.6 Consider Problem 12.3 and apply the modified secant method for what follows. Show Eqs. (12.3.48)–(12.3.50). Show that the linear problem defined by Eq. (12.3.45) is reduced to

$$\frac{\overline{\overline{\varepsilon}}^{(1)}_{eq}}{\overline{\overline{\varepsilon}}^{(2)}_{eq}} = \left(\frac{c_2}{c_1}\frac{\dfrac{\partial\overline{\mu}}{\partial\mu_1}}{\dfrac{\partial\overline{\mu}}{\partial\mu_2}}\right)^{1/2}, \quad \text{with} \quad \frac{\mu_1}{\mu_2} = \frac{\sigma_0^{(1)}}{\sigma_0^{(2)}}\left(\frac{\overline{\overline{\varepsilon}}^{(1)}_{eq}}{\overline{\overline{\varepsilon}}^{(2)}_{eq}}\right)^{m-1}$$

with a lower bound Hashin–Shtrikman estimation of $\overline{\mu}$, show that

$$\frac{\overline{\overline{\varepsilon}}^{(1)}_{eq}}{\overline{\overline{\varepsilon}}^{(2)}_{eq}} = \left\{\left[1 + \frac{2}{5}\left(\frac{\mu_1}{\mu_2} - 1\right)\right]^2 + \frac{6}{25}c_1\left(\frac{\mu_1}{\mu_2} - 1\right)^2\right\}^{1/2}.$$

Show that the flow stress of the composite with rigid inclusions is expressed by

$$\overline{\sigma}_0 = \overline{\sigma}_0^{(2)}\frac{(1 + \tfrac{3}{2}c_1)^{m+1/2}}{(c_2)^m}.$$

12.7 Consider Problem 12.4 and show that the nonlinear problem using the modified method writes

$$\bar{\bar{\varepsilon}}_{\text{eq}}^{(2)} = \left[\frac{4}{9c_1} \bar{\varepsilon}_m^2 + \frac{1}{1 + \frac{2}{3}c_1} (\bar{\varepsilon}_{\text{eq}})^2 \right]^{1/2}.$$

Rewrite the overall constitutive laws for the spherical and deviatoric parts and check the following equality:

$$\frac{\partial^2 W(\bar{\varepsilon}_{\text{eq}}, \bar{\varepsilon}_m)}{\partial \bar{\varepsilon}_{\text{eq}}\, \partial \bar{\varepsilon}_m} = \frac{\partial \bar{\sigma}_m}{\partial \bar{\varepsilon}_{\text{eq}}} = \frac{\partial^2 W(\bar{\varepsilon}_{\text{eq}}, \bar{\varepsilon}_m)}{\partial \bar{\varepsilon}_m\, \partial \bar{\varepsilon}_{\text{eq}}} = \frac{\partial \bar{\sigma}_{\text{eq}}}{\partial \bar{\varepsilon}_m}$$

SUGGESTED READINGS

Castañeda, P. (1991). The Effective Mechanical Properties of Nonlinear Isotropic Composites Materials, *J. Mech. Phys. Solids,* Vol. 39, pp. 45–71.

Castañeda, P. (1992). New Variational Principles in Plasticity and Their Applications to Composite Materials, *J. Mech. Phys. Solids,* Vol. 40, pp. 1757–1788.

Castañeda, P. (1996). Exact Second-Order Estimates for the Effective Mechanical Properties of Nonlinear Isotropic Composite Materials, *J. Mech. Phys. Solids,* Vol. 44, pp. 827–862.

Castañeda, P. and P. Suquet. (1995). On the Effective Mechanical Behavior of Weakly Inhomogenous Nonlinear Materials, *Eur. J. Mech. A/Solids,* Vol. 14, pp. 205–236.

Castañeda, P. and J. R. Willis. (1988). On the Overall Properties of Nonlinearly Viscous Composites, *Proc. R. Soc. Lond. A,* Vol. 416, pp. 217–244.

Castañeda, P. and M. Zaidman. (1994). Constitutive Models for Porous Materials with Evolving Microstructure, *J. Mech. Phys. Solids,* Vol. 42, pp. 1459–1497.

Gilormini, P. (1995). A Shortcoming of the Classical Nonlinear Extension of the Self Consistent Model, *C. R. Acad. Sci. Paris, Série IIb,* Vol. 320, pp. 115–122.

Hutchinson, J. W. (1976). Bounds and Self-Consistent Estimates for Creep of Polycrystalline Materials, *Proc. R. Soc. Lond.,* Vol. A 348, pp. 101–127.

Michel, J. C. and P. Suquet. (1992). The Constitutive Law of Nonlinear Viscous and Porous Material, *J. Mech. Phys. Solids,* Vol. 40, pp. 783–812.

Suquet, P. (1993). Overall Potentials and Extremal Surfaces of Power Law or Ideally Plastic Composites, *J. Mech. Phys. Solids Struct.,* Vol. 41, pp. 981–1002.

Suquet, P. and P. Castañeda. (1993). Small-Contrast Perturbation Expansions for the Effective Properties of Nonlinear Composites, *C.R. Accad. Sci. Paris, Série II,* Vol. 317, pp. 1515–1522.

Qiu, Y. P. and G. G. Weng. (1992). A Theory of Plasticity for Porous Materials and Particle-Reinforced Composites, *J. Appl. Mech.,* Vol. 59, pp. 261–268.

13

MICROMECHANICS OF MARTENSITIC TRANSFORMATION IN SOLIDS

Martensitic transformation in solids has received considerable attention in recent years because of its broad potential applications. Such phase change can be induced by the application of stress as well as by temperature change. The original interest in martensitic phase transformation came from the special mechanical properties of the transformation product, namely the strength of martensite and the corresponding thermomechanical treatments to produce adequate behavior for engineering applications.

Various strain mechanisms enter the behavior of such materials. Most of them are ductile materials, and, therefore, in addition to the plastic flow, a microstructure evolution accompanying the phase change leads to complex strain mechanisms ensuring basically the coexistence of the two phases. These strain mechanisms result from an accommodation process, which enhances ductility at a particular strength level by means of "transformation induced plasticity," known as the TRIP effect. The classical definition given to this phenomenon corresponds to the irreversible strain observed even if the specimen is loaded by a stress state less than the yield stress of the softer phase.

In such materials, the overall behavior depends strongly on the temperature since the martensitic phase transformation kinetics is related with temperature through the so-called chemical energy. Furthermore, since martensitic phase transformation occurs without diffusion through a cooperative shear movement of atoms, it is recognized that the applied as well as the internal stresses assist the transformation. The role of plastic strain on the progress of martensitic transformation is much more complicated. Martensitic phase transformation occurs on cooling without applied stress at M^s temperature. Above M^s, the critical stress to undergo martensitic phase transformation increases linearly with temperature up to the M_s^σ temperature defined as the maximum temperature at which martensitic transformation occurs by elastic stress. In this temperature range, martensitic transformation is defined as the stress-assisted transformation. At temperatures above M_s^σ, significant plastic flow precedes the transformation, and an additional contribution to transformation arises from the production of new nucleation sites by plastic deformation. In this temperature regime where the transformation critical stress decreases significantly, the phase change is defined as the plastic-strain-induced transformation. The temperature dependence of the critical stress for causing martensitic transformation is schematically represented in Figure 13.1 where the M^d temperature is introduced to reflect the stability of austenitic phase.

From a micromechanics point of view, materials with phase transformation can be considered as two-phase composite materials where the "reinforcement" phase evolves. In addition to the usual thermoelastic properties, which may be assumed homogenous for most of these materials, the inelastic behavior of austenite–martensite two-phase material appears strongly heterogeneous. According to the diversity of active mechanisms and the evolving microstructure, various coupling have to be described. From a kinematical point of view, a volume element experiences a plastic strain in its austenitic state, followed by the instantaneous transformation strain according to discrete

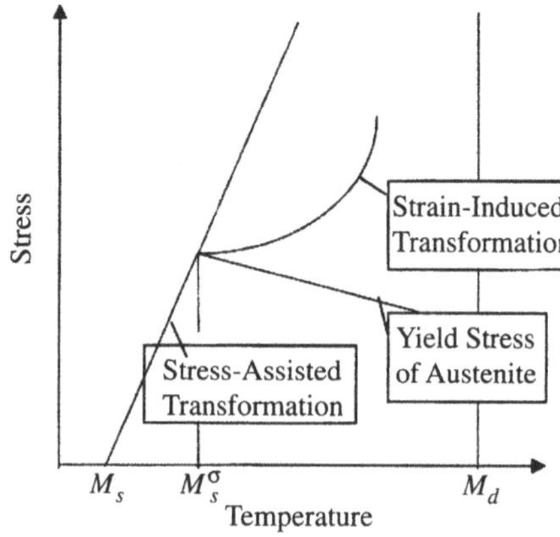

Figure 13.1 Schematic representation of critical stress for martensitic transformation as function of temperature.

values of the possible set of transformation strains and plastic flow in the martensitic state. This phenomenon is also known as dynamic softening. Beside the chemical part of the free energy, the elastic part is due to applied stresses, as well as due to internal stresses generated by the incompatibilities of the total inelastic strain field (plastic + transformation). These couplings render possible a good combination between ductility and strength in materials with phase transformation. In fact any tendency to increase ductility in conventional materials will decrease the strength. For example, a good combination between strength and ductility is a required property to design lightweight structures in automotive bodies and also to improve body safety.

To reproduce the various features of materials exhibiting martensitic phase transformation is not a simple task. Since the problem has to be considered from a multiscale point of view taking into account the microstructure evolution accompanying martensitic transformation, only partial solutions have been developed during the last decades.

Basically, modeling the behavior of such materials raises, naturally, the following problems:

1. To define adequate variables describing the microstructure evolution during the phase transformation. For example, the volume fraction f of martensite or a set of volume fractions f^I ($I = 1, 2, \ldots$) may constitute an appropriate internal variable. If it is the case, the evolution law of this variable should be specified. It may be formally written as

$$\dot{f} = (\ )\dot{\sigma} + (\ )\dot{T}$$

in terms of control variables, stress and temperature, for example. It should be noticed that any attempt to perform a systematic solution to this problem has to be accomplished through a thermodynamic study to derive energetic criteria for phase transformation.

2. To define the appropriate scale to describe the transformation strain or TRIP strain, accompanying the phase transformation. As this inelastic strain results from complex accommodation processes, which occur at the microscale, morphological considerations of the transformation product have to be taken into account. At this scale, the definition of the TRIP strain may leads to a set of transformation strains $\boldsymbol{\varepsilon}^{trI}$ ($I = 1, 2, \ldots$), which depends on the possible crystallographic orientations of martensite related to the concept of martensitic variants.

It may be possible to adopt a more macroscopic description by introducing a kind of metallurgical variable $\overline{\boldsymbol{\varepsilon}}^{\text{tr}}$ (σ,T) to deal with the TRIP strain. However, such a description may lead to a model with limited predictive capabilities.

3. To adopt an appropriate nonlinear homogenization scheme for a well-defined representative volume element (RVE). The choice of an appropriate RVE is also conditioned per the choice of internal and metallurgical variables describing the microstructure evolution and the morphological aspects of the transformation. If the problem is described by the macroscopic variables f and $\overline{\boldsymbol{\varepsilon}}^{\text{tr}}$ (σ,T), the RVE simply represents a two-phase austenite–martensite composite material. For such a case, simple isotropic secant approach (see Chapter 12) may be sufficient to describe the overall behavior. However, a multivariant description of the transformation product may require the definition of various RVE. For example, an RVE representing a single crystal with a certain number of martensitic variants (different orientations), surrounded by an austenitic matrix, is required to approximate the stress and strain distribution in each phase. Another RVE can also be defined to account for the polycrystalline structure of these materials.

Complete descriptions of the above interrelated problems are still under investigation both experimentally and theoretically. In particular, refined microstructural analyses are needed to understand various mechanisms at different scales. As the problem of multiscale modeling is still challenging for these materials, the reader will find in this chapter an introduction to the different microscale mechanisms with certain guidance in terms of micromechanical modeling. We pay particular attention to the problem of coexistence of two phases and resulting discontinuities in terms of stress, strain, and mechanical properties. For such a purpose we describe in details the concept of energy momentum, originally introduced by Eshelby. We show how this framework can be adapted to derive energetic criteria for phase transformation at different scales.

13.1 PHASE TRANSFORMATION MECHANISMS AT DIFFERENT SCALES

The elementary mechanism accompanying martensitic transformation lies in a nondiffusion process at the atomic level, which transforms the parent crystallographic lattice (austenite) into a new crystallographic lattice of the transformation product (martensite); see Figure 13.2. This

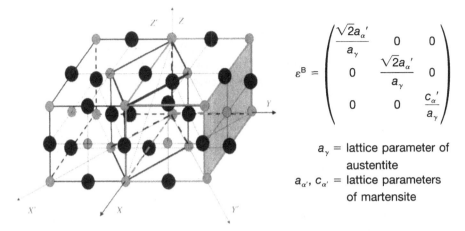

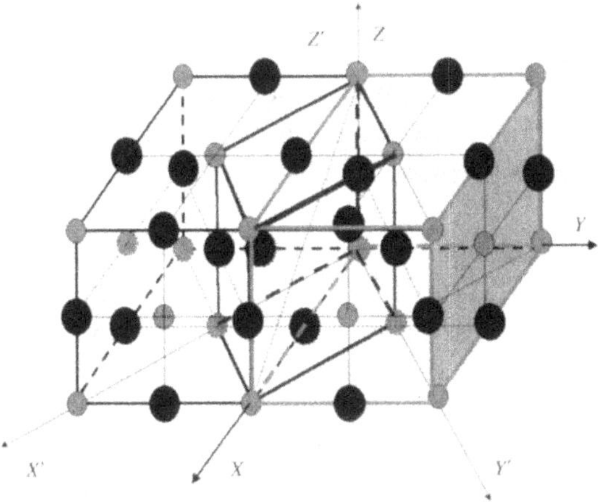

$$\varepsilon^B = \begin{pmatrix} \frac{\sqrt{2}a_{\alpha}'}{a_{\gamma}} & 0 & 0 \\ 0 & \frac{\sqrt{2}a_{\alpha}'}{a_{\gamma}} & 0 \\ 0 & 0 & \frac{c_{\alpha}'}{a_{\gamma}} \end{pmatrix}$$

a_{γ} = lattice parameter of austentite

$a_{\alpha'}, c_{\alpha'}$ = lattice parameters of martensite

Figure 13.2 Bain strain resulting from lattice change during phase transformation.

solid–solid phase transformation results from purely geometrical features in a local lattice distortion called the Bain strain.

The Bain strain $\boldsymbol{\varepsilon}^B$ describing the geometrical transformation from austenite to martensite is incompatible in nature and therefore requires an accommodation procedure that, in general, affects the transformed regions. This accommodation step depends on the mechanical properties of the parent and product phases and leads to a typical length, shape, and spatial distribution of the product of phase transformation.

The concept of the incompatibility accompanying the Bain strain means that the lattice change will generate elastic strain both in parent and product phases and, as a result, internal stresses. These stresses may be sufficient to initiate relaxation mechanisms with certain diversity depending on the mechanical properties of the considered material, which are controlled by the chemical composition, thermomechanical processes, and other parameters. The relaxation mechanisms may be activated as follows:

- Inelastic deformations by plastic shear or twinning in parent and product phases
- A typical resulting morphology of the transformed domains, which consists, in general, in a plate and lath shapes
- Formation of autoaccommodating groups of martensite

Indeed, a complete determination of such an accommodating step relies on a fine description of the typical microstructure of the transformation product. Therefore, it could only be accomplished through approximate ways leading to the crystallographic theories of martensitic transformation. One of the well-known theories is the method of Wechsler, Liebermann, and Read (1953), which describes the accommodation step by an inelastic strain preserving the parent phase lattice and leading to the concept of habit plane or interface between austenite and martensite.

This theory is based on the concept of inelastic compatibility of a local transformation strain $\boldsymbol{\varepsilon}^{\text{tr}}(\mathbf{r})$ assumed to be uniform within an elementary transformed volume V^I. In other words, the methodology assumes a given morphology of the transformed region and deals with the compatibility conditions through two different ways: one is based on a concept of energy minimization using the elementary Eshelby's inclusion, the other deals with the problem of stress and strain discontinuities across the interface S^I of the considered domain, where the compatibility conditions rely on the coherency of this interface.

Eshelby's Inclusion Method

Through the basic concept of Eshelby's inclusion problem, the transformed domain V^I is assumed to be ellipsoidal in shape and experiences an eigenstrain corresponding to the lattice distortion described by the Bain strain $\boldsymbol{\varepsilon}^{\text{B}}$. If we denote by $\mathbf{L}$ the elastic stiffness tensor of the infinite medium and by V its volume, the free energy density reads

$$W = \frac{1}{2}\frac{V^I}{V}\,\boldsymbol{\varepsilon}^{\text{B}}:\mathbf{L}:(\mathbf{I} - \mathbf{S}^I):\boldsymbol{\varepsilon}^{\text{B}}, \tag{13.1.1}$$

where $\mathbf{S}^I$ is the Eshelby tensor.

If the lattice change do not generate any elastic deformation and therefore any increase of the elastic free energy, the compatibility conditions should fulfill the following eigenvalue problem:

$$(\mathbf{I} - \mathbf{S}^I):\boldsymbol{\varepsilon}^{\text{B}} = 0. \tag{13.1.2}$$

In general, with the purely geometric lattice change described by $\boldsymbol{\varepsilon}^{\text{B}}$, Eq. (13.1.2) has no solution. Therefore, an additional inelastic mechanism is required to satisfy the eigenvalue problem raised by (13.1.2). One of the accommodation mechanisms corresponds to a plastic shear or twinning within the transformed domain by preserving its crystallographic lattice; see Figure 13.3. The resulting strain is called "lattice

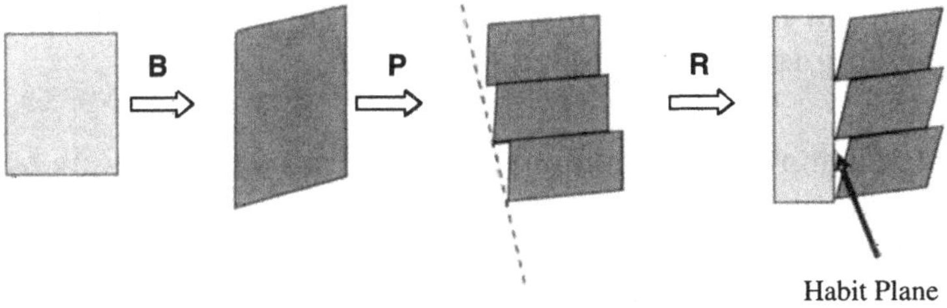

Figure 13.3 Schematic representation of the accommodation process by twinning or slip in martensite by lattice invariant strain (B, Bain strain; P, plastic deformation or twinning; R, rotation).

invariant strain" and denoted here by $\boldsymbol{\varepsilon}^{\text{inv}}$. From this analysis, the transformation strain $\boldsymbol{\varepsilon}^{\text{tr}} = \boldsymbol{\varepsilon}^{\text{B}} + \boldsymbol{\varepsilon}^{\text{inv}}$ is compatible so that the conditions

$$(\mathbf{I} - \mathbf{S}^I):\boldsymbol{\varepsilon}^{\text{tr}} = 0 \tag{13.1.3}$$

is satisfied if

$$\det(\boldsymbol{\varepsilon}^{\text{tr}}) = 0 \text{ (determinant).} \tag{13.1.4}$$

Therefore, a complete determination of $\boldsymbol{\varepsilon}^{\text{tr}}$ from (13.1.4) can be performed following the next steps:

1. Plastic shear or twinning is assumed to occur along a slip system of the transformed region with a magnitude γ so that the lattice invariant strain reads

$$\boldsymbol{\varepsilon}^{\text{inv}} = \gamma \mathbf{R}, \tag{13.1.5}$$

where $\mathbf{R}$ defines the orientation tensor of the slip or twin system defined by its unit normal $\tilde{\mathbf{n}}$ and shear direction $\tilde{\mathbf{m}}$ as

$$R_{ij} = \tfrac{1}{2}(\tilde{m}_i \tilde{n}_j + \tilde{m}_j \tilde{n}_i). \tag{13.1.6}$$

2. The second step defines an interface called the habit plane between the two phases, so that the transformation strain is considered to be predominately shear along this plane. That is

$$\varepsilon_{ij}^{\text{tr}} = \tfrac{1}{2} g(m_i n_j + m_j n_i), \tag{13.1.7}$$

where **m** is the shear direction and **n** the unit normal to the habit plane; g is the magnitude of the shear.

As stated above, the accommodation process occurs mainly by shear, however, when phase transformation is accompanied by substantial plastic flow, a volumic component is measured, which ranges from 1 to 4%. In other words, the two vectors **m** and **n** are not necessarily perpendicular (Fig. 13.4).

From Eq. (13.1.7) and for a given $\boldsymbol{\varepsilon}^B$ and **R**, the compatibility equations (13.1.3) and (13.1.4) determine g and the vectors **m** and **n**. This ends the second step; see Problem 13.1.

Method Using Jump Conditions

For this analysis, the compatibility conditions lie in the definition of a coherent interface or habit plane S^I between the parent and product phases with **n** being the unit normal of the surface. This hypothesis corresponds to the continuity of displacement ($[\mathbf{u}] = 0$) and/or velocity ($[\mathbf{v}] = 0$) fields as well as the stress vector ($[\boldsymbol{\sigma}] \cdot \mathbf{n} = 0$), where $[\mathbf{x}] = \mathbf{x}^+ - \mathbf{x}^-$ denotes the jump of **x** across the interface. In the present case, the "positive" side belongs to the parent phase (austenite), whereas the "negative" one corresponds to the transformed regions (martensite).

It is interesting to note that, in the case of shape memory alloys where the transformation occurs mainly without a plastic deformation and under specific conditions, the analysis can also be adapted to martensite–martensite interfaces.

The field equations expressed at the interface can be reduced to the following compatibility conditions; see Problem 13.2:

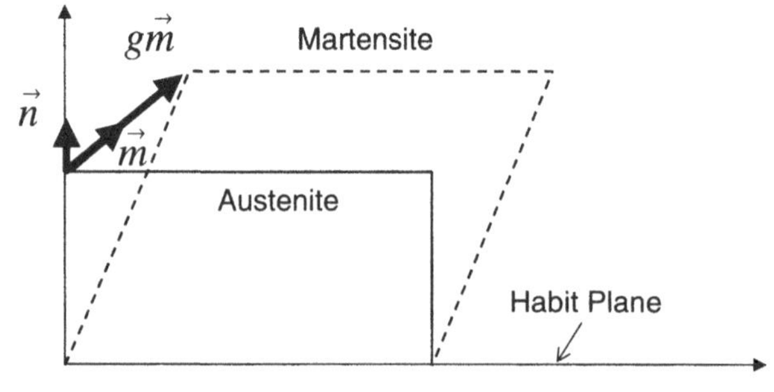

Figure 13.4 Schematic representation of Eq. (13.1.7).

$$[\varepsilon_{ij}] = \varepsilon_{ij}^{+} - \varepsilon_{ij}^{-} = \tfrac{1}{2}(\lambda_i n_j + \lambda_j n_i), \qquad (13.1.8)$$

where the total strain at each side of the interface $\boldsymbol{\varepsilon}^{+,-} = \boldsymbol{\varepsilon}^{e+,e-} + \boldsymbol{\varepsilon}^{tr+,tr-}$ consists of an elastic part $\boldsymbol{\varepsilon}^{e}$ and a transformation contribution $\boldsymbol{\varepsilon}^{tr}$ resulting from the accommodation process accompanying the phase change; $\boldsymbol{\lambda}$ is a vector to be determined.

For simplicity, the behavior from each side of the interface is described by linear elastic constitutive law, such that

$$\boldsymbol{\sigma}^{+,-} = \mathbf{L}^{+,-}:(\boldsymbol{\varepsilon}^{+,-} - \boldsymbol{\varepsilon}^{tr+,tr-}). \qquad (13.1.9)$$

If we further introduce a reference homogeneous elastic medium $\mathbf{L}$ such that

$$\mathbf{L}^{+,-} = \mathbf{L} + (\mathbf{L}^{+,-} - \mathbf{L}) = \mathbf{L} + \Delta\mathbf{L}^{+,-},$$

Eq. (13.1.9) is equivalent to

$$[\boldsymbol{\sigma}] = \mathbf{L}:[\boldsymbol{\varepsilon}] - \mathbf{L}:[\boldsymbol{\varepsilon}^{tr}] + [\Delta\mathbf{L}:\boldsymbol{\varepsilon}^{e}], \qquad (13.1.10)$$

or by taking into account the condition $[\boldsymbol{\sigma}] \cdot \mathbf{n} = 0$, it results that

$$\mathbf{L}:[\boldsymbol{\varepsilon}]:\mathbf{n} = \mathbf{L}:[\boldsymbol{\varepsilon}^{tr}]:\mathbf{n} - [\Delta\mathbf{L} : \boldsymbol{\varepsilon}^{e}]:\mathbf{n}. \qquad (13.1.11)$$

Substituting (13.1.8) into (13.1.11) leads to the following equations whose resolution gives the components of the vector $\boldsymbol{\lambda}$. That is

$$\Im_{ik}\lambda_k = \aleph_i, \qquad (13.1.12)$$

where $\Im_{ik} = L_{ijkl}n_j n_l$ and $\aleph_i = L_{ijkl}\,[\varepsilon_{kl}^{tr}]n_j - [\Delta L_{ijkl}\varepsilon_{kl}^{e}]n_j$.

Clearly, Eq. (13.1.11) shows that a compatible phase transformation (e.g., $\boldsymbol{\varepsilon}^{e+,e-} = 0$) requires heterogeneous transformation strain, for example, $[\boldsymbol{\varepsilon}^{tr}] \neq 0$. Therefore, it results from (13.1.11) that a compatible expression of the transformation strain reads

$$[\varepsilon_{ij}^{tr}] = [\varepsilon_{ij}] = \tfrac{1}{2}(\lambda_i n_j + \lambda_j n_i). \qquad (13.1.13)$$

Direct applications of compatibility conditions (13.1.13) can be generated as follows:

1. The Bain strain described in Figure 13.2 from purely geometrical considerations does not fulfill (13.1.13). In fact, if from the austenitic side we state that $\boldsymbol{\varepsilon}^{tr+} = 0$ and from the martensitic side $\boldsymbol{\varepsilon}^{tr-} = \boldsymbol{\varepsilon}^{B}$, Eq. (13.1.13) gives

$$[\varepsilon_{ij}^{tr}] = \varepsilon_{ij}^{B} \neq \tfrac{1}{2}(\lambda_i n_j + \lambda_j n_i). \qquad (13.1.14)$$

2. The accommodation process described in the previous section, which relies on a slip or twinning in the transformed regions [see (13.1.5)], fulfills the compatibility conditions (13.1.13), such that if one assumes that $\boldsymbol{\varepsilon}^{tr+} = 0$ and $\boldsymbol{\varepsilon}^{tr} = \boldsymbol{\varepsilon}^{B} + \boldsymbol{\varepsilon}^{inv}$, the compatibility conditions writes

$$[\varepsilon_{ij}^{tr}] = -(\varepsilon_{ij}^{B} + \varepsilon_{ij}^{inv}) = \tfrac{1}{2}(\lambda_i n_j + \lambda_j n_i), \qquad (13.1.15)$$

   which is equivalent to Eq. (13.1.7).

More recently, mathematical-based approaches have been proposed to predict other compatible morphologies during phase transformation. For details, the reader could refer to the Suggested Readings at the end of this chapter.

Notion of Martensitic Variants

As discussed in previous sections, the accommodation process accompanying the lattice change during phase transformation leads to a typical morphology of the transformation product, which implicitly results in a heterogeneous transformation strain $\boldsymbol{\varepsilon}^{tr}(\mathbf{r})$. In the previous analysis, we adopt an accommodation process through a shear or twinning by defining a habit plane between austenite and martensite leading to expressions (13.1.7) and (13.1.15) of the transformation strain $\boldsymbol{\varepsilon}^{tr}$. From purely crystallographic considerations and due to high symmetries of the parent phase, 24 couples of $\mathbf{n}$ and $\mathbf{m}$ are possible to define a habit plane leading to the notion of martensitic variant.

From this concept, each possible martensitic variant is described by its crystallographic orientation, morphology, and transformation strain, which are implicitly interrelated by the accommodation process of the lattice change.

Kinematics at Single-Crystal Level

The purpose of this section is to discuss the different strain mechanisms during phase transformation at the single-crystal level. This step is required to build micromechanical tools for the first scale transition linking the appropriate features of transformed regions to constitutive equations of single crystals.

At this level, the concept of martensitic variants is taken into account by assuming that the transformation strain is piecewise uniform within elementary transformed regions corresponding to the different martensitic variants, so that one can write

$$\boldsymbol{\varepsilon}^{\mathrm{tr}}(r) = \sum_{I=1}^{N} \boldsymbol{\varepsilon}^{\mathrm{tr}I}\, \theta^{I}(r), \tag{13.1.16}$$

where $\theta^{I}(r)$ are the Heaviside step functions for the different transformed domains or variants I and N their number. The transformation strain $\boldsymbol{\varepsilon}^{\mathrm{tr}I}$ associated with a martensitic variant I is described by (13.1.7) as

$$\varepsilon_{ij}^{\mathrm{tr}I} = \tfrac{1}{2} g\,(m_i^I n_j^I + m_j^I n_i^I) = g R_{ij}^I, \tag{13.1.17}$$

where $\mathbf{R}^I$ is the orientation tensor of the variant I.

For a single crystal (taken as an RVE) with volume V subjected at its external boundary ∂V to a displacement $u_i = \bar{\varepsilon}_{ij} x_j$, and undergoing an inelastic strain field $\varepsilon^{\mathrm{tr}}(r) + \varepsilon^{p}(r)$ within V, the total macroscopic $\bar{\boldsymbol{\varepsilon}}$ strain is

$$\bar{\boldsymbol{\varepsilon}} = \frac{1}{V}\int_V (\boldsymbol{\varepsilon}^{e}(\mathbf{r}) + \boldsymbol{\varepsilon}^{\mathrm{tr}}(\mathbf{r}) + \boldsymbol{\varepsilon}^{p}(\mathbf{r})]\, dV = \langle \boldsymbol{\varepsilon}^{e}(\mathbf{r}) + \boldsymbol{\varepsilon}^{\mathrm{tr}}(\mathbf{r}) + \boldsymbol{\varepsilon}^{p}(\mathbf{r})\rangle. \tag{13.1.18}$$

Note that the elastic strain field $\boldsymbol{\varepsilon}^{e}(r)$ arises from $\bar{\boldsymbol{\varepsilon}}$ and the incompatibilities of the fields $\boldsymbol{\varepsilon}^{\mathrm{tr}}(r)$ and $\boldsymbol{\varepsilon}^{p}(r)$, where $\boldsymbol{\varepsilon}^{p}(r)$ denote the plastic strain field.

In general, it is difficult to distinguish experimentally between the plastic contribution and the transformation one at the macroscopic level. However, in terms of modeling, we will be able to make this distinction if we adopt a crystallographic description of phase transformation.

For homogeneous elasticity with its elastic stiffness $\mathbf{L}$, one obtains directly the elastic part $\boldsymbol{\varepsilon}^e$ of $\bar{\boldsymbol{\varepsilon}}$ as follows:

$$\bar{\boldsymbol{\varepsilon}} = \mathbf{L}^{-1} : \bar{\boldsymbol{\sigma}} + \bar{\boldsymbol{\varepsilon}}^{\mathrm{tp}}, \tag{13.1.19}$$

where $\bar{\boldsymbol{\sigma}} = \langle \boldsymbol{\sigma}(\mathbf{r}) \rangle$, and $\bar{\boldsymbol{\varepsilon}}^{\mathrm{tp}} = \langle \boldsymbol{\varepsilon}^{\mathrm{tr}}(\mathbf{r}) + \boldsymbol{\varepsilon}^{\mathrm{p}}(\mathbf{r}) \rangle$ corresponds to the macroscopic inelastic strain arising from the phase transformation and the plastic flow. At this stage, one can distinguish between two cases where the physical aspects of martensitic transformation change significantly:

1. In the case of shape memory alloys, the plastic contribution is neglected and the transformation strain becomes

$$\bar{\boldsymbol{\varepsilon}}^{\mathrm{tr}} = \frac{1}{V} \int_V \boldsymbol{\varepsilon}^{\mathrm{tr}}(\mathbf{r})\, dV. \tag{13.1.20}$$

Due to typical properties of the strain field $\boldsymbol{\varepsilon}^{\mathrm{tr}}(r)$, one can differenciate between two simplified forms for $\bar{\boldsymbol{\varepsilon}}^{\mathrm{tr}}$:

- The microstructure of variants as well as the fact that $\boldsymbol{\varepsilon}^{\mathrm{tr}}(r)$ is piecewise uniform, this leads to

$$\bar{\boldsymbol{\varepsilon}}^{\mathrm{tr}} = \frac{1}{V} \sum_I \int_{V^I} \boldsymbol{\varepsilon}^{\mathrm{tr}I}\, dV = \sum_I \boldsymbol{\varepsilon}^{\mathrm{tr}I} f^I, \tag{13.1.21}$$

where $\varepsilon^{\mathrm{tr}I}$ are material constants, and f^I are the internal variables subjected to the following constraints:

$$0 \le \sum_I f^I \le 1, \qquad 0 \le f^I \le 1. \tag{13.1.22}$$

- Since the transformation strain vanishes within the austenitic phase, thus (13.1.21) can be written as

$$\bar{\boldsymbol{\varepsilon}}^{\mathrm{tr}} = \frac{1}{V} \int_{V^M} \boldsymbol{\varepsilon}^{\mathrm{tr}}(\mathbf{r})\, dV = \frac{V^M}{V} \left\{ \frac{1}{V^M} \int_{V^M} \boldsymbol{\varepsilon}^{\mathrm{tr}}(\mathbf{r})\, dV \right\} = f \tilde{\boldsymbol{\varepsilon}}^{\mathrm{tr}}, \tag{13.1.23}$$

where $f = \Sigma_I f^I$ is the total volume fraction of martensite, and $\tilde{\boldsymbol{\varepsilon}}^{\mathrm{tr}}$ being the average transformation strain over the total volume V^M of martensite, which is an unknown to be determined from solving the deformation problem.

2. The case of iron base alloys where the phase transformation is accompanied by substantial plastic flow in the parent phase. In this case, it is more critical to distinguish at the macroscopic level between a macroscopic transformation strain [in the sense of Eq. (13.1.23)] and a macroscopic plastic strain. This basically results from the fact that the transformation part contains a plastic contribution resulting from the accommodation process discussed above. An incremental formulation with a multivariant description is more appropriate for this class of materials (see below).

Incremental Formulation: Concept of Moving Boundaries

In most heterogeneous materials, the mechanical properties may be assumed piecewise uniform at appropriate scales. These materials are made up of different phases and/or grains separated by interfaces called grain or phase boundaries. The interfaces are in general stationary with respect to particles and often considered as perfect, at least at low temperatures. This hypothesis corresponds to the continuity of displacement ($[\mathbf{u}] = 0$) and/or velocity ($[\mathbf{v}] = 0$) fields as well as the stress vector ($[\sigma] \cdot \mathbf{n} = 0$). The continuity assumptions extended to equivalent forms for the volume lead to the usual localization and homogenization relations, from which the classical scale transition methods are developed. The overall behavior of the RVE is then deduced from the microstructure and the local behavior (basically the purpose of previous chapters).

In various situations where inelastic strains result from discrete physical mechanisms such as twinning, martensitic transformation, or recrystallization, the previous hypotheses are no longer valid because of evolving microstructures or moving boundaries whose velocities are different from those of the particles. Consequently, the strain field and/or mechanical properties exhibit discontinuities across moving boundaries and therefore:

- Additional terms in localization and homogenization relations have to be introduced in the framework of local thermodynamics and micromechanics to account for the moving boundaries.
- Driving forces, germination, and growing laws describing the evolving microstructure need to be determined.

These are discussed next.

Kinematics of Evolving Microstructure

In materials undergoing martensitic phase transformation, the parent and product phases coexist with the same chemical composition but differ in their lattices, volumes, and shapes. As a result, a complex accommodation process (discussed in the previous section) occurs leading to a typical morphology of the transformation product with multiple interfaces S or boundaries within the transformed region itself, and with the parent phase. During a loading increment (stress or temperature), two scenarios are possible:

- The material experiences phase transformation with a complex movement of interfaces between parent and product phases.
- Under particular conditions, reorganization of preexisting transformed regions is possible by a typical movement of interfaces within these regions.

For example, it is interesting to link, from kinematics purpose, the macroscopic strain increment $\dot{\bar{\boldsymbol{\varepsilon}}}$ acting on the boundary of an RVE to the velocity field. The velocity $\mathbf{v}$ of particles belonging to the external boundary of the RVE is assumed as

$$v_i = \dot{\bar{\varepsilon}}_{ij} x_j = \frac{d\bar{\varepsilon}_{ij}}{dt} x_j. \tag{13.1.24}$$

From the definition of macroscopic strain

$$\bar{\varepsilon}_{ij} = \frac{1}{V} \int_V \varepsilon_{ij}(\mathbf{r})\, dV.$$

The strain increment is given by

$$\dot{\bar{\varepsilon}}_{ij} = \frac{d}{dt}\left(\frac{1}{V}\right) \int_V \varepsilon_{ij}(\mathbf{r})\, dV + \frac{1}{V}\frac{\partial}{\partial t}\left[\int_V \varepsilon_{ij}(\mathbf{r})\, dV\right]. \tag{13.1.25}$$

By neglecting the change in volume V of the RVE, (13.1.25) is reduced to

$$\dot{\bar{\varepsilon}}_{ij} = \frac{1}{V}\frac{\partial}{\partial t}\left[\int_V \varepsilon_{ij}(\mathbf{r})\, dV\right]. \qquad (13.1.26)$$

The strain field is discontinuous along the different moving interfaces, the time derivative in (13.1.26) reads

$$\dot{\bar{\varepsilon}}_{ij} = \frac{1}{V}\int_V \frac{\partial \varepsilon_{ij}(\mathbf{r})}{\partial t}\, dV - \frac{1}{V}\int_S [\varepsilon_{ij}(\mathbf{r})] w_\alpha n_\alpha\, dS, \qquad (13.1.27)$$

or

$$\dot{\bar{\varepsilon}}_{ij} = \frac{1}{V}\int_V \left[\frac{\partial \varepsilon_{ij}(\mathbf{r})}{\partial t} - [\varepsilon_{ij}(\mathbf{r})]\, w_\alpha n_\alpha \delta(S)\right] dV, \qquad (13.1.28)$$

where $\delta(S) = \int_S \delta(\mathbf{r} - \mathbf{r}')\, dS'$ is the Dirac delta function. $[\varepsilon] = \varepsilon^+ - \varepsilon^-$ is the jump of the local strain field; $\mathbf{w}$ is the velocity of the interface at a material point, where the outward unit normal to the interface is denoted by $\mathbf{n}$; and $w_\alpha n_\alpha$ is a scalar describing the normal velocity of the interface.

Applications

For example, the time derivative of (13.1.20) shows different mechanisms associated with martensitic transformation in shape memory alloys, which are not revealed through a simple "static" comparison between the actual configuration and the reference one (austenite). By taking into account the discontinuities $[\boldsymbol{\varepsilon}^{tr}(\mathbf{r})] = \boldsymbol{\varepsilon}^{tr+} - \boldsymbol{\varepsilon}^{tr-}$ of the strain field $\boldsymbol{\varepsilon}^{tr}(r)$ along the moving boundaries, one has

$$\dot{\bar{\boldsymbol{\varepsilon}}}^{tr} = \frac{1}{V}\int_V \dot{\boldsymbol{\varepsilon}}^{tr}(\mathbf{r})\, dV - \frac{1}{V}\int_S [\boldsymbol{\varepsilon}^{tr}(\mathbf{r})] w_\alpha n_\alpha\, dS, \qquad (13.1.29)$$

and since the concept of stress-free or eigenstrain is adopted for the transformation strain, so that $\dot{\boldsymbol{\varepsilon}}^{tr}(\mathbf{r}) = 0$. It then follows that

$$\dot{\bar{\boldsymbol{\varepsilon}}}^{tr} = -\frac{1}{V}\int_S [\boldsymbol{\varepsilon}^{tr}(\mathbf{r})] w_\alpha n_\alpha\, dS, \qquad (13.1.30)$$

where S represents all of the moving boundaries (austenite–martensite as well as boundaries between the martensitic variants), and $\omega_\alpha n_\alpha$ is the normal velocity of the interface.

The "crystalline" description of the martensitic transformation (analogous to crystal plasticity) comes directly from (13.1.30); see Problem 13.3:

$$\dot{\overline{\boldsymbol{\varepsilon}}}^{\mathrm{tr}} = \sum_I \boldsymbol{\varepsilon}^{\mathrm{tr}I} \dot{f}^I, \tag{13.1.31}$$

where in (13.1.31) the distinction between transformation and exchange between existing variants is not explicitly shown.

The different mechanisms are more explicitly described when one expands the time derivative of (13.1.23):

$$\dot{\overline{\boldsymbol{\varepsilon}}}^{\mathrm{tr}} = \dot{f}\tilde{\boldsymbol{\varepsilon}}^{\mathrm{tr}} + f\dot{\tilde{\boldsymbol{\varepsilon}}}^{\mathrm{tr}}. \tag{13.1.32}$$

In fact, the analysis of (13.1.32) allows us to distinguish the following phenomena:

- Transformation ($\dot{f} \neq 0$) without deformation ($\tilde{\boldsymbol{\varepsilon}}^{\mathrm{tr}} = 0$, $\dot{\tilde{\boldsymbol{\varepsilon}}}^{\mathrm{tr}} = 0$).
- Transformation ($\dot{f} \neq 0$) with deformation ($\tilde{\boldsymbol{\varepsilon}}^{\mathrm{tr}} \neq 0$) but without reorientation ($\dot{\tilde{\boldsymbol{\varepsilon}}}^{\mathrm{tr}} = 0$).
- Deformation by reorientation ($\dot{\tilde{\boldsymbol{\varepsilon}}}^{\mathrm{tr}} \neq 0$) without transformation ($\dot{f} = 0$).
- Transformation, deformation, and reorientation ($\dot{f} \neq 0$, $\tilde{\boldsymbol{\varepsilon}}^{\mathrm{tr}} \neq 0$, $\dot{\tilde{\boldsymbol{\varepsilon}}}^{\mathrm{tr}} \neq 0$).

In the case of iron-based alloys, the average inelastic strain of the single crystal is given by

$$\overline{\boldsymbol{\varepsilon}}^{\mathrm{tp}} = \frac{1}{V}\int_V [\boldsymbol{\varepsilon}^{\mathrm{tr}}(\mathbf{r}) + \boldsymbol{\varepsilon}^{\mathrm{p}}(\mathbf{r})]\, dV, \tag{13.1.33}$$

where $\boldsymbol{\varepsilon}^{\mathrm{p}}(\mathbf{r})$ describes simultaneously the plastic strain of an elementary volume element at austenitic and martensitic states. When the plastic flow becomes significant during phase transformation, the reorientation and inverse transformation mechanisms can be disregarded. The progression of the transformation can be described by the instantaneous growth of new plates or laths in the austenitic phase. The time derivative of (13.1.33) leads to

$$\dot{\bar{\boldsymbol{\varepsilon}}}^{\text{tp}} = \frac{1}{V}\int_V \dot{\boldsymbol{\varepsilon}}^{\text{tp}}(\mathbf{r})\,dV - \frac{1}{V}\int_S [\boldsymbol{\varepsilon}^{\text{tp}}(\mathbf{r})] w_\alpha n_\alpha\,dS. \qquad (13.1.34)$$

The volume integral in (13.1.34) describes the plastic flow in residual austenite and preexisting martensitic variants, it could be written as the following form involving the average plastic strain rate $\dot{\bar{\boldsymbol{\varepsilon}}}^{pA}$ in austenite and the average one over a martensitic variant. That is

$$\frac{1}{V}\int_V \dot{\boldsymbol{\varepsilon}}^{\text{tp}}(\mathbf{r})\,dV = (1 - f)\dot{\bar{\boldsymbol{\varepsilon}}}^{pA} + \sum_I f^I \dot{\bar{\boldsymbol{\varepsilon}}}^{M_I}. \qquad (13.1.35)$$

The surface integral is much more complicated. It requires the determination of the inelastic strain jump at each point of the moving interface, which can be expressed as

$$[\boldsymbol{\varepsilon}^{\text{tp}}(\mathbf{r})] = \boldsymbol{\varepsilon}^{p+} - (\boldsymbol{\varepsilon}^{p-} + \boldsymbol{\varepsilon}^{\text{tr}_I}) \approx -\boldsymbol{\varepsilon}^{\text{tr}_I}. \qquad (13.1.36)$$

Such a jump may be assumed to result only from the accommodation process leading to the definition of compatible transformation strain $\boldsymbol{\varepsilon}^{\text{tr}_I}$. However, as can be seen from (13.1.36), a heterogeneous plastic strain $\boldsymbol{\varepsilon}^{p+,-} \neq 0$ may occur during the phase transformation. Its determination is not a simple task and may require complex numerical calculations. Since the transformation strain $\boldsymbol{\varepsilon}^{\text{tr}_I}$ contains a plastic contribution corresponding to lattice invariant strain, the contribution of $\boldsymbol{\varepsilon}^{p+,-}$ can be neglected. Therefore, the surface integral reads

$$-\frac{1}{V}\int_S [\boldsymbol{\varepsilon}^{\text{tp}}(\mathbf{r})] w_\alpha n_\alpha\,dS = \sum_I \boldsymbol{\varepsilon}^{\text{tr}_I} \dot{f}^I. \qquad (13.1.37)$$

Substituting (13.1.35) and (13.1.37) into (13.1.34) yields

$$\dot{\bar{\boldsymbol{\varepsilon}}}^{\text{tp}} = (1 - f)\dot{\bar{\boldsymbol{\varepsilon}}}^{pA} + \sum_I f^I \dot{\bar{\boldsymbol{\varepsilon}}}^{M_I} + \sum_I \boldsymbol{\varepsilon}^{\text{tr}_I} \dot{f}^I. \qquad (13.1.38)$$

In (13.1.38) one can distinguish the following contributions:

- The first term describes the average plastic flow in the residual austenitic phase.
- The second term corresponds to the plastic flow in the preexisting martensitic phase.
- The last term expresses the formation of new plates or laths.

At the single-crystal level, the plastic flow of residual austenite may be described through the crystallographic slip by introducing the slip systems of FCC lattices. The plastic strain increment is then expressed by (see Chapter 11)

$$\dot{\bar{\boldsymbol{\varepsilon}}}^{pA} = \sum_{g} \mathbf{P}^{g} \dot{\gamma}^{g}, \tag{13.1.39}$$

where $\mathbf{P}^g$ is the Schmidt tensor and $\dot{\gamma}^g$ the shear rate on slip system g.

However, it is more difficult to track the crystallographic slip in martensitic variants. This is due to the large number of variables (24 × 24 variables, 24 possible variants, for each 24 possible slip system). Due to the typical fine microstructure of the transformation product and their typical crystallographic orientation, the plastic deformation of martensite may have no significant contribution to the overall flow of the RVE. This statement is also supported by the fact that martensite is a hard phase. Equation (13.1.38) is then reduced to

$$\dot{\bar{\boldsymbol{\varepsilon}}}^{\mathrm{tp}} = (1 - f) \sum_{g} \mathbf{P}^{g} \dot{\gamma}^{g} + \sum_{I} \boldsymbol{\varepsilon}^{\mathrm{tr}_I} \dot{f}^{I}. \tag{13.1.40}$$

Equation (13.1.40) contains a classical crystal plasticity problem with internal variables, namely the volume fraction of each possible martensitic variant. To come out with a constitutive law of the RVE representing the single crystal, evolution laws of these internal variables have to be specified. In general, this step requires additional approximations and can be accomplished through two different ways:

1. To establish a thermodynamic framework to obtain the driving forces for the activation of the internal variables. This method is more appropriate to derive energetic criteria for the phase transformation.
2. To assume evolution laws based on experimental observations without any link with thermodynamics. This method does not require the determination of driving forces; however, its predictive capacities are limited.

In next sections we present an example displaying the first methodology. This is possible by adopting the original work of Eshelby's momentum tensor.

Eshelby's Energy Momentum Tensor

Eshelby introduced an important concept to deal with the configuration of a solid. It corresponds to the energy momentum tensor, which is another way to combine the field equations with energetic considerations. Let us consider a homogeneous material without any incompatibility in which the elastic energy density $\omega(\mathbf{r},\boldsymbol{\beta})$ only depends on the elastic distortion β. Therefore, we can perform the following:

$$\omega_{,i}\,(\mathbf{r},\boldsymbol{\beta}) = \frac{\partial \omega}{\partial x_i} = \frac{\partial \omega}{\partial \beta_{jk}}\frac{\partial \beta_{jk}}{\partial x_i}, \tag{13.1.41}$$

where the stress tensor reads

$$\sigma_{jk} = \frac{\partial \omega}{\partial \beta_{jk}}, \tag{13.1.42}$$

and $\beta_{jk} = u_{j,k}$, and $\mathbf{u}$ is the displacement vector.

From the equilibrium equation $\sigma_{ik,k} = 0$, one can easily show that

$$\omega_{,i}\,(\mathbf{r},\boldsymbol{\beta}) = [\sigma_{jk}\,(\mathbf{r})u_{j,i}\,(\mathbf{r})]_{,k}. \tag{13.1.43}$$

Thus, (13.1.43) is equivalent to

$$[\omega(\mathbf{r},\boldsymbol{\beta})\delta_{ik} - \sigma_{jk}\,(\mathbf{r})u_{j,i}\,(\mathbf{r})]_{,k} = 0. \tag{13.1.44}$$

Eshelby's energy momentum tensor denoted here by $\boldsymbol{\Omega}$ is expressed by

$$\Omega_{ik} = \omega\delta_{ik} - \sigma_{kj}u_{j,i}. \tag{13.1.45}$$

Therefore, the energy momentum tensor satisfies the following equation resulting from (13.1.44):

$$\Omega_{ik,k} = 0 \qquad 13.1.46)$$

Let us consider now the case of martensitic transformation where the accommodation process leads to a nonhomogeneous transformation strain field $\boldsymbol{\varepsilon}^{tr}(\mathbf{r})$ undergoing a jump $[\boldsymbol{\varepsilon}^{tr}(\mathbf{r})]$ across an interface between the parent and product phases. Let us define the following local quantity $F_i = [\Omega_{ik}]n_k$, which can be expressed from (13.1.45) as

$$F_i = [\omega\delta_{ik} - \sigma_{jk}u_{j,i}]n_k = [\omega]n_i - [\sigma_{jk}u_{j,i}]n_k. \qquad (13.1.47)$$

From the diagonal symmetries of the elastic constant $\mathbf{L}$ and the decomposition of the total distortion $\boldsymbol{\beta}$ into an elastic part $\boldsymbol{\beta}^e$ and a transformation contribution $\boldsymbol{\beta}^{tr}$, one obtains for the jump of elastic energy (see Problem 13.4)

$$[\omega] = \tfrac{1}{2}(\sigma_{ij}^{+} + \sigma_{ij}^{-})][\beta_{ij}^{e}]. \qquad (13.1.48)$$

On the other hand, the continuity of the traction vector, $[\sigma]\cdot\mathbf{n} = 0$, leads to

$$[\sigma_{jk}u_{j,i}]n_k = \tfrac{1}{2}(\sigma_{jk}^{+} + \sigma_{jk}^{-})[u_{j,i}]n_k = \tfrac{1}{2}(\sigma_{jk}^{+} + \sigma_{jk}^{-})[\beta_{ji}]n_k. \quad (13.1.49)$$

Substituting (13.1.48) and (13.1.49) into (13.1.47), one has

$$F_i = \tfrac{1}{2}(\sigma_{jk}^{+} + \sigma_{jk}^{-})\{[\beta_{jk}^{e}]n_i - [\beta_{ji}]n_k\}, \qquad (13.1.50)$$

and by taking into account the compatibility equation (13.1.8), it results

$$[\beta_{ji}]n_k = [u_{j,i}]n_k = \lambda_j n_i n_k = \lambda_j n_k n_i = [\beta_{jk}]n_i. \qquad (13.1.51)$$

Finally with (13.1.51), it follows that

$$F_i = [\Omega_{ik}]n_k = \tfrac{1}{2}(\sigma_{jk}^{+} + \sigma_{jk}^{-})[\varepsilon_{jk}^{tr}]n_i. \qquad (13.1.52)$$

If one follows a complete thermodynamic study—see Cherkaoui et al. (1998)—one can show that the scalar

$$F = F_i w_i \qquad (13.1.53)$$

is the mechanical driving force if one considers the normal velocity of the interface $w_\alpha n_\alpha$ as the internal variable describing the microstructure evolution. This variable is, in general, not appropriate to derive constitutive equations of materials with evolving microstructure. It requires time-consuming and complex computational tools. By means of additional hypotheses taking into account the typical morphology of martensite, an alternative description is possible. It could be considered as another scale of description, which adopts the volume fractions of martensite as internal variables.

13.2 APPLICATION: THERMODYNAMIC FORCES AND CONSTITUTIVE EQUATIONS FOR SINGLE CRYSTALS

Thermodynamic Driving Forces

Recently, Cherkaoui et al. (1998) extended Eshelby's (1970) pioneering work on the energy momentum tensor to derive the thermodynamic driving force acting on a moving boundary point between the product and the parent phases. That is

$$F = -\tfrac{1}{2}(\boldsymbol{\sigma}^{+} + \boldsymbol{\sigma}^{-}):[\boldsymbol{\varepsilon}^{\mathrm{tp}}] + [\varphi], \tag{13.2.1}$$

where $[\boldsymbol{\varepsilon}^{\mathrm{tp}}]$ and $[\varphi]$ denote the jumps of the inelastic strain and the chemical energy, respectively; $\boldsymbol{\sigma}^{+}$ and $\boldsymbol{\sigma}^{-}$ correspond to the limiting values of the local stress field for each side of the moving interface, which can be linked with the help of interface operators $\mathbf{Q}(\mathbf{L}, \mathbf{n})$,

$$\boldsymbol{\sigma}^{+} = \boldsymbol{\sigma}^{-} - \mathbf{Q}(\mathbf{L}, \mathbf{n}) : [\boldsymbol{\varepsilon}^{\mathrm{tp}}], \tag{13.2.2}$$

depending on the unit normal to interface and elastic constant of the material. With (13.2.2), (13.2.1) can be rewritten as

$$F = [\tfrac{1}{2}\mathbf{Q}(\mathbf{L}, \mathbf{n}) - \boldsymbol{\sigma}^{-}]:[\boldsymbol{\varepsilon}^{\mathrm{tp}}] + [\varphi]. \tag{13.2.3}$$

As noticed above, the introduction of (13.2.3) as a thermodynamic driving force assumes that the interface normal velocity is taken as the internal variable characterizing the phase transition. It is also noticed that equation (13.2.3) requires the determination of the local stress and strain fields in each point of the moving interface. Therefore, the determination of (13.2.3) for any topology of the interface could be obtained only through strong numerical calculations. To go a step further in the scale transition, an alternative relation of (13.2.3) is obtained by introducing the concept of an ellipsoidal growing. The former hypothesis assumes that the moving interface represents an ellipsoidal growing domain. This is in accord with the microstructure developed during martensitic transformation corresponding to the formation of micro-domains of a typical morphology of plate or lath.

Furthermore, as we discussed in the previous kinematic study, the martensitic transformation progresses by the nucleation and instantaneous growth of new domains. This is related to a high interface velocity between the parent and product phases, leading consequently to the following expression for the local jump $[\boldsymbol{\varepsilon}^{\mathrm{tp}}]$ of the inelastic strain

$$[\boldsymbol{\varepsilon}^{\mathrm{tp}}]^I = \boldsymbol{\varepsilon}^{\mathrm{p}+} - (\boldsymbol{\varepsilon}^{\mathrm{p}-} + \boldsymbol{\varepsilon}^{\mathrm{tr}I}) = -\boldsymbol{\varepsilon}^{\mathrm{tr}I}. \tag{13.2.4}$$

In other words, the instantaneous growth of a martensitic domain allows the continuity of the plastic strain over its boundary. In connection with (13.1.40), it implies that the volume fractions f^I of each possible martensitic variant are the internal variables characterizing the phase transition at the mesoscale, the integration of (13.2.3) along the ellipsoidal domain interface, and thanks to (13.2.4) one obtains the driving force F^I for the "flow" $\dot{f}^I$ of a martensitic variant due to the formation of a new domain. That is,

$$F^I = \boldsymbol{\sigma}^-:\boldsymbol{\varepsilon}^{\mathrm{tr}I} - B(T - T^0) + \tfrac{1}{2}\boldsymbol{\varepsilon}^{\mathrm{tr}I}:\mathbf{L}:(\mathbf{I} - \mathbf{S}^I):\boldsymbol{\varepsilon}^{\mathrm{tr}I} + \tfrac{1}{2}\boldsymbol{\varepsilon}^{\mathrm{tr}I}:\mathbf{L}:\frac{\dot{\mathbf{S}}^I}{\dot{f}^I}:\boldsymbol{\varepsilon}^{\mathrm{tr}I} f^I, \tag{13.2.5}$$

where $\mathbf{I}$ is the identity tensor and $\mathbf{S}^I$ is Eshelby's tensor depending on the aspect ratio as well as on the orientation of the martensitic micro-domain; $\dot{\mathbf{S}}^I$ corresponds to the variation of $\mathbf{S}^I$ due to the ellipsoidal growth. For the sake of simplicity, a homothetic growth is assumed giving $\dot{\mathbf{S}}^I = 0$. The term $B(T - T^0)$ is a linear approximation of the chemical energy jump $[\varphi]$. B is a material constant and T^0 the equilibrium temperature. It should be noticed that the derivation of (13.2.5) requires the hypothesis of a uniform stress field $\boldsymbol{\sigma}^-$ within the growing ellipsoidal domain. Details on the derivation of (13.2.5) as well as on the notion of ellipsoidal growing can be found in Cherkaoui et al. (1998).

Equation (13.2.5) gives the thermodynamic driving force for nucleation and growth of martensitic microdomains belonging to different martensitic variants; it requires the knowledge of the stress field inside the growing domains. In addition to the overall applied stress, this stress field contains several contributions due to different couplings between plasticity and phase transformation at the microscale. In the following, several ways for the determination of internal stresses are discussed from a micromechanics point of view:

1. The simplest way is to neglect any source of internal stresses. In such conditions, Patel and Cohen (1953) have formulated the energy term resulting only from the interaction of applied stress with transformation strains. This corresponds in (13.2.5) to the term $\overline{\boldsymbol{\sigma}} : \boldsymbol{\varepsilon}^{\mathrm{tr}_I} - B(T - T^0)$, where $\boldsymbol{\sigma}^-$ is assumed equal to the applied stress $\overline{\boldsymbol{\sigma}}$, and all the other contributions are neglected.
2. It is possible to assume an ellipsoidal growth of a martensitic microdomain inside a homogeneous stress field corresponding to the average stress over the austenitic phase. At the current configuration of the RVE, if we denote by $\overline{\boldsymbol{\varepsilon}}^{PA}$ the average plastic strain in the austenitic phase, the interactions between plasticity and martensitic phase transformation are taken into account through the average stress $\overline{\boldsymbol{\sigma}}^A$ in the austenitic phase; see Exercise Problem 13.5:

$$\overline{\boldsymbol{\sigma}}^{A} = \overline{\boldsymbol{\sigma}} - \sum_{I=1}^{N} f^I \mathbf{L} : (\mathbf{I} - \mathbf{S}^I) : (\overline{\boldsymbol{\varepsilon}}^{PA} - \boldsymbol{\varepsilon}^{\mathrm{tr}_I}), \qquad (13.2.6)$$

where N is the number of active martensitic variants at the current configuration of the RVE and f^I their volume fractions.

With the instantaneous growth hypothesis, the stress inside a growing microdomain $\boldsymbol{\sigma}^-$ belonging to a martensitic variant I is related to $\overline{\boldsymbol{\sigma}}^A$ by the following simple form (see Problem 13.6):

$$\boldsymbol{\sigma}^{-} = \overline{\boldsymbol{\sigma}}^{A} - \mathbf{L} : (\mathbf{I} - \mathbf{S}^I) : \boldsymbol{\varepsilon}^{\mathrm{tr}_I}. \qquad (13.2.7)$$

With (13.2.7), (13.2.3) leads to

$$F^I = \overline{\boldsymbol{\sigma}}^{A} : \boldsymbol{\varepsilon}^{\mathrm{tr}_I} - B(T - T^0) - \tfrac{1}{2}\boldsymbol{\varepsilon}^{\mathrm{tr}_I} : \mathbf{L} : (\mathbf{I} - \mathbf{S}^I) : \boldsymbol{\varepsilon}^{\mathrm{tr}_I}. \qquad (13.2.8)$$

In the thermodynamic force (13.2.8), one can distinguish two contributions:

- A long-range internal stress effect through the term $\overline{\boldsymbol{\sigma}}^{A} : \boldsymbol{\varepsilon}^{\mathrm{tr}_I} - B(T - T^0)$. The stress $\overline{\boldsymbol{\sigma}}^{A}$ contains the effects of plastic strain of both phases, as well as of transformation strains undergone by the preexisting martensitic variants. Depending on the applied and internal stresses, $\overline{\boldsymbol{\sigma}}^A$ plays the role of variant selection.

- A self-internal stresses effect through the term $\frac{1}{2}\boldsymbol{\varepsilon}^{trI} : \mathbf{L} : (\mathbf{I} - \mathbf{S}^I) : \boldsymbol{\varepsilon}^{trI}$ depending essentially on the morphology of the growing microdomain.

The effect of the internal stress field (emerging from the plastic flow) on the martensitic phase transformation is known as the strain-induced martensitic phase transformation phenomenon. Classically, the plastic flow at the grain level is described by a homogeneous plastic strain through the plastic slip on crystallographic glide systems. Intragranular stresses arising from this description correspond to the second-order internal stresses. However, for ductile materials undergoing martensitic phase transformation, the role of plastic strain in phase transition is more complicated. In fact, the martensitic plates nucleate at dislocation pile-ups, dislocation dipole, or intersection of slip bands. In such a situation, the description of the strain-induced martensitic transformation is insufficient through a homogeneous plastic strain. In other words, a third-order stress field emerging from the heterogeneity of plastic strain has to be taken into account. This can be performed by the decomposition of the local austenitic plastic strain $\boldsymbol{\varepsilon}^p(\mathbf{r})$ into:

- A uniform part $\overline{\boldsymbol{\varepsilon}}^{pA}$ corresponding to its average over the whole volume of the austenitic grain (this part is described by the crystallographic slip).
- A fluctuating part $\Delta\boldsymbol{\varepsilon}^p(\mathbf{r})$ taking into account the heterogeneity of the plastic strain due to various configurations of inelastic defects.

To take into account the fluctuations of the plastic strain, (13.2.7) giving the stress inside a growing microdomain belonging to a martensitic variant I is extended to

$$\boldsymbol{\sigma}^- = \overline{\boldsymbol{\sigma}}^A - \mathbf{L} : (\mathbf{I} - \mathbf{S}^I) : \boldsymbol{\varepsilon}^{trI} + \boldsymbol{\sigma}^p(\mathbf{r}), \qquad (13.2.9)$$

where $\boldsymbol{\sigma}^p(\mathbf{r})$ is a third-order stress field resulting from $\Delta\boldsymbol{\varepsilon}^p(\mathbf{r}) \cdot \boldsymbol{\sigma}^p(\mathbf{r})$ is evaluated using an inclusion problem, where we assume that the martensitic domain nucleates and grows at a plastic defect with volume V^p and undergoing a plastic heterogeneity with strength $\Delta\boldsymbol{\varepsilon}^p$. The initial shape of the martensitic nucleus coincides with one of the plastic defects. In such conditions, the thermodynamic force (13.2.8) is extended to the following form [see Cherkaoui et al. (2000)]:

$$F^I = \overline{\boldsymbol{\sigma}}^A : \boldsymbol{\varepsilon}^{\text{tr}_I} - B(T - T^0) - \tfrac{1}{2}\boldsymbol{\varepsilon}^{\text{tr}_I} : \mathbf{L} : (\mathbf{I} - \mathbf{S}^I) : \boldsymbol{\varepsilon}^{\text{tr}_I} + \kappa g H^{I_n}\gamma^n, \tag{13.2.10}$$

where

$$H^{I_n} = \mathbf{R}^I : \mathbf{L} : (\mathbf{S}^I - \mathbf{S}^{p_n}) : \mathbf{P}^{A_n}, \tag{13.2.11}$$

$\mathbf{S}^{p_n}$ is Eshelby's tensor depending on the morphology of the plastic "defect" assumed to have the initial shape of the martensitic nucleus; g and $\mathbf{R}^I$ are given in Eq. (13.1.17); γ^n is the plastic slip in the residual austenitic phase; $\mathbf{P}^{A_n}$ is the corresponding Schmidt tensor; and κ is a model parameter.

For the plastic flow in residual austenitic phase, the associated driving forces are the resolved shear stresses on slip systems given by (see Chapter 11)

$$\tau^{A_n} = \overline{\boldsymbol{\sigma}}^A : \mathbf{P}^{A_n}. \tag{13.2.12}$$

As shown by equation (13.2.12), the plastic flow is affected by the phase transformation through the average stress $\overline{\boldsymbol{\sigma}}^A$. This constitutes in addition to the transformation strain the mechanisms of ductility enhancement in these materials.

Critical Forces and Constitutive Equations of a Single Crystal

From an energetic point of view, the formation of a martensitic domain is allowed, if the associated driving force F^I overcomes a transformation barrier F^c. This critical force stems from the energy necessary to rebuild the crystal lattice at the front of the developed domain. F^c can be considered as a material parameter.

Concerning the plasticity of the austenitic phase and since the predominant mechanisms are the plastic slips on the crystallographic glide systems, the usual hardening matrix H_A^{gh} is introduced describing the self and the latent hardening in FCC metals; see Chapter 11. In such conditions, the critical force τ_A^{cg} to achieve for plastic slip on a glide system g is expressed as

$$\tau_A^{c_g} = \tau_A^{c_0} + H_A^{gh}\gamma^h, \tag{13.2.13}$$

where $\tau_A^{c_0}$ is the initial critical shear stress and is identical for all the slip systems.

When the thermodynamic driving forces reach their critical values, the consistency rule leads to the evolution laws for $\dot{f}^I$ and $\dot{\gamma}^n$. From (13.2.10) and (13.2.12), the consistency rule applied to the transformation type of internal variables gives the following expression:

$$\dot{\overline{\boldsymbol{\sigma}}}:\mathbf{R}^I + g\dot{f}^J\mathbf{L}:(\mathbf{I} - \mathbf{S}^J):\mathbf{R}^I : \mathbf{R}^J - f^J\mathbf{L}:(\mathbf{I} - \mathbf{S}^J):\mathbf{R}^I:\mathbf{P}^{A_n}\dot{\gamma}^n$$
$$+ \kappa H^{I_n}\dot{\gamma}^n - \frac{B}{g}\dot{T} = 0. \qquad (13.2.14)$$

In the same way, one obtains from (13.2.13) and (13.2.12) the following relation for the plastic type of internal variables:

$$\dot{\overline{\boldsymbol{\sigma}}}:\mathbf{P}^{A_g} + g\dot{f}^J\mathbf{L}:(\mathbf{I} - \mathbf{S}^J):\mathbf{P}^{A_g} : \mathbf{R}^J$$
$$- f^J\mathbf{L}:(\mathbf{I} - \mathbf{S}^J):\mathbf{P}^{A_g} : \mathbf{P}^{A^n}\dot{\gamma}^n - H_A^{gn}\dot{\gamma}^n = 0. \qquad (13.2.15)$$

which is added to Eq. (13.2.14) to build a nonlinear system whose solution gives the evolution laws of the internal variables X_i characterizing the martensitic phase transformation and the plastic flow of the parent phase. This could be formally written as

$$\dot{X}_i = A^i_{kl}\,(\boldsymbol{\sigma},T,\mathbf{X})\dot{\overline{\sigma}}_{kl} + B^i\,(\boldsymbol{\sigma},T,\mathbf{X})\dot{T}. \qquad (13.2.16)$$

Equation (13.2.16) can be also expressed formally as function of internal variables X_i as follows:

$$\dot{\overline{\varepsilon}}^{\text{tp}}_{ij} = M^k_{ij}\,(\mathbf{X})\dot{X}_k. \qquad (13.2.17)$$

Combining (13.2.16) and (13.2.17), the inelastic response of an austenitic single crystal is expressed as

$$\dot{\overline{\varepsilon}}^{\text{tp}}_{ij} = M^k_{ij}\,(\mathbf{X})\,A^k_{mn}\,(\boldsymbol{\sigma},T,\mathbf{X})\dot{\overline{\sigma}}_{mn} + M^k_{ij}\,(\mathbf{X})B^k\,(\boldsymbol{\sigma},T,\mathbf{X})\dot{T}. \qquad (13.2.18)$$

By adding the elastic response $\dot{\boldsymbol{\varepsilon}}^e = L^{-1} : \dot{\overline{\boldsymbol{\sigma}}}$, one obtains the constitutive equation of an austenitic single crystal undergoing martensitic phase transformation coupled with plasticity. That is

$$\dot{\varepsilon}_{ij} = [L^{-1}_{ijmn} + M^k_{ij}(\mathbf{X})\, A^k_{mn}(\boldsymbol{\sigma},T,\mathbf{X})]\dot{\sigma}_{mn}$$

$$+ M^k_{ij}(\mathbf{X})\, B^k(\boldsymbol{\sigma},T,\mathbf{X})\dot{T}. \tag{13.2.19}$$

Equation (13.2.19) can also be formally written as

$$\dot{\varepsilon}_{ij} = (\mathbf{L})^{-1}\dot{\sigma}_{mn} - \mathbf{m}\dot{T}, \tag{13.2.20}$$

where **L** and **m** are the average tangent elastoplastic and thermal moduli of a single crystal, respectively.

Equation (13.2.20) is similar to a nonhomogeneous thermoelastic problems where both elastic modulus and thermal expansion coefficients fluctuate in a given composite material. In our case, it corresponds to a polycrystalline aggregate of single crystals with given crystallographic orientations with their constitutive laws described by (13.2.20). Here the thermal contribution arises only from the chemical energy related to phase transformation.

Under purely mechanical loading conditions ($\dot{T} = 0$), Eq. (13.2.20) is reduced to the incremental formulation adopted by Hill for his self-consistent model; see Chapter 11. An extension of Hill's approach to account for the "thermal" fluctuations of local fields is not straightforward. In the next section, we recall the field equations within a thermomechanical framework. The analysis leads to a thermomechanical integral equation. A self-consistent approximation of this equation reduces to Hill's model when a purely mechanical loading is applied.

13.3 OVERALL BEHAVIOR OF POLYCRYSTALLINE MATERIALS WITH PHASE TRANSFORMATION

In the previous section, the constitutive equation of an austenitic single crystal has been established within micromechanics and thermodynamics frameworks. This accomplishes the first transition method, which takes into account different features of martensitic transformation and its couplings with plastic flow. Interactions between crystals have to be undertaken to describe the thermomechanical behavior of a polycrystalline aggregate, where each constituent is described by the constitutive equation (13.2.20).

For such a purpose, let us consider a representative volume element of the polycrystalline aggregate with volume V, subjected on its boundary ∂V to a velocity field $\mathbf{v}(\mathbf{r})$ such that

$$\nu_i(\mathbf{r}) = \dot{u}_i(\mathbf{r}) = \dot{\bar{\varepsilon}}_{ij}x_j \quad \text{if} \quad \mathbf{r}(x_i) \in \partial V, \tag{13.3.1}$$

where $\bar{\boldsymbol{\varepsilon}}$ is the macroscopic applied strain.

The field equations

$$\dot{\sigma}_{ij,j}(\mathbf{r}) = 0, \quad \dot{\varepsilon}_{ij}(\mathbf{r}) = \tfrac{1}{2}[\dot{u}_{i,j}(\mathbf{r}) + \dot{u}_{j,i}(\mathbf{r})],$$
$$\dot{\sigma}_{ij}(\mathbf{r}) = L_{ijkl}(\mathbf{r})\dot{\varepsilon}_{kl}(\mathbf{r}) - m_{ij}(\mathbf{r})\dot{T} \tag{13.3.2}$$

can be combined to solve the problem through the solution of a thermomechanical integral equation; see Problem 13.7:

$$\dot{\varepsilon}_{mn}(\mathbf{r}) = \dot{\bar{\varepsilon}}_{mn} + \int_{V'} \Gamma^0_{mnij}(\mathbf{r} - \mathbf{r}')[\Delta L_{ijkl}(\mathbf{r}')\dot{\varepsilon}_{kl}(\mathbf{r}') - \Delta m_{ij}(\mathbf{r}')\dot{T}]\, dV'. \tag{13.3.3}$$

To derive (13.3.3), a reference homogeneous medium with properties $\mathbf{L}^0$ and $\mathbf{M}^0$, and with the same boundary conditions (13.3.1), is introduced such that

$$L_{ijkl}(\mathbf{r}) = L^0_{ijkl} + \Delta L_{ijkl}(\mathbf{r}), \qquad m_{ijkl}(\mathbf{r}) = M^0_{ijkl} + \Delta m_{ijkl}(\mathbf{r}), \tag{13.3.4}$$

where $\Gamma^0_{mnij}(\mathbf{r} - \mathbf{r}')$ is the modified Green's function, expressed by

$$\Gamma^0_{mnij}(\mathbf{r} - \mathbf{r}') = -\tfrac{1}{2}[G^0_{ki,kl}(\mathbf{r} - \mathbf{r}') + G^0_{kj,li}(\mathbf{r} - \mathbf{r}')] \tag{13.3.5}$$

in terms of Green's functions $\mathbf{G}^0(\mathbf{r} - \mathbf{r}')$ related to the reference homogeneous medium. Originally, the self-consistent mean field theory has its great interest in the properties of the modified Green tensor Γ^0, which can be divided for any homogeneous medium with tangent moduli $\mathbf{L}^0$ into a local part Γ^{loc} and nonlocal part Γ^{nloc} such as

$$\Gamma^0_{ijkl}(\mathbf{r}) = \Gamma^{\text{loc}}_{ijkl}\delta(\mathbf{r}) + \Gamma^{\text{nloc}}_{ijkl}(\mathbf{r}). \tag{13.3.6}$$

Substituting (13.3.6) into (13.3.3), and using the properties of the Dirac function $\delta(\mathbf{r})$, the integral equation becomes,

$$\dot{\varepsilon}_{mn}(\mathbf{r}) = \dot{\bar{\varepsilon}}_{mn} + \Gamma^{\mathrm{loc}}_{mnij}[\Delta L_{ijkl}(\mathbf{r})\dot{\varepsilon}_{kl}(\mathbf{r}) - \Delta m_{ij}(\mathbf{r})\dot{T}]$$
$$+ \int_{V'} \Gamma^{\mathrm{nloc}}_{mnij}(\mathbf{r} - \mathbf{r}') \quad [\Delta L_{ijkl}(\mathbf{r}')\dot{\varepsilon}_{kl}(\mathbf{r}') - \Delta m_{ij}(\mathbf{r}')\dot{T}]\, dV', \tag{13.3.7}$$

where the integral form in (13.3.7) is generally difficult to estimate due to high and stochastic fluctuations of the field $\Delta\mathbf{L}(\mathbf{r}'):\dot{\varepsilon}(\mathbf{r}') - \Delta\mathbf{m}(\mathbf{r}')\dot{T}$. To overcome this difficulty, the self-consistent mean field theory for elastic materials chooses a reference medium $(\mathbf{L}^0, \mathbf{M}^0)$ so that the mean value of the field $\Delta\mathbf{L}(\mathbf{r}'):\dot{\boldsymbol{\varepsilon}}(\mathbf{r}') - \Delta\mathbf{m}(\mathbf{r}')\dot{T}$ vanishes and therefore the integral in (13.3.7) can be neglected. This condition of vanishing mean value of the fluctuating field is also known as the self-consistency condition. In fact, this condition writes

$$\int_{V'} [\Delta\mathbf{L}(\mathbf{r}'):\dot{\boldsymbol{\varepsilon}}(\mathbf{r}') - \Delta\mathbf{m}(\mathbf{r}')\dot{T})\, dV' = 0. \tag{13.3.8}$$

On the other hand, one can show in a straightforward manner that Eq. (13.3.8) leads to the following macroscopic behavior (see Problem 13.8):

$$\dot{\bar{\sigma}}_{ij} = L^0_{ijkl}\dot{\bar{\varepsilon}}_{kl} - M^0_{ij}\dot{T}. \tag{13.3.9}$$

Expression (13.3.9) shows an interesting property that consists in the typical choice of the reference medium $(\mathbf{L}^0, \mathbf{M}^0)$ to fulfill the consistency condition stated above. Clearly, it follows from (13.3.9) that the properties of the reference medium should be the effective properties of the considered composite, that is, $\mathbf{L}^0 = \overline{\mathbf{L}}$ and $\mathbf{M}^0 = \overline{\mathbf{M}}$.

Under the self-consistent approximation, Eq. (13.3.7) is reduced to

$$\dot{\varepsilon}_{mn}(\mathbf{r}) = \dot{\bar{\varepsilon}}_{mn} + \Gamma^{\mathrm{loc}}_{mnij}[(L_{ijkl}(\mathbf{r}) - \overline{L}_{ijkl}]\dot{\varepsilon}_{kl}(\mathbf{r})$$
$$- [m_{ij}(\mathbf{r}) - \overline{M}_{ij}]\dot{T}. \tag{13.3.10}$$

Here Γ^{loc} is calculated with respect to the effective properties $\overline{\mathbf{L}}$ of the polycrystalline material.

Equation (13.3.10) may be reorganized as follows:

$$\dot{\varepsilon}_{ij}(\mathbf{r}) = A_{ijkl}(\mathbf{r})\dot{\bar{\varepsilon}}_{kl} - a_{ij}(\mathbf{r})\dot{T}, \tag{13.3.11}$$

where the concentration tensors **A** and **a** are expressed by

$$\begin{aligned} A_{mnkl}(\mathbf{r}) &= \{I_{mnkl} - \Gamma^{\mathrm{loc}}_{mnij}[L_{ijkl}(\mathbf{r}) - \overline{L}_{ijkl}]\}^{-1}, \\ a_{kl}(\mathbf{r}) &= A_{klmn}(\mathbf{r})\Gamma^{\mathrm{loc}}_{mnij}[m_{ij}(\mathbf{r}) - \overline{M}_{ij}]. \end{aligned} \tag{13.2.12}$$

Note that Eq. (13.3.11) solves the concentration or localization problem, which allows us to determine the effective properties as

$$\overline{\mathbf{L}} = \langle \mathbf{L}(\mathbf{r}):\mathbf{A}(\mathbf{r})\rangle_V, \tag{13.3.13}$$

$$\overline{\mathbf{M}} = \langle \mathbf{L}(\mathbf{r}):\mathbf{a}(\mathbf{r}) + \mathbf{m}(\mathbf{r})\rangle_V, \tag{13.3.14}$$

where the following properties have been used:

$$\dot{\bar{\boldsymbol{\sigma}}} = \langle \dot{\boldsymbol{\sigma}}(\mathbf{r})\rangle_V, \qquad \dot{\bar{\boldsymbol{\varepsilon}}} = \langle \dot{\boldsymbol{\varepsilon}}(\mathbf{r})\rangle_V, \qquad \dot{\bar{\boldsymbol{\sigma}}} = \overline{\mathbf{L}}:\dot{\bar{\boldsymbol{\varepsilon}}} - \overline{\mathbf{M}}\dot{T}.$$

Note that the mean field self-consistent model developed by Hill and extensively discussed in Chapter 11 can be described by the localization problem (13.3.11) in the case of purely mechanical load ($\dot{T} = 0$). In fact, Eshelby's elementary inclusion problem used by Hill to solve the concentration problem is deduced from (13.3.11) by assuming piecewise uniform tangent modulus, so that

$$\dot{\varepsilon}^r_{ij} = A^r_{ijkl}\dot{\bar{\varepsilon}}_{kl} \quad \text{with} \quad A^r_{ijkl} = [I_{mnkl} - \Gamma^{\mathrm{loc}}_{mnij}(L^r_{ijkl} - \overline{L}_{ijkl})]^{-1}. \tag{13.3.15}$$

In Eq. (13.3.15), one can readily show the relationship between Γ^{loc} and the Hill's tensor **H** and as a result the equivalency between Eq. (13.3.13) and (11.1.61) given in Chapter 11; see Problem 13.9.

Finally, to adopt the above framework for polycrystalline materials, the representative volume element is assumed to be an aggregate of N single crystals with homogeneous properties ($\mathbf{L}^g$, $\mathbf{m}^g$) ($g = 1, \ldots, N$) and volume fractions f^g. The integral forms (13.3.13) and (13.3.14) are reduced to the following discrete expressions:

$$\overline{\mathbf{L}} = \langle \mathbf{L}(\mathbf{r}):\mathbf{A}(\mathbf{r})\rangle_V = \sum_{g=1}^{N} f^g \mathbf{L}^g:\mathbf{A}^g, \tag{13.3.16}$$

$$\overline{\mathbf{M}} = \langle \mathbf{L}(\mathbf{r}):\mathbf{a}(\mathbf{r}) + \mathbf{m}(\mathbf{r})\rangle_V = \sum_{g=1}^{N} f^g(\mathbf{L}^g:\mathbf{a}^g + \mathbf{m}^g), \tag{13.3.17}$$

where

$$\mathbf{A}^g = [\mathbf{I} - \Gamma^{\text{loc}}(\overline{\mathbf{L}}):(\mathbf{L}^g - \overline{\mathbf{L}})]^{-1}, \tag{13.3.18}$$

$$\mathbf{a}^g = \mathbf{A}^g : \Gamma^{\text{loc}}(\overline{\mathbf{L}}):(\mathbf{m}^g - \overline{\mathbf{M}}). \tag{13.3.19}$$

As discussed in Chapter 11, the implementation of Eq. (13.3.16) to (13.3.19) required an iterative numerical scheme. Note that the tangent moduli ($\mathbf{L}^g$, $\mathbf{m}^g$) ($g = 1, \ldots, N$) are deduced from Eq. (13.2.19), which requires the resolution of a nonlinear system involving the various internal variables related to phase transformation and plastic flow in each considered single crystal. Indeed, the present problem dealing with the behavior of polycrystalline materials with phase transformation is much more complicated than a purely plastic problem. Numerical implementation of the present framework and applications to real polycrystalline materials can be found in Cherkaoui et al. (2000).

PROBLEMS

13.1 Let us assume that the accommodation process accompanying the Bain strain occurs by plastic deformation $\boldsymbol{\varepsilon}^p$ on a slip system with unit normal $\tilde{\mathbf{n}}$ and shear direction $\tilde{\mathbf{m}}$. We denote by γ, the magnitude of shear.

(a) If $\tilde{\mathbf{n}}$ is of (1,0,1) type and $\tilde{\mathbf{m}}$ parallel to (1,0,−1) direction, provide the expression of $\boldsymbol{\varepsilon}^p$.

(b) If the Bain strain is expressed by

$$\boldsymbol{\varepsilon}^{\mathrm{B}} = \begin{pmatrix} a & 0 & 0 \\ 0 & a & 0 \\ 0 & 0 & c \end{pmatrix} \text{ with } a = 0.132,\ c = -0.199$$

(a and b are lattice parameters)

express the compatibility condition and determine γ.

(c) Determine **n** and **m** in Eq. (13.1.7).

13.2 Prove the compatibility conditions (13.1.8).

13.3 Let us consider a RVE with volume V and that consists of a two-phase austenite–martensite composite. We denote by $f = V^M/V$ the volume fraction of martensite. Let us assume that the existing martensite is described by an ellipsoidal domain with boundary or interface S. If the phase transformation occurs by the ellipsoidal growth of this domain:

(a) Show that the volume fraction change is expressed by (indications: use the half-axes of an ellipsoid)

$$\dot{f} = \frac{\dot{V}^M}{V} = \frac{1}{V}\int_S w_\alpha n_\alpha \, dS.$$

(b) If the transformation strain is assumed to be piecewise uniform, so that

$$\boldsymbol{\varepsilon}^{\text{tr}}(r) = \sum_{I=1}^{N} \boldsymbol{\varepsilon}^{\text{tr}I}\theta^I(r),$$

prove Eq. (13.1.31).

13.4 Prove Eq. (13.1.48).

13.5 Consider a RVE with an austenitic matrix containing N martensitic variants. Each variant I has experienced an eigen strain $\boldsymbol{\varepsilon}^{\text{tr}I}$ during the phase transformation. Denote by $\overline{\boldsymbol{\sigma}}^A$ the average stress in the matrix and by $\overline{\boldsymbol{\sigma}}^{M_I}$ the one in a martensitic variant. By using Kroner's inclusion problem where the infinite medium represents the austenitic matrix:

(a) Show that $\overline{\boldsymbol{\sigma}}^{M_I} = \overline{\boldsymbol{\sigma}}^A + \mathbf{L}:(\mathbf{I} - \mathbf{S}^I):(\overline{\boldsymbol{\varepsilon}}^{p_A} - \boldsymbol{\varepsilon}^{\text{tr}I})$.

(b) Prove Eq. (13.3.20).

13.6 Consider an infinite medium representing an austenitic matrix with prescribed $\overline{\boldsymbol{\sigma}}^A$ and $\overline{\boldsymbol{\varepsilon}}^{p_A}$. Assume that a martensitic ellipsoidal domain growth instantaneously in this matrix. If we denote by $\boldsymbol{\sigma}^-$ the stress within this domain: Show that

$$\boldsymbol{\sigma}^- = \overline{\boldsymbol{\sigma}}^A - \mathbf{L}:(\mathbf{I} - \mathbf{S}^I) : \boldsymbol{\varepsilon}^{\text{tr}I}.$$

13.7 Using Green's functions for infinite medium, prove the integral equation (13.3.21).

13.8 Prove Eq. (13.3.22).

13.9 Provide the relation between $\boldsymbol{\Gamma}^{loc}$ and Hill's tensor and prove the equivalency between Eq. (13.3.13) and Eq. (11.1.61) given in Chapter 11.

REFERENCES

Cherkaoui, M., M. Berveiller, and X. Lemoine (2000) Couplings Between Plasticity and Martensitic Phase Transformation: Overall Behavior of Polycrystalline TRIP Steels, *Int. J. Plasticity*, Vol. 16, pp. 1215–1241.

Cherkaoui, M., M. Berveiller, and H. Sabar (1998). Micromechanical Modeling of Martensitic Transformation Induced Plasticity (TRIP) in Austenitic Single Crystals, *Int. J. Plasticity,* Vol. 14, pp. 597–626.

Eshelby, J. D. (1970). Energy Relations and the Energy-Momentum Tensor in Continuum Mechanics, In *Inelastic Behavior of Solids,* M. F. Kanninen, W. F. Alder, A. R. Rosenfield, and R. I. Joffee, eds., McGraw-Hill, New York, pp. 77–115.

Patel, J. R. and M. Cohen (1953). Criterion for the Action of Applied Stress in the Martensitic Transformation, *Acta Metall.,* Vol. 1, pp. 531–538.

Wechsler, M. S., D. S. Lieberman, and T. A. Read (1953). On the Theory of the Formation of Martensite, *AIME Trans. J. Metals,* Vol. 197, pp. 1503–1515.

SUGGESTED READINGS

Diani, J. M., and D. M. Parks (1998). Effects of Strain State on the Kinetics of Strain Induced Martensite in Steels, *J. Mech. Phys. Solids*, Vol. 46, pp. 1613–1635.

Fischer, F. D. and S. M. Schlögl (1995). The Influence of Material Anisotropy on Transformation Induced Plasticity in Steel Subject to Martensitic Transformation, *Mech. Mater.,* Vol. 21, pp. 1–23.

Franciosi, P., M. Berveiller, and A. Zaoui (1980). Latent Hardening in Copper and Aluminium Single Crystal, *Acta Met.*, Vol. 28, pp. 273.

Gautier, E. and A. Simon (1988). In *Phase Transformation,* Vol. 87, G. W. Lorimer, ed., Institute of Metals, London, pp. 285–287.

Greenwood, G. W. and R. H. Johnson (1965). The Deformation of Metals under Small Stresses During Phase Transformation, *Proc. Roy. Soc.,* Vol. A283, pp. 403.

Leblond, J. B., J. Devaux, and J. C. Devaux (1989) Mathematical Modelling of Transformation Plasticity in Steels, I: Case of Ideal-Plastic Phases, *Int. J. Plasticity*, Vol. 5, pp. 551–572.

Magee, C. L. (1966). *Transformation Kinetics, Microplasticity and Aging of Martensite in Fe-31 Ni,* Ph.D. Thesis, Carnegie Institute of Technology, Pittsburgh.

Olson, G. B. and M. Cohen (1975). Kinetics of Strain-Induced Martensitic Nucleation, *Metall. Trans. A,* Vol. 6A, pp. 791.

Stringfellow, R. G., D. M. Parks, and G. B. Olson (1992). A Constitutive Model for Transformation Plasticity Accompanying Strain-Induced Martensitic Transformation in Metastable Austenitic Steels, *Acta Metall. Mater.*, Vol. 40, pp. 1703–1716.

INDEX

Printed in the United States
By Bookmasters